THE *Passionate* APE

Bad Sex, Strong Love, and Human Evolution

Craig Hagstrom

RIVERFOREST PRESS
BAINBRIDGE ISLAND · WASHINGTON

The Passionate Ape

Cover design and illustration by Lightbourne, © 2001 Lightbourne
Cover art © 2001 Olivia Schemanski

RiverForest Press, P.O. Box 11694,
Bainbridge Island, Washington 98110-5694

First Printing April 2001

Publisher's Cataloging-in-Publication Data
Hagstrom, Craig, 1948-
    The Passionate Ape / by Craig Hagstrom
p. 395 cm. 28
    Includes bibliographical references and index.
ISBN 0-9702626-5-5
1. Human evolution
2. Human origins
3. Love
4. Anthropology

599.938   00-191637

For my parents,
who took us to the pasture
to see how calves are born

Grudging Acknowledgements

My wife Sherry has survived this project since November 1989, and supported us for a year while I worked on the book full time. She read several drafts, and listened while I tried to explain each new idea that popped into my head. These ideas often announced themselves by my leaping up running to find pencil and paper at occasionally inopportune moments. No one should have to put up with all this.

I also thank my sister Sharon Smith, a relentless supporter and tireless reader. But when we all get together for dinner, and Sherry and Sharon have a glass of wine or so, the helpful criticism is sometimes overwhelming, and one does begin to have second thoughts about so much assistance.

Theodore A. Rees Cheney helped immensely, not only by his editing of the draft, but also through ideas I gained by reading his work. Material from his creative writing course sparked the greatest improvements in this book. He is worth much more than he was paid.

Elaine Morgan offered several suggestions after reading an early draft. For her ideas at that time, and for the inspiration she provided in her own books, I am very grateful.

Warning and Disclaimer

This book does not give advice. It does not advise on legal, ethical, medical, psychological or other matters. It does not permit, excuse or condone any behavior forbidden by law or by local custom. It recommends lawful, honorable and responsible public and private conduct. The publisher and author do not provide legal, medical or other professional services. If you require an expert opinion, seek the services of a properly qualified and licensed professional.

This book discusses a controversial hypothesis. To help people see themselves in a new way, the book uses selected ideas to set up viewpoints. The concepts selected might not be commonly accepted and might even be hotly contested. To keep it in perspective, you may mentally precede each sentence with "The author thinks …". Of course, you could do that with any book.

This book is an exercise in imagining. Considering the number of new ideas presented here, at least some of them will probably turn out to be wrong. If invalid ones could be reliably identified before publication, they would have been omitted. There are too many hypotheses to surround each one with the customary parenthetical remarks. Since this is not a scholarly work for a technical readership, the book states ideas more emphatically than technical books normally would, but that does not mean that the ideas are proven or accepted.

The purpose of this book is to educate and to entertain. Neither the author nor the publisher shall have any responsibility or liability to any person or entity with respect to any loss or damage caused, or alleged to have been caused, directly or indirectly, by information or ideas contained in this book.

If you do not wish to be bound by the above, you may return this book to your bookseller for a full refund. Your bookseller may require proof of purchase, and may require that returned books be in substantially the same condition as when purchased.

Preface

Our hearts wither without passion like orchids without rain. Yet love can also bring danger; passion and parachuting are both fatal in some cases, and hazardous often. Love's fulfillment and disappointments direct our course, as we weigh the devotion we have against the excitement and danger of passion that we might attain. We strive to reach love as if, dying of thirst, we see in the distance an explosive oasis.

Humans have powerful passions but frail relationships, and love's greatest impact is in its absence. To lose someone you love deeply feels more painful than death. I learned this lesson too young, and spent some decades trying to understand why love can hurt. While on that mental journey, I accidentally found a possible explanation for how love evolved in the first place.

My father died on a September Sunday in the summer I turned thirteen. He was an airline pilot, taking off from Chicago when a control cable parted, and his plane crashed a few seconds after lifting into the air. My father's departures and arrivals had measured out my days as a child. We relished each return like schoolchildren waiting for Friday afternoon. Looking back, it seems our first sentence as babies was to ask when he would come home. The cycles of his presence clocked our lives.

I remember my mother waiting in the car while I ran to meet my father in the crew-dispatch office near the passenger terminal. Now a quieter area, those buildings were once the center of activity. Red-leather benches filled a room smaller than the seating area of a single modern airport gate. One unguarded door opened to the asphalt where people walked out to climb stairs into the planes. I ran through

"

the crew office and up to a counter (I couldn't see over the top) asking breathlessly if the plane had landed. The clerk leaned over a bit to see me, and grinned while I bounced impatiently for his answer. I stood confused by his silence, as my father walked up behind me and smiling touched my shoulder. I remember holding him and being held in return, his stubble on my cheek, the smell of him.

He often claimed to nap, lying on the living room floor. From the time we began to crawl we stalked and assaulted him. We climbed upon him to wrestle, testing our strength, warming ourselves in his glow. Now I understand that my father lay on the floor not to nap but as an invitation to us. He showed us his love without words, yet so clearly that I can weep forty years later remembering him.

Standing in my parents' bedroom we heard my mother tell us his plane had crashed. Instantly my life shattered. I do not know the person I would have been if my father had lived; that future vanished as his plane broke apart, leaving instead the person I am.

It astonished me to see my love for my father become agony from the mere knowledge that he would not come home. I had taken his love for granted as any child will, and never considered the loss of it. Love arose, I would have guessed, to make you feel good when close to the person you loved. I was shocked to see how a feeling once taken for granted became unrelenting pain merely by knowing he would not be back.

As we filed from my parents' bedroom a fist of pain clenched me, leaving me breathless in suffocating loss. Each of us held our grief as a mother holds a dead child, not knowing what to do. None of us could make it better; none could let it go. You cannot run from a pain inside, so I stood still as if I were looking down at my heart lying like a dog in the road. I wanted numbness but felt clawed open. My body lived day to day glassy-eyed, while my heart tried to grasp its new situation.

For the next decade pain echoed through me like thunder in a cavern. Clearly I had misunderstood love; only a very bad thing could hurt so much. Love gave moderate pleasure but immense pain, and so I swore an oath in rage and tears: I would not be touched and hurt any more. Better to live behind walls so thick none could touch me than to feel that pain again. Better to parry with words than to embrace, to build an island than a bridge, to stay alone.

I pretended to go on bravely without my father, but it was a sham. From the corner of my eye I watched friends all unaware of the deadly snare I had uncovered. They behaved as if love were a good thing, happy in their ignorance. Why, I wondered, did only I see through a deceitful emotion? Had the disastrous death of my father laid bare something better left unrecognized?

I had always been a tinkerer, disassembling things to see how they worked. Now I was fascinated by this brutal emotion which had injured me so badly. As my friends began to date, to test adulthood's emotional contours, I studied them to see what made it all function while keeping them blissfully unaware. My life centered on exploring emotion from a safe distance, weighing acquaintances' behavior, seeing how the world worked for them but not for me.

We had hardly buried my father before I began adopting another. Just down the road lived our attorney and his family, his daughter one of my classmates. He was huge above the waist with legs shriveled by polio, and he drew me like a magnet. On a fishing trip I pulled his sweater over me and could cross my legs without stretching the fabric. I pulled him and his family over me for another warmth and made myself small enough to be no trouble.

I saw no danger; I sought only closer contact with people who were our neighbors and friends long before my father died. With him gone and my mother busy supporting us, my newly-adopted father took a larger place than he held before.

A few years later he died of cancer and I had to grieve again. My oath of isolation had completely failed to protect me and I raged at my lack of resolve. Without ever noticing the process, I had done exactly what I swore never to do, and had to pay again for caring. From life's most brutal lesson I apparently learned nothing at all.

I couldn't understand why we had love when other animals seemed to do well enough without it. I couldn't see why it attached so tenaciously, or why it approached with such stealthy tread. Its awesome impact suggests that we need it, and apparently can't control it. I've loved several times since then because I couldn't help it, finally making an uneasy truce with an unruly emotion that never is far from my mind. Only when I began this book did I start to see passion's contours.

In the 1960's, Robert Ardrey depicted our ancestors as hunter-gatherers living on African savannas, a killer species from an otherwise mild family. I read all his books, but the engrossing picture he painted never felt quite real. The Mighty Hunter didn't resemble anyone I could imagine. Desmond Morris wrote The Naked Ape *in 1967, calling humans the sexiest apes, saying that the value of a good sex life shaped our evolution. That sounded like a terrific goal, but it didn't seem to match lives of people I knew, with relationships barely hanging together.*

In 1982 I found a used copy of The Descent of Woman *by Elaine Morgan, like a breath of fresh air. She suggested that we evolved for some time as semi-aquatic apes, beachcombers during a drought, and that our physical and mental evolution were affected by aquatic needs. Her work continued a school of thought that has been pondered as long as the savannah hypothesis, but with less publicity. She discussed the conversion to frontal sex, the birth of rape and love, and an array of tantalizing evidence. Our subcutaneous fat, hairlessness, finger webbing and other features are easily recognized as traits of other aquatics.*

If we were studying humans today for the first time, I suspect we'd quickly accept this marine hypothesis, but mainstream anthropology has not yet embraced the aquatic ape. The scientific method is an excellent tool for stepwise progress, but makes an awkward compass for changing a theoretical course. Schools of thought form "notches" where careers grow, and scientists are reluctant to abandon a familiar notch, preferring instead to fit conflicting evidence into the accustomed frame. Changing a scientist's mind requires much more rigorous proof than moving him into his first notch.

This mindset, coupled with an early misunderstanding of what the aquatic ape hypothesis implied, soured the relationship between anthropology and the aquatic ape. Seriously discussing the aquatic hypothesis could damage an anthropologist's professional standing today, even if he found the evidence worth exploring. This has left exploration of the aquatic ideas to enthusiastic amateurs, with mixed results.

Of all the tricky paths where scientists (and amateurs) tread, emotions may be the most treacherous. Anthropologists typically portray human love as a sophisticated emotion, an accidental product of our large and powerful brains. Only humans (this approach holds) can talk, compare their views on the world, and know each other well enough to fall in love. Pair bonds are good enough for ape-men plodding across the African plains, while love came more recently, rising above mere need to give us rapture.

But love doesn't act sophisticated, it acts in your guts, like fear and hate. The clinical bond that Morris describes doesn't feel like passion I have known. The glow of idealized rapture has little to do with the hollowness of knowing that my father was torn and scattered over a cornfield. My finding a replacement didn't feel like the product of a sophisticated brain. It felt like hunger.

As a hobby I absorbed anthropology but my real focus was other business, exploring love as a geologist explores forces below the ground. Stripping away the greenery and gravel we can find the roots of mountains; probing through the Sahara's sand we can detect the beds of rivers that once flowed there. And at the bottom lies the oldest rock of all, marked by elementary force. I wanted to uncover passion's bedrock, but had no clue where to look.

I started this book as hobby work, to show how the aquatic hypothesis might answer some interesting anthropological questions. My goal was to write about twenty pages in a couple weeks, documenting some ideas on beards, metabolism and anatomy. The book wasn't intended to be about passion, but about the anatomy of an aquatic ape. As the pages stacked up, fear and doubt popped up as recurring problems for our ancestors to surmount. As I weighed alternate evolutionary solutions, the most feasible lubricant for aquatic relationships kept turning out to be love, or rather its tepid predecessor.

With some hesitation, I gripped my nemesis. If my personal history shaped my aquatic thesis, then any answers I found might be illusory. In science and life, we find what we go looking for; if I look for passion, I'll find it in fossils. On the other hand, if passion's evolution were truly our formative event, then I had the perfect schooling to see what it did to us. Perhaps the evolution of human emotions can be best seen by a traumatized amateur anthropologist. Perhaps my passion-based evolutionary scenario is correct, available for seeing by anyone whose "credentials" place them at the necessary vantage point. If so, I would be derelict not to report what I found, though my credentials are also my weakness.

The concept of love as an evolved emotion, not an accident, arising to serve a function and compel specific behavior, stunned me. If love evolved then we should see its creating force at work in love's expression. If love evolved as compulsion then we should see its function most clearly when it compels, as we see speed's value when an animal runs. If overcoming fear and doubt led to passion, then it is passion surmounting fear, not passion satiated and at peace, that shows what passion can do. Love evolved not to make happiness but to cause pain when unfulfilled, not to make togetherness pleasurable but to make absence hurt.

In this book I'll try to describe why love evolved this way. Love's obsessive focussing first evolved in aquatic women, because men in the water had little to compete with. A woman with a tendency to obsession had a better chance of being pleased with a man she encountered, because her susceptibility magnified an individual male and raised his status in her eyes. This small advantage launched us on a trajectory our cousins never needed because their females had clearer proofs of male adequacy.

Love evolved not to make you feel good when you hold your lover, but to make you feel bad when apart. My pain at my father's death was trivial in evolutionary terms, the accidental result of close primate family structure and the human need for intimacy. My grief did not show passion misaligned; it expressed its function in pure form. Love exists to make us surmount obstacles to stop the pain. Bad luck made my father's death an unsolvable problem for those left behind because he was dead instead of merely away.

Absent or dead, the beloved still lives in the mind and fills our empty space. I wept for my father not because he was nonexistent but because he was too real to ignore and yet not near. Forty years his broken body has rested under a sunny slope overlooking the river and the airport. Yet if I reflect, just for a moment, I can call him clearly to mind again, warm and alive. I am four years old again as he lifts me high up, turning as we go, his happy son in orbit.

Human evolution's key lies in passion, because love fosters changes in the passionate mind. A human can obsessively focus on another, and can obsessively focus on a cure for cancer, or inventing airplanes, or any other goal. Mental traits we evolved as courting tools became the traits that set us apart from our cousins, and give shape and meaning to our lives.

I fell too early into an emotional desert. Through good luck and the kindness of others, I emerged relatively unscathed. While out there digging for water, I found a treasure, the sand-covered traces of love's first evolution, the unexpected gem which I happened to know something about. By comparing humans to other apes, I found a way to understand why passion and obsession were necessary for us but not for them, and why love lets humans prevail while our cousins hide in the trees.

This book is the treasure map's first sketch.

Table of Contents

Table of Illustrations

The Bonobo

The bonobo is the least-known of the Great Apes. It is often called the "pygmy chimpanzee", but this suggests simply a smaller version of the basic design, like a Shetland pony with respect to horses. In fact, the bonobo is a species with status equal to the chimp's, and not a smaller animal at all. In some respects it might be useful to include it in this discussion, because bonobo sexuality is highly unusual and might offer some lessons.

I have intentionally omitted all reference to it here, and speak of the Great Apes as including only humans, gorillas, orangutans and chimpanzees. I do this mostly to simplify phrasing, because this book is intended for lay readers who are unlikely to be familiar with the bonobo. To include it would require some background equivalent to that for the other apes, and a discussion distinguishing the bonobo from chimps, with the danger of reader confusion over the "pygmy" term. The marginal benefit does not seem to be worth either the investment or the risk of distraction from the main thesis.

Chapter 1 - Waves In The Gene Pool

Imagine the first love, ever in the world. Picture the first lovers.

After forest-dwelling apes, but before hairless hunters on the African plains, walked humans who first felt passion. After life in the trees, but before the stone-tipped spear, lived those who loved each other for the very first time. And you can see them, if you try.

This ancient prize still shapes us today, each in our own way. With a little thought we can recreate the world, guess at the date, suggest the place where it all began. Without removing the magic, we can uncover its pragmatic source. Human passion marks a detour we took, half-hidden in the woods. We passed by a stone cairn where two paths cross, a meeting whose meaning leaves me breathless.

Make no mistake. Mine is no self-congratulatory fable; humanity is not the pinnacle of evolution. In the boot camp of our souls, where weakness meant death, passion ensured our survival. Rapture is as vital for humans as flying for hawks. We evolved love to its necessary level, as teeth or talons suit their need. We live with passion uneasily, but live with it we must; it rarely gives us peace of mind, but it does give us children.

Long before we were hunters we evolved to be passionate. The earliest wanderers on the plains knew much about hunting, but even more they knew love. Nothing, they realized, could compare to finding the one who listened with the most attentive ear, whose embrace was closest and warmest, whose eyes looked back most directly into one's own.

Something leads us to fall in love, though when not in love we may marvel at our own lack of sense. Something overcomes vast differences between men and women, making a pair of lovers from a pair of strangers. We did not evolve in order to marry, but marriage is consistently how we assuage our hungers. Though far from perfect, monogamy based on love satisfies better than any other way of life. Many can imagine no other course.

For all its power, passion is only a building block of a much larger structure. Humanity's strength is not in how we forge weapons, but in how we form bonds of iron from hearts of flesh. The real benefit comes not from our happiness, but from the evolutionary changes wrought in our brains by passion. If humans long survive, passion will be the reason, while chimps, gorillas and orangs could easily die in the next hundred or thousand years.

The thinnest filament of chance saved us. Impossible to predict but visible in retrospect, a cosmic accident gave us the power to overcome our flaws. But for this slimmest stroke of fortune we would long ago have gone extinct, as gorillas, chimps, and orangs may soon all die. And the reason they now dwindle is part of the reason we will probably continue.

Passion defines us. Its arrival marks our birth; it's the key to who we are today. A human can learn to hunt, or to work in a factory, or do any number of things, but in all human groups, in all our cultures, we feel the same when we love. It cannot be taught or learned, though cruelty can subdue and warp it. No other creature feels what we feel.

Come with me back to the beginning. Set aside the hunting myths you have heard your whole life, and listen to a different story of our birth. I want to show how we fell in love, and what it means.

Parallel Primates

My recital will be hard to follow if you take human superiority for granted. We've heard the litany of human advantages: Big brains make us smarter; hunting gave us skills to conquer the world; human love is a superior bond of intelligent brains, creating a family unique in the world. Myths, all of them.

True, we had hunting bands, but hunting did not form our character. Men and older children sought small game, or scavenged the kills of other predators. Women stayed nearer home with younger children, and gathered vegetables and seeds. We lived as hunters not because we evolved to be hunters, but because our preceding evolution made hunting and gathering the most convenient way to survive. As a tall child plays basketball, we became hunters because it was easy. The Mighty Hunter myth was our evolution's *product,* not its *cause.*

When we began hunting, we already walked upright. Those who first colonized the African plains already paired in stable monogamous relationships. The very first hunter-gatherers had close families; they chanted together at work, sang songs in the evenings, and with spoken words planned the next day. Mothers and fathers together raised slow-growing young, and the parents considered themselves married. Early searchers for game might have been smarter than modern humans. This intelligent, speaking, singing, loving animal resulted from an earlier epic.

Humans are not superior, but carry enormous wounds from evolution's wrong turns. We cling to an evolutionary dead-end of the primate order; our near relatives dwindle year by year because we share deeply-rooted flaws.

Our closest primate relatives (chimps, gorillas and orangs) all have so many points of similarity to us that we cannot tell which is our nearest brother. Yet in our hearts we could be from different planets. Painfully timid, these three cannot even protect their own territories. While other primates gleefully attack intruders (perhaps from a safe height), our cousins retreat; their timidity condemns them. Heartsickness rots our closest relatives, and its nature will later illuminate our good fortune.

This timidity dismayed early primatologists, who had trouble finding a human parallel in a cowardly haystack. Baboons were the first large primates studied in detail, not because they are the most instructive but because they will face humans at close range. African farmers must frequently repel baboon invasions, not those of chimps, gorillas or orangutans. Baboons attack intruders of other species, including humans, often fatally. They quickly acclimate to human presence, so when we needed a human model we used the primates we could find, though we knew they were not our nearest relations.

Chimps by contrast are difficult to study because of their shyness, taking months to become acclimated to human company. In personal combat they rarely touch each other; chimps seriously wound others only when males attack smaller females, or when gangs of males overwhelm lone victims. Where other mammals will butt heads and occasionally kill each other, chimps fight by waving branches near their rivals. Not an impressive model for the Mighty Hunter.

Gorillas are no more inspiring, though in early contacts they terrified humans. Only in extreme desperation will a gorilla face an attacker to protect his family. Plagued by timidity, gorilla males regularly inbreed rather than leave the family to compete for mates. They cling to tiny pockets, isolated places, retreating instantly at any hint of danger. This largest of primates is famous (and loved) for his gentleness. But what led this evolution toward mildness? Why be so gentle, when a little brutality could gain so much for such a giant?

The orangutan is strong enough to destroy a crocodile's jaws, but its shyness makes it the least-studied of the great apes. Researchers following this red ape have found a solitary and drifting animal. Male orangs avoid each other by sound alone and may never meet other males. Females spend silent hours in trees, hiding from males nearby. With perhaps more strength per pound than any other ape, the orang may also be the most timid.

A vast mental gap separates humans from other apes. When a chimp hears a new noise, he usually freezes in alarm and checks his escape route (though chimps in groups will stalk interesting sounds). Gorillas and orangs move deeper into the trees. When a human hears the same, he goes to see what it might be. Some quality in the human mind makes us confident of ourselves as no other primate is confident. It is not mere *intelligence* that makes us brave and them timid. Each of us knows smart people with no confidence at all, and bold idiots.

What makes humans ready to sacrifice the lives of their young men to defend their country, while an orang will not protect even a patch of brush? Before we had aircraft or rifles or sharpened rocks, what made humans, the weakest of apes, into the bravest?

Human Evolution Suite

I have a complex story to tell. Assembling this puzzle was the most fun I ever had with my head, though many parts made me angry and some made me weep. I wish it were simpler, but it's only fair to warn you it's not.

We prefer to explain ourselves with simple stories. The Mighty Hunter myth is a favorite, stating that important and unique human features came from the needs of bipedal hunters on the African plains. To hunt game we needed to hold weapons and walk upright; we lost hair to keep cool in the chase, evolved big brains to learn tricks, language to plan the hunt. We had so much to learn that we lengthened our childhood to learn it, and evolved stronger bonds to nurture slow-growing young.

Super-Sex evolved to weld together our hunting bands. Frontal sex made closer mates; female orgasms made enthusiastic partners. A woman's bare skin lured a man, her breasts echoed the curve of her buttocks, and invited his touch. Women's constant sexuality became part of the male-female bargain. Men, we are told, evolved the largest penis in the primate world (relative to body size) to keep wives eagerly awaiting their mates' return. Great fun, all this, but not reality.

Our hunter-gatherer myth resembles a simple tune, satisfying enough when whistled but unable to occupy an orchestra for long. We have about eight million years of evolution to cover; a single mythic episode cannot contain all we endured.

Human evolution is not a tune but a suite — a series of melodies, each different and telling a different story. Movements of our suite have a sequence; the final movement sounds wrong unless we've heard the preceding ones. Each movement has a beauty all its own. These will continue, beyond our control, whether or not we see the process. None of the movements follows *inevitably* from the one before, but none can be out of sequence.

Humans evolved through stages; new needs often conflicted with prior stages' evolutionary results. Near the end of our suite I will argue that large brains signal stupidity, not intelligence. That probably sounds silly now, because the preceding movements are omitted. But when you hear the final melody, I hope it makes sense after all.

Only our lucky survival connects the melodies of our suite. Nothing in the earliest prehumans made our appearance inevitable, and at each stage many died without issue. If all had died we wouldn't be worrying about it, but that doesn't make our presence predestined. Looking back we can find patterns, and draw some lessons from the track. I hope to show you the path we took, but a map would help.

Our suite contains five major movements, covered in ten chapters. In some of those, males evolved most-rapidly; in others, females were evolution's pivot. Some steps were almost simultaneous; other steps might have had intervals of a million years.

Human Evolution Suite	
Movements	**Chapters**
The Aquatic Ape	1. Introduce the theory, life on land
	2. Life in the water, body changes
Emotional Growth	3. Frontal sex orgasm loss
	4. Courting evolution, tepid love
New Sexual Focus	5. Return to land, advantage of disinterest
	6. Women's passion and self-repression
Intellectual Decline	7. Obtuse men and the courting collapse
	8. Targeting girls, evolving the caricature
The Mighty Hunter	9. Neoteny snowballs, illusion of orphancy
	10. Religion appears, security and confidence

We meet our ancestors as they enter the Indian Ocean, to escape the Pliocene drought. Aquatic life led to frontal sex and a reoriented vagina. This cost women the automatic orgasm, and they evolved emotions as replacements. We moved back to land as the drought lifted, and women's emotions grew to powerful passion. Men were still promiscuous, and sidestepped feminine passions by sexually targeting naive women. Women seeking tolerable mates created a selective pressure favoring dumber men. As men's intelligence declined, they strengthened their preference for young women. Men's sexual focus caused our maturation to slow, causing brains to grow. This increased the shyness we share with the other apes, who are going extinct because of it. But instincts we evolved for mating accidentally overcame our fear, and the Mighty Hunter story began.

Our story is not simple. We have been through many hard times and narrow escapes. This book presents a radical view of our past, yet a wealth of evidence substantiates what I will say.

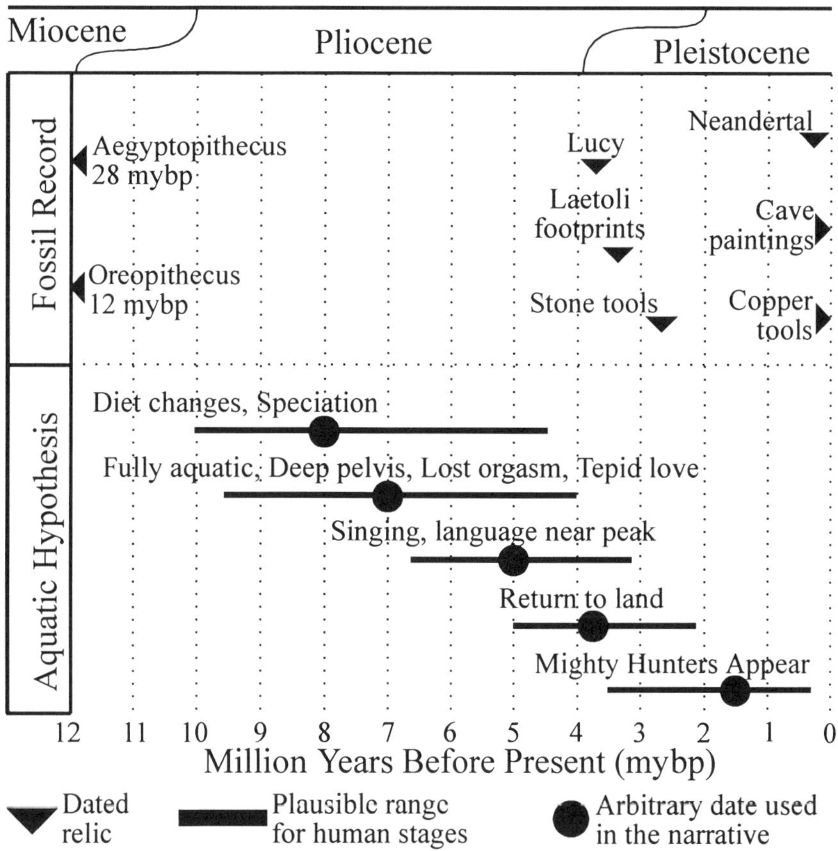

Figure 1: The Calendar

The calendar bottom half brackets date ranges for my discussion, and within the ranges I've picked arbitrary dates so I can simplify the phrasing. The top half of the figure shows generally-accepted dates, giving the aquatic scenario some perspective.

Sex Is A Mess

Through all the human evolution suite, like a funereal theme, drones sexual misalignment. In multi-million-year cycles, men and women have worked against each other with long-term and often tragic results.

To hear this musical lament, we must discard our habitual praise of human sexuality. If humans epitomize evolutionary perfection, then human sexuality must naturally be perfect. No other animal cares for its mate's pleasure; no other tends the contents of its mate's heart. No other spends so much time thinking about sex.

Our view of *female* sexuality, in particular, owes much to self-serving myths, partly because many gauge male expertise by counting female orgasms. Women don't orgasm all the time, so if our sexuality is superior then clearly no other female can climax at all. Such logic compels us to find unique reasons why women *should* orgasm while other females should have no need to. During copulation, the placid demeanor of other species' females once seemed to reinforce this view. Each woman's orgasm became another proof of human sexuality's transcendance.

But we mistook placid for inorgasmic. We now have data on female orgasms in elephants, cows, chinchillas and snails. Housecats are famous for their orgasms; female primates climax reliably and visibly. Orgasms are prevalent in widely-disparate orders, measurable wherever we look, not newly-evolved to keep women interested. Suddenly women do not lead orgasmic evolution, but trail behind. Women are remarkable not for having orgasms, but for having so few.

It's time we started over. Let's begin with what we already know.

Human sexuality is a mess. Life's greatest challenge is finding a mate. We spend our lives searching for a tolerable relationship; many never find one. The stress of this pursuit causes many people to regularly throw up before a date. This highest and best form of reproduction, lauded by anthropologists, for many of us turns out to be our lowest and worst memories.

Instead of wondering why women *started* to orgasm, ask why they are *stopping*. Only about 30% of women can orgasm from coitus, but if all other female primates can, then our female ancestors could, too. Rather than arguing that women evolved to suit men, curves and skin inviting touch, ask why women evolved inorgasmic bodies. Forget human sexual superiority, and ask why women are interested in sex at all, with such a miserly payoff.

Lack of emotion impresses us in other animals' coupling. Instead of arguing that this proves the advantage of human bonding, ask why no other animal needs emotion to make sex work. What happened to us, that sex is so often intolerable unless it comes with love? And what is this love, that it can lure us out on dates even if we must first vomit?

Human sexuality is the awkward but vital baggage of a naked ape. We have frail relationships but powerful passions. It was love, not tools, that defined the human condition and set our course. From the start of recorded history sex has shown no improvement, and I think the same problems have been with us for perhaps four million years. This *had* to affect us; modern humans were formed by the need to find and then tolerate mates.

Human sexual evolution reflects obsession and misjudgment, women's fears and men's confusion. Our story pivots on vanishing female orgasms and love's compensations. Often harsh, always directionless, natural forces repeatedly pinned us in tight corners from which only a lucky few emerged to face the next test.

Human relationships are patchworks of instinctive and learned behaviors, now so distant from the primate mainstream that (by comparison) homosexuality seems a trivial aberration. Our promiscuous ancestors' casual sexuality was shed to save our species; successive waves of change left layers of emotional debris. Women paid the higher price, but the battle scarred us all.

Still, we are remarkable. As each sex evolved for its own needs, the other evolved to cope. The fortuitous result was a powerful and instinctive fear that might have destroyed us. Yet along the way we happened to evolve unrelated abilities, which combined in the end to rescue us from our hidden terrors. We lie buried so deeply in our own dysfunction we cannot see it, any more than a fish can see water. Like astronauts, we must gain some distance to see the curve of this emotional sphere.

Human supremacy (if it exists) is not in our weapons or intellect, but in our ability to conjure from thin air the emotional support we desperately need. Our sexuality wounded us deeply, and in the end saved us.

Primordial Population

I want to introduce some of your relatives — not very heroic, but yours just the same. We begin with the common ancestor of chimps, gorillas, orangs and humans, living twelve to fifteen million years ago in East African forests. We come from the "superfamily" Hominoidea, encompassing apes and humans. Our Miocene root strongly resembled chimps, but we can add some gorilla and orang to imagine this animal from whom we all descend.

Sometime between twelve and six million years ago, drought and starvation forced us into the sea near what is now East Africa. We lived in the ocean for from two to six million years, while evolutionary forces reshaped our bodies and minds. Our aquatic evolution left its mark in our hairless bodies, subcutaneous fat and upright posture. I'll show how our time in the water also gave us language, love, and eventually religion.

Quiet bays were our first home and larder, and as we evolved to be aquatic we ventured into the open ocean. But because we had insurmountable physical handicaps (legacies of our primate heritage) we never felt fully at home in water. When the drought eased and food became plentiful again, some of us returned to the land.

Evolution, however, tends not to reverse its course. If conditions return to some earlier state, random mutations are unlikely to simply *subtract* any recently evolved and now-unneeded features. Instead, new forms and behaviors evolve to suit the new conditions. Whales did not re-evolve gills to replace lungs when they entered the sea, but evolved breath-holding and sealed nostrils. When we left the water we *might* have become chimplike again, but the odds are always against such a solution, and we did not.

Our sea-changed bodies were awkward on land, and more changes appeared in the returnees. Modern humans are the sum of both changes, transformed by two epic moves. We can explain all human features by these transitions; Mighty Hunter theories do not encompass us so neatly.

Our language and courage resulted from life in the water and the move back to land, woven into a primate heritage. Rape and marriage also came from this ancient double move. Instincts that had always worked for primates ceased to work for us in the water. New instincts evolved to replace them, we carried those back to land, and then modified the result.

In 1929, Professor Sir Alister Hardy envisioned the aquatic ape for the first time. He credited Wood Jones' book <u>Man's Place Among the Mammals</u> as the source of his inspiration. Hardy pondered the idea for thirty years before he spoke of it on a BBC program in 1960 and wrote of it in two later articles. It caused a stir at the time, and then was forgotten. In 1967, Desmond Morris devoted a page or so of <u>The Naked Ape</u> to the idea, and then went on to more exciting stuff. Elaine Morgan developed Hardy's suggestion further, and wrote <u>The Descent of Woman</u> in 1972, <u>The Aquatic Ape</u> in 1982, and others.

In recent decades we've learned much about primate interactions. When one learns a new way to feed, for example, mimicking spreads the solution through the troop; a primate troop is smarter than any one member. How rapidly ideas spread depends largely on the status of the first to learn it. If the dominant (or "alpha") male learns a new trick, the whole troop may use it within hours; if a child learns it, the rest may not adopt it for years. This least follower (the "omega" male or female) is not normally a source of inspiration.

A bunch of anthropologists, like any troop of primates, learns fastest if the alpha anthropologist shows the way. As yet no L. S. B. Leakey or Ernst Mayr has stepped forward to shepherd the aquatic ape idea. Though Elaine Morgan has done yeoman service, the theory's credentialled champion is not yet visible. Hence this incremental contribution from what one might charitably call an omega anthropologist.

For twenty years I've tried to see aquatic apes, envisioning life in water as an extension of normal primate life on land. As aquatic apes grew clearer in my mind, their behaviors and emotions started to feel more human. I want to describe connections I found to modern life, unexpected things of value.

We can start by sketching a primate population before any human uniqueness appeared. We can mentally transport them through an aquatic phase and describe how they might react. Each act, each change should be believable in the primate context, one of many possible courses any primate might take. As it happens, one of these paths leads to us.

Life On Land

We squatted perhaps two feet tall, a dozen million years ago. Brown hair covered our tailless bodies, with faces mostly bare. Males were a bit larger than females, with larger canine teeth. Males played more roughly, and frequent injuries often caused females to avoid males except for sex. Both sexes tended toward baldness with age. Though we were not exactly like chimps, no primate makes a better model.

We knucklewalked on all fours, though we could stand more erect to look about or walk a few steps. Jane Goodall reported a chimp who habitually walked upright to hold a crippled arm off the ground. In dominance displays, males rushed back and forth on two legs while waving branches for show, the branch as much for balance as for threat.

Small groups scattered through the forests of East Africa, two to ten of us per square mile. Groups broke and reformed as individuals wandered sometimes together, often apart. These were errands, not random moves, each member knowing where he wanted to go and what to do. As days passed, each person repeatedly met known companions in familiar places; each kept company with old friends and newcomers. Think of a large family drawn to different attractions at an amusement park.

Repeated meetings demand intelligence for recognizing each other, knowing rank, judging mood. A male who was your buddy yesterday may have just lost a fight, and be feeling prickly. A female who scorned you last week might now be warmer. Primates are socially aware because primate societies are complex. Where social groups are also mobile and fluid, even higher social skills are needed. Protocol was everything.

In good times we spent only a few hours a day eating, like diners at a long salad bar, though travelling between courses might take all day. In lean times, finding food was our major occupation. Primates have no instinctively targeted food, though we prefer sweet and fruity tastes, and meat. Within any area, learning and habit guided the choice of foods. Each fed himself, though a mother would stop her infant from eating something unhealthful. Pregnant females had no disadvantage, since groups moved slowly, and pursued prey not from need but for sport.

Adults shared food only when males captured young pigs, baboons or small monkeys. Chimps do stalk game (but only in irregular binges) and Goodall says chimp teams will warily stalk any unidentified noise. Those who catch prey are not obliged to share, but other adults and youngsters will wheedle pieces if they can.

Our forebears fed with hands and teeth. Of the hundred or more foods they ate, none required any tool more sophisticated than a twig to extract termites from a mound, or a sponge of crumpled leaves for soaking drinking water from tree hollows. The most complex dining acts were precision hammering of some nuts, and termite-fishing.

Dominance

Dominance and status held their attention more than did feeding. Primates (and most other mammals) recognize dominance in others, and most females prefer dominant mates. Critical to primate society and mating choices, status is a powerful and instinctive cue.

Display, not combat, is an ape's usual path to dominance, because our near relatives are easily impressed. Chimps often lose their nerve in a fight without ever being touched, over-awed by mere show. Young chimps may howl in terror at the approach of an adult who has never even struck them; adults often take care to calm thc youngsters, patting them reassuringly. Loneliness can plague some adult males, when others become too nervous to stay near them. Deadly attacks are known, but most combat is by bluff.

Imagine a barroom brawl in a Western epic. Fists hammer faces with pile-driving impact, the battle wavers unconvincingly, and finally our hero emerges victorious, leaving the villain underfoot. Stirring stuff, this, and though few of us ever witness or endure such a fight, we apparently enjoy simulations enough to keep movie-makers in business.

Now imagine a different fight. Black Bart bellows in rage and slams his barstool on the floor; our hero Lance Sterling thumps his stool in response. After smashing a bottle, Bart jumps on the pieces. Lance answers by running back and forth a couple of feet from Bart, stomping loudly. In desperation Bart drop-kicks a spittoon, but Lance counters by hurling a chair clean out the front door. Black Bart begins to weep, at which point Lance pats him on the back and they share milk and cookies.

The first version sells more tickets, but we began with something like the second case, where neither combatant touches the other. Our near relatives still fight that way; in our part of the primate spectrum, single combat is rare. Only when a gang of males attacks a lone individual, or an adult assaults an infant, is there real danger.

Chimps fight not with claws and teeth but by extravagant displays of noise and motion. Like the second description of barroom brawlers, they rush back and forth waving branches and thumping whatever is loudest. (One chimp boosted his status when he discovered how loud five-gallon cans could be.) Without ever touching their opponents, they rage and bluster until one loses his nerve, at which point they groom and comfort each other in the chimp equivalent of sharing milk and cookies.

The key is to *impress*, not to frighten. Threats suppress interaction, and onlookers move away, while performance promotes interaction and draws the audience in.   Putting on a show means more to us than bloodshed. We each hold our place in the world by how well we do, how much money we make, not by how many we kill. Like our near relatives and our primordial population, we applaud a good display but shun real violence.

How we compete reflects how ape minds evolved. A baboon troop fears very little; an ape troop fears nearly everything. I'll later outline evolutionary origins for the fear, but in brief our ancestors lived in a world they found large and frightening, filled with invisible dangers.

Individuals typically responded by finding a friend they could depend on. When you're lonely and nervous it helps to have someone near, even if they're a bit nervous, too.

For apes, dominance comes from bluster and friendship, and the woven threads of friendships give shape and assurance to life. Primates may be unique among animals in recognizing these links. Geese may mate for life, but I doubt they *understand* what a bond is, or is not. But primates know who is with whom, and how they're getting on. When any two interact, they create a new thing — a relationship — and only a primate can see it.

Sex, Sex, Sex

Sex provided pleasant entertainment for our ancestral population, in spirit closer to square dances than lovers' trysts. Groups of males clustered near any estrous female, copulating in turn, dozing and munching like hairy polygamous picnickers. Modern gang rapes differ from this primordial party mostly because a modern woman views with horror the coital tournament her female predecessor eagerly sought.

Estrus is most species' sexual trigger, a female erotic hunger and attractiveness to males, timed to her ovulation. She emits odors to notify males of her state; for some, garish displays of swollen tissue attract his eye. Many species have no apparent estrus, all females being ready at the same time and all males somehow knowing it. Salmon, grunion and horseshoe crabs come to mind. But where females in a population vary in their fertility, estrus is the common pattern, signaled by odor, anatomy, and behavior.

Male behavior and female estrus are inseparable. Odor and swelling evolved because males noted it; male awareness evolved because it led to fertile females. Female attributes and male recognition evolve together, varying by species but always leading males to ready females. The genetic palette of female features works only because male awareness belongs to the same species. Estrus is a property of a *species*, not of females.

Estrus varies widely in primates. Chimps (the model for our common ancestor) display extravagant labial swellings and distinctive odors. Other species have less obvious signs. The orang displays no visible swelling, but females become more-demanding at the middle of their cycle. Male orangs are more aggressive when near such a female; apparently the male recognizes something human observers don't.

Female displays reach their peak in multi-male troops, and the females often make distinctive copulatory noises. Both probably evolved because they got male attention, and promoted male competition useful to females. In monogamous species coupling is a quiet affair; women who make noise during sex betray our promiscuous heritage.

Female cycles synchronized our ancestors' sexuality. At regular intervals the uterus grows a fresh layer of soft and receptive tissue for an egg's implantation. Hormone levels change; female sexual tensions rise as external genitals swell and lubricate. Labia impede penile entry when flaccid but become firm when swollen, defining the vaginal opening as inflating an inner tube shapes an open center, a lubricated extension of the vaginal barrel. Sexual scent trails behind her like the right perfume on a pretty woman, and she will couple with any male in sight.

And sight is often the key. Scent doesn't linger in branches, and keen eyesight finds fruit or enemies more-reliably. Fifteen million years ago, after twenty million years of arboreal evolution, our sense of smell fell far behind that of most land animals. Today, after our aquatic interval and the human sexual collapse, our smell sensitivity has degraded far more. When my story begins, our smell sensitivity was far better than a modern human's.

Our precursors traveled in family-based clusters, or in groups of adult males who might be brothers or friends. Females in stationary species might advertise estrus with subtle signals at close range; our more mobile lineage probably called long-distance, as do chimps. A chimp female's swollen genitals stand out like a beacon in the dark fur of her rump, as she leans over to winkle out termites perhaps, oblivious to her effect on distant males. Goodall reports male chimps racing to a female when they spot her pink flash.

Females can't smell their own estrus odor, though they recognize their swelling and associate it with male attention. Males, however, can't get excited *without* odor; visual signals attract attention, but odor triggers desire. Goodall reports males sticking their fingers into the vulva of swollen females, then sniffing their fingers. For our ancestors, if she didn't smell right she just wasn't sexy.

No man knows estrus' appeal. The last man who felt it died several million years ago. Modern women feel a distant echo of swelling passions, but can't make men see it; indeed, women's inability to announce estrus was crucial to our evolution. We might compare estrus to food. Like a meal so tasty we continue eating after we're full, estrus fires sexual interest to the point of exhaustion. As food's taste and texture are understood without explanation, so we once knew without words which female was ready *right now*. As our mouths water when the odor of our meal hits us, so a ripe female's odor raised every penis in the area. What modern men feel when ogling a pretty girl is a pale shadow of our original hunger.

For our primordial population some ten million years ago, sex worked like a well-greased machine. As with any animal, copulation used angles and motions that gave impregnation and pleasure. For dogs or sheep, the four-footed female stance and classic male approach both stifle experimentation. There isn't much a wildebeest can fiddle with, even if bright enough to think of something new. Primates by contrast are sexual artists, experimenting with frontal sex when younger, with oral sex and manipulating the partner's genitals.

Sex for our ancestors was about as intimate as shaking hands. A male chimp makes what Goodall calls a "sweeping" gesture with his arm, and the female immediately crouches to present her raised backside. (The male's freedom to do this may depend on his status and the presence of higher-ranking males.) A female may initiate coitus with a reluctant male, by presenting to him until he shows interest. The female's appetite is often far ahead of the male's; she may complete her estrus cycle bearing bite marks from males grown exhausted and irritated by her constant importunings.

Female Orgasms

Female enthusiasm owed much to reliable orgasms — one or more per copulation. Female orgasms are well documented in several primate species and many mammals, with lab measurements and field observations closely paralleling women's responses.

Orgasms in males and females depend on similar anatomy, and seem virtually identical in sensation. In both sexes, fluid-filled bulbs expand during sexual stimulation. In women these bulbs straddle the vaginal opening; in men they are in the same location, inside the body near the base of the penis. When stimulation reaches a threshold, they explosively relax and trigger rhythmic contractions. Orgasm can be achieved without a clitoris or penis — those merely provide the sensation needed to reach the threshold. Men who have suffered penile amputation can still have orgasms.

In males, copulation stimulates the penis directly. In females, coition may stimulate the clitoris, the vaginal walls or deeper structures, or surface skin. As long as something causes bulbar expansion, she can climax. For both sexes, orgasms are an immediate physical reward for animals oblivious to the long-term benefits of reproduction.

For Miocene females, orgasmic stimulation came primarily from deep-touch nerves (the G-spot) stroked by the penile head, along with clitoral stimulation. The deep-touch response comes from bladder nerves whose stroking gives erotic sensations. The bladder lies near the vagina's inner end, with the urinary duct sandwiched between the uterus and pubic bone. In rear-entry sex, the penis' underside massaged bladder and urethral nerves against the pubic bone, giving reliable orgasms.

Women have largely lost G-spot stimulation, and are evolving new eroticisms to replace it. Many can orgasm from labial/clitoral friction, or from the penis massaging the pubococcygeal muscle. Some can climax from listening to music, or from whole-body activities like skiing. Someday women may be so sensitive to other stimuli that orgasms are again reliable. But today orgasms remain hard for women to achieve, and our history pivots on what women did when they lost this dependable reward.

A modern woman's lovemaking is much like other primates' complete estrus cycle. Estrus begins with labial swelling and vaginal lubrication over several days, as blood engorges the genitals. A plateau period follows, when the female is instantly ready for sex and may repeatedly solicit any male. Finally her body returns to its normal state; swollen genitalia return their fluid to the bloodstream, and the female returns to a more placid existence until her next cycle.

We retained the anatomy but lost the cyclic pattern. Swelling and lubrication, once automatic at intervals, we now evoke under partial control. Small relaxations after each orgasm were once followed a few days later by massive detumescence as estrus ended; now each coupling (possibly multi-orgasmic) triggers all the detumescence and the entire clitoral complex drains. Estrus in our ancestors was a long sensual tide, with pleasant waves of repeated coital orgasms. We condensed and heightened this; a tidal wave replaced a tide.

Women have been mislaid in the last few million years. What other females take for granted we must work for, and help each other to achieve. In no other species does male skill limit female satisfaction; in no other is she regularly denied a climax. By losing estrus, women lost the brink-of-orgasm readiness of our cousin primates. Vaginal reorientation cost her the G-spot, and coitus often must last a long time for penile massaging alone to cause an orgasm. Her emotional limits usually demand that she satisfy her sexual needs through a single male, though one crummy erection can rarely supply all she could use. We have come a very long way from our hairy promiscuous picnickers.

Even when not actively soliciting sex, female chimps willingly copulate with adolescent males, or infants, obliging them twenty times or more in a single session. Goodall reports an infant male attempting to reach an estrous female about the time he was learning to walk. If you imagine a woman laughingly stripping to let four- and five-year-old boys poke at her, while their mothers look on approvingly, you have a sense of our Miocene sexuality.

By our standards, our forebears were wildly promiscuous. Chimp females may have strong preferences for sex partners, yet they will also mate with many different males up to 50 times a day. A Barbary macaque female copulates with every sexually mature male in the troop, averaging 17 minutes between encounters.

Nonreproductive sex is common in primates, and easy sexual access may help smooth the waters for these highly social animals. But our promiscuity probably owes more to the infanticide that many primates share. A primate child faces no greater danger than attack by an adult male, and Sarah Hrdy suggests a female would gain by sowing male confusion. If a female mates with all males, none of them knows her child is not his own, and therefore safe to kill.

Females *evolved* promiscuity rather than planned it. Any more-insatiable females would have coupled with more males, lost fewer young to infanticide, and left a growing proportion of equally insatiable female young. Modern multi-orgasmic women owe more to ancestral infanticide than to the need for a rewarding pair bond.

Transient impotence is the male insurance against infanticide, ensuring that one female couldn't couple with the same male too many times. A male who could sexually satisfy a female would monopolize her and exclude other males, showing those males their non-paternity and so signing the death-warrant for his own young. Transient impotence benefits not only the individual male but also his breeding population. Where transient impotence prevails, paternity will be consistently unclear and infanticide lower. Insatiable females and inadequate males will tend to out-reproduce competing populations where males are made of firmer stuff.

All these clues point to highly sexual ancestors for humans, as promiscuous as any other primate. By trading coital orgasms for emotional stability, women illustrate how far we have separated from our forebears.

So we return to our cheerful hairy picnickers, for whom sex is a pleasant community entertainment. An enthusiastic female flits from male to male, wearing them all out eventually. Her preference for dominant ones and her affection for male friends are important to her, but the female will use every erect penis she can find. Young males are interrupted only when the alpha males are fresh enough to feel like pushing them aside, but in the end everyone gets plenty. For females sex is like going to the bathroom, automatic and reliable.

The Social Matrix

Our primordial fathers and mothers lived in different mental worlds. Females wandered with other mothers, or alone. Each cared for one or two youngsters, cuddled them, played baby games. Through this quiet world of maternal care tumbled noisy gaggles of mature males, yelling at each other and hurling barstools out the door, followed by an adoring retinue of younger males. Think of young toughs chasing each other through the children's section of a park, frightening the little ones and upsetting the mothers.

Males compete noisily for brief advantage; females form society's stable core. Males come and go, and the strongest today may sire more infants, but the female invests *years* in her family and her place. Risking more if she fights, gaining little from any single confrontation, females tend to compete carefully, in ways so different from male competition that early researchers missed them entirely. By cooperating with some and displacing others, female lineages can become powerful dynasties with far more impact than any one male can have.

Females compete often through skillful mothering. Treetop evolution forces closeness on primates; mother and infant must remain in constant contact to prevent falls. The resulting passionate desire for physical contact was primate society's foundation. Touching's constant reassurance in infancy (and the mother's need to touch) fostered a high tolerance for company. Benefits of cooperation with other females (who could hold the baby for a minute) helped foster sisterhoods that led to dynasties and finally to primate society.

Children can be stupid when parents are good; curiosity is dangerous without a mother's monitoring. Inquisitiveness evolves freely in primates because primate mothers already needed to be excellent — it costs nothing extra to guard against thoughtless inquisitiveness. A mother's constant presence also allows infants' emotional needs to grow; if reassurance were nonexistent, infants could not evolve to require it.

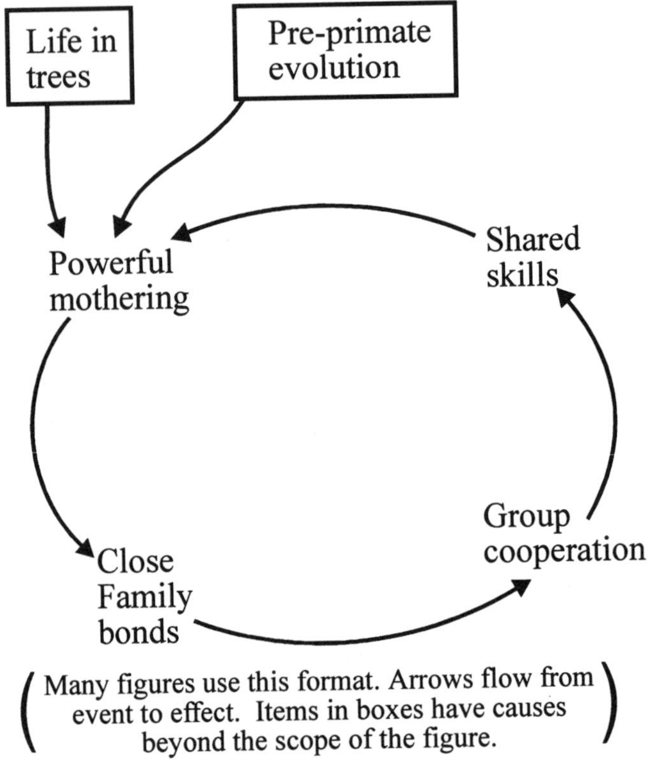

Figure 2: Mothering Skills

Primate success owes much to this mutually beneficial behavioral spiral. High mothering skills made mothers a reliable source of comfort and cuddling, and the young grow into interdependent adults. Groups of such adults are cooperative and flexible, pooling intelligence and resources, leading to primates' general success. In such groups, females can develop and share mothering skills and time, and so complete the loop.

We can't describe a primate outside its troop context; a lone primate is not fully functional. Loose or tight, group allegiances define individual identity, while group pressures delimit personal freedom. We share this with other orders. The porpoise and killer whale spend their entire lives in a family setting; to isolate either will drive it insane, as humans become insane in solitary confinement.

Primate groups are unique, and probably the world's most complex. High intelligence and intimate family bonds create subtle interactions of individuals and alliances. The singular primate ability to manipulate others' bodies creates tactile rewards. A porpoise may lead a family group but cannot pick fleas from its companions; no porpoise knows the intimacy coming from such attention. Ardrey (1966) calls this combination of vulnerable body, powerful brain and close society the key to primate success.

The group has a life beyond any individual. Elders and their relationships are a constant presence, the mold where an infant's mind is cast. More than just having good parents, to the infant's mind this feels as basic as food or air. Infants grow in an atmosphere of constant personal contact and cooperation.

(Yet infanticide looms wherever an infant can meet a male not its father. But infants don't know their peril, and the male's attack comes quickly. Rarely will an infant survive to be more careful next time. So childhood seems a time of playful and loving interaction. Infanticide is as invisible to the infant as an oncoming car is invisible to the turtle on the road.)

Adult interactions and personalities form the infant's world, as your family was already functioning before you arrived. A very young child doesn't realize that his parents were once separate; this umbrella relationship seems eternal. For any primate infant, adults have been there forever.

For females in particular, security and stability are vital. In most primates (including our near relatives) females often remain with their natal troop for life. Males may remain with female groups for a decade, so females are likely to have known their sexual partners for many years if not from birth. Promiscuity may evolve to forestall infanticide, but a female's casual sexuality may also owe much to her familiarity with all her suitors.

Primate troops follow one or more alpha members. In some species, leaders discipline harshly; in others, they guide more genially. Males fill the alpha role for most primates, though prosimians often have female leaders. Long-term female lineages may have more effect on the group's success, but *apparent* power rests with the one who picks the daily route, whose rising ends the community nap. A respected leader, male or female, is a "super-parent" for the group.

Apes care relatively little about group membership, though territorial defense and attacks on strangers do happen. Measured against other primates, ape groups accept new members more easily because social boundaries are lax. Again, chimps will be a model for our ancestors.

Chimps cluster in groups of four to eight, though a chimp's acquaintances may include 40 or 50. A chimp's circle of friends may never be seen all in one place; many such circles may overlap. Like your friends at work not knowing your hunting pals, a wandering chimp might meet individuals who may rarely or never meet each other. Yet within this range chimps form teams, recognize boundaries, and occasionally wage war.

Chimps cluster in casual associations with common goals. Females gather in nursery groups, sharing child-rearing duties. These contain perhaps fifteen members, often mostly children. Males collect in groups of two to eight (fishing buddies, if you will) but sometimes travel alone. Juvenile males follow groups of older males (too young to be their equals), or form temporary groups of their own. The largest chimp aggregations are of adult males and childless females. Goodall reports the entire male population collecting around an estrous female, and remaining together for a week. Lacking strict membership rules, these collections might tolerate a stranger — or might not.

Clusters meet as they move, and meetings cause great excitement and pleasure. Adult males embrace each other; males kiss the hands of females. Chimp greetings include bowing, holding hands and patting each others' face or genitals. Goodall suggests these greeting rituals (more complex in chimps than in most primates) evolved because there are so many such encounters. Greetings are enthusiastic and noisy, and go on for some time.

Yet they split again when their excitement dies. Individuals stay with one or the other group, or wander on alone. Chimps behave as if they expected each meeting to fulfill some important need; they move apart as if they realize their error. A young child visiting relatives shortly after Christmas can easily be confused and think more presents are forthcoming. When he realizes his mistake, early excitement can quickly turn to a lack of interest. Chimps act like that.

Chimps and gorillas owe slight allegiance to groups, and may accept strangers where other primates would attack. Orangs care even less than do chimps; orang families eat in adjacent trees without ever acknowledging each other's presence. Flexible and fickle human allegiances closely parallel the loose and vague associations of our nearest relatives.

In The Old Country

Immigration is a horizon of the imagination. Life in the old country, whatever old country it was, appears timeless and vague to the immigrant's grandchildren. The common predecessor of chimps, gorillas, orangs and humans lies beyond such a horizon. The chimpanzee stayed closest in form to our common precursor, and our cousins' minds resemble each other well enough that we can guess how they thought and behaved. But we don't know what happened, back in the old country, that made our ancestor first evolve apart from other primates, that made them leave.

Some have called the chimp-like Miocene ape a "generalized" primate, as if evolution were somehow uncertain how to place it. But our ancestor evolved specializations exactly as it needed to, and the uncertainty lies in our minds. It ate a broad diet of plants and vulnerable animals, and had the hands and body suited to gathering what it needed. Large for a primate (like a baboon), it lived on the ground and adapted for walking. This terrestrial physique evolved on an older arboreal foundation, which included agility and prehensile feet and hands.

Its most important facet, in the old country, was how it lived. Just as immigrants bring traditions with them, we carry in us the traces of our ancestor's behavior, and we share these with chimpanzees, gorillas and orangutans. Because we four are very alike, we can partly reconstruct our predecessor's body, mind and lifestyle.

Our ancestor's vital difference came not from being a relatively large primate, but from being a slow-growing one. All the great apes (gorillas, chimps, orangs and humans) grow slowly, so our ancestor grew slowly too. Between fifteen and twenty million years ago, for unknown reasons and along with other changes, our common ancestor parted from other primates by slowing its maturation.

This may have given a more-cooperative society; slow-growing animals tend to be less aggressive, and live in closer groups. Or slow growth may have helped them retain curiosity later into life, and most anthropologists credit much of our success to curiosity. Or a random mutation may have slowed their growth, just as other mutations occur by chance, and it happened that such slowing did our ancestors no immediate harm and so the trait persisted.

Our Miocene ancestor probably spent a third of its life growing to adulthood. It bequeathed that trait to its descendants, so the great apes are not only the largest primates but also the slowest to mature. Our common ancestor was placid and shy, living in cooperative groups based mostly on female family bonds. Though there were fights, group life revolved around politics more than force. Though some children died of infanticide and clumsiness and snakebite, mothers were competent and supportive, and families were close.

Few left the old country. Chimps remain closest in form to our Miocene ancestor, and we may think of them as the ones who did not emigrate. Gorillas and orangs stayed closer to the ancestral form than we did, the equivalent of moving to slightly colder or warmer climates. But for a heartsickness that plagues them now, they could have gone on forever. The emigration of a few had no effect on those who remained behind; but for a drought none would have left at all.

Chapter 2 - Aquatic Transition

Pliocene dryness ended this congenial life ten or twelve million years ago. When drought parched the land we parted from other apes, and took refuge in Africa's eastern coastal waters. We faced the same ocean other mammals have faced, and broad themes repeated in us. Like whales we gained body fat and lost hair; like seals we developed finger webbing.

Our transition may have taken several million years, countless generations of tiny changes. None knew they were evolving, any more than you feel giddy at the foot of height we've gained in the last few hundred thousand years. But our being primates shaped the details of our aquatic evolution, just as the whale and manatee differ now because their ancestors differed when they lived on land. Our primate heritage limited our aquatic form.

Three primates left fossils suggesting an aquatic life, showing remarkable similarities in their evolution. About twelve million years ago *Oreopithecus* became an aquatic swamp-dweller in what is now Italy; those swamps' coal deposits bear its remains. Its relation to us remains unclear, after decades of debate. *Paranthropus,* contemporaneous with our own lineage in eastern Africa, was less closely related to us than are chimpanzees. The third, *Australopithecus,* might be our direct ancestor. It left fossils up to six million years old, including bones we named Lucy, and we will refer to *Australopithecus* many times.

These three (*Oreopithecus*, *Paranthropus* and *Australopithecus*) share significant features. They all have bowl-shaped pelves and extended iliac crests (the ridge where your belt rides). All have *gluteus maximus* attachments moved well to the rear (i.e., buttocks). They have reinforced ileo-femoral ligament anchors, preventing hyper-extension of the leg at the hip. All have relatively small *acetabula* (hip joints), indicating low weight-bearing ability, and extended ischia (the bones you sit on) to provide kicking leverage. They lack the *acetabulo-cristal buttress* present in modern humans, which supports the hip joint in walking. *Australopithecus* shows knee condyles (weight-bearing surfaces) that strongly imply hyper-extension of the knee, useful for a swimmer.

These three suggest that primates can regularly evolve a common suite of features, and aquatic life provides an environment making such features adaptive. Walking upright on land provides different selective pressures that might give similar results, and these aquatic and terrestrial viewpoints are hotly debated. I think that we (and at least some other primates) had an aquatic past; this book explores how it might have happened for us.

Our own aquatic transition is hidden in dry centuries. Over life's half-billion years on earth, vast ages left no fossil record. At any time, few places in the world — volcanic ash or settling mud — can preserve bones until they fossilize. Fossil deposits are snapshots, recording those brief moments when a certain place had just the right conditions. In dessicated East Africa there were few such places and times.

A Movable Feast

We find our hairy picnickers twelve million years ago looking for food. Many primates eat a varied diet, and may travel far to find it; they can change diet rapidly when conditions permit. As the least territorial of primates, our ancestors' feeding routes were set more by habit than boundary, guided mostly by their knowledge of the area's foods.

We'll adopt an imaginary group as our family for the aquatic transition. They lived in low country, some miles inland from the Indian Ocean. We first see them on their daily food searches, twenty individuals rambling in clusters of three or five.

In their rounds they ignored many plants, both inland and near the beach. Some tasted bad, some were poisonous or difficult to digest, or harder to peel. Youngsters would occasionally experiment, while older members stuck to accustomed foods. As long as a snack wasn't actually dangerous, a mother wouldn't interfere with her child's dietary investigations.

A hundred thousand years after we adopted our imaginary lineage, the Pliocene dryness had begun. Food routes still covered the same country, but in some years trees had less fruit, in other years none. Near the beach and the brackish swamps, foods scorned five thousand generations earlier had become part of their diet. With more reliable (if less delectable) dinners available near the coast, they had alternatives when the inland crops failed. They rarely set foot in the water, but often rinsed food in it for the salty taste. Children toyed with critters scuttling over the beach. For each of them, this was the way life had always been.

Another hundred thousand years passed; the Pliocene drought deepened. Forests shrank, grasslands replaced them. With fewer rewards, travelling inland was no longer a daily event. Once every few days was enough to gather the little that grew. Much of their time was spent on the beaches collecting crabs and mussels. Having grown up playing with them, they learned to eat them. Crabs take little skill to catch, and easily replaced the termites and grubs our family had once harvested.

Swimming became common. Adults waded in to chase an escaping crab, perhaps, but children swam for fun. They paddled in tidal pools, holding their breath and looking around underwater, often catching snacks. As generations passed they became better at finding food this way, retaining the behavior into adulthood.

Another hundred thousand years passed, another five thousand generations, and inland feeding routes were gone. The Pliocene drought erased forests; open grasslands replaced fruit trees. Baboons, lions and wildebeest roamed a prairie where apes once wandered in shade. Our family now lived in a thin strip of trees along the coast, near salty tidal swamps. For each, this was how it had always been.

The beach and shallow water became their lives' focal point. They walked a few miles along the shore, harvesting favorite oyster beds or choice seaweed. They occasionally had peaceful meetings with similar groups on either side. With no movement inland and none out to sea, their range was a one-dimensional strip of coastline.

They gathered all their food in the water or on the beach, and each spent hours in the water every day. Soon after rising, each went swimming to catch a slow fish, or picked up a crab at the water's edge. Adults returned to squat on the shingle, grooming each other, copulating as opportunities arose, until hunger sent them back to the water. Children went into the water more often than adults and stayed longer, playing games with each other and chasing their food.

When wading, new mothers stayed in shallower water with infants clinging to their fur. At the age when young chimps learn to walk, babies learned to swim. Paddling at first, with a hand or foot still holding Mom, they soon splashed in small circles around her. By the time they were a year old, they could hold their breath, see underwater, and swim from one adult to another. For each, water was as much home as land had ever been. Adults took waves like old sailors roll with their ship. They dozed standing neck-deep, keeping balance with the same unconscious awareness that now keeps us from rolling out of bed.

Another hundred thousand years, and we were fully aquatic. We slept treading water, or standing on shallow bottoms in quiet bays. Children slept holding their parents or other adults, as a sleeping child holds a blanket. No one went up on the beach anymore. Our diet came from only aquatic foods; nutrients we once got from termites and fruit we now got from clams and kelp.

Pliocene Long Division

Like a bulldozer taking out the middle of a wall, the Pliocene drought plowed the center from the forests of Eastern Africa. As dryness slowly changed vast forests into plains, isolated food resources became refuges for our ancestors and their relatives.

Drought effects were uneven. Local climates preserved small forests, surrounded by savannah or desert. Century by century, green islands shrank; inhabitants had neither the foresight nor the means to walk a hundred miles to the next woodland. When forests died, inhabitants died, and the Pliocene survivors were those already in favorable sites when the drought began.

The four ape species (humans, gorillas, chimps and orangs) are the Pliocene's children. A small portion of our ancestral stock stayed in what did remain of Miocene forests, and survives today as chimps, closely resembling our common root. In foothills were other Miocene apes, and as drying progressed they moved higher into mist-watered heights where gorillas still live today. In cooler air larger bodies gave survival advantages, not for muscle power but for warmth. Other apes straggled along riverbanks, sustained by scant greenery. Crowding led to starvation, unsociable apes ate better, and so misanthropy evolved. These survive today as orangs, the most solitary of apes. Perhaps dozens of populations took other paths, but none are now with us.

Habitat	Strategy	Example
Forest remnants	Retain Miocene behavior where conditions permit	Chimpanzee
Savanna	No survivors	None
Foothills	Move higher to feed on mist-watered vegetation, gain body size to retain heat. Return to lowlands as conditions permit.	Gorilla
Riverbanks	Spread along rivers, evolve social distaste to encourage spacing. Expand into jungle where conditions permit.	Orangutan
Seacoast	Become aquatic, return to land as conditions permit.	Human

Our forebears lived near seacoasts. As the drought deepened, we retreated to the beach and looked for clams, and then moved into the water. We weren't superior; our parents just happened to live in one of the survivable places.

Our salty tears suggest that we evolved in *salty* water. Fresh water animals have no means to flush salt from the body, but saltwater animals do. Marine loggerheads and terrapins weep salt tears, as do saltwater crocodiles; freshwater crocodiles do not. Sea snakes (but not terrestrial snakes) weep salt. Saltwater lizards weep salt; terrestrial ones don't. Sea birds flush highly-concentrated salt water from their nasal glands.

Mammals show the same pattern. Beaver and freshwater manatee have no salt-expelling glands. Elephants weep salt tears (and some other evidence suggests they evolved in the sea), while pinnipeds have two sets of glands to purge salt. The glands used vary by species, but all have salt-expelling organs near their eyes. These mammals also expel saltwater when under stress; seals and elephants weep like humans when frightened, confined, or deprived of their young.

Our tears are not very salty now; Morgan thinks we had only begun to evolve such glands. I suspect it has more to do with the need to *conserve* salt once we moved back to land, where salt was hard to find. Our tears may once have been so concentrated that we needed no fresh water at all, just as seals may drink from the sea.

By the middle of the Pliocene we lived in salt water, and would not see our ape relatives for a long, long time. Perhaps the drought took a half-million years to squeeze apes (and other species) into separate enclaves. In our slow retreat we abandoned the land for protected bays along the Indian Ocean coasts. Time in the water began as excursions, but eventually became our full-time life for several million years. In this giant tub we find our forebears floating naked together without benefit of marriage. Think of southern California without telephones.

Initial Transformation

Food brought us to the beach, and diet guided our initial transformation. No fruits on the beach had tough skins to nip and break, we had no nuts to crack with molars. On the beach and in the shallows, powerful jaws had little value. Worse, food could be hazardous there. Cracking mussels in the teeth risks cuts from the shards. A bit of nut shell or twig wasn't dangerous, but ingested clamshell splinters could injure. In the forests we had eaten small animals whole — grubs, termites, ants and caterpillars. On the beach all the best foods had to be cracked open, and the containers could hurt.

Primates love to crack hard things to get the soft insides. After catching a monkey, a chimp shatters the victim's skull and eats the brain. Chimps use stone hammers skillfully on hard nuts, and our ancestors used the same techniques and shifted the target as they began to evolve in the shallows.

We instinctively bite what we can't break with our fingers. An early aquatic ape whose bite happened to be weaker would resort to hammer stones more quickly, reducing the likelihood of injury from shards. So life on the beach gave a survival advantage to jaw weakness, not power, and humans began evolving the weakest bite of any ape.

Once cracked open, shellfish demand a precise scissoring bite to sever the innards. Like fence posts on each side of our jaws, large canine teeth prevent sideways movement of incisors. Adults had once needed powerful jaws and large canines for dominance contests; now they were a hindrance.

Any who could cope better with marine foods gained nutritional advantage. Seaweed-munchers and carrion-scavengers quickly lost out to others living on oysters and cracked crab. Sucking became a valuable skill, but sucking isn't something adult hominoids do well. Infants' mouths are built for sucking, and during our early beach evolution children had an advantage over adults. As each grew older, his decreasing lip mobility and growing canine teeth became dual hindrances.

The smaller an individual's canines, the better he could use incisors; those with mobile lips got more nutrition from marine foods. Over generations, smaller teeth and flexible lips consistently paid off in healthier, better-fed adults. So canine teeth shrank because big canines were in the way, while adults retained infantile lips to better suck the innards of hapless crustaceans.

These both reflect "neoteny", retaining juvenile features in an adult body. Neoteny often correlates to slow growth, and the great apes (including gorillas, chimps and orangs) are the slowest-growing primates. (All apes show some juvenile features in adults, and I'll return to this later.) As we entered the sea, the value of childish mouths began to push humans further toward neoteny, giving supple lips and smaller canine teeth, slowing our maturation even more.

Thumbrise, Thumbset

Our hands were evolution's second focus as we adapted to shallow-water foods, splashing after trapped fish. When we touched something, half-seen in the roiled water, we scooped with curled fingers, thumb next to index finger. Often it got away. Those who could trap slippery prey *between* thumb and fingers had a more secure grip and a more certain dinner. Variations in fishing success gave long-term differences in survival. Evolving to grab at fast-moving and slippery prey, our fingers splayed out to act as a cage, the thumb rotating (becoming "opposed") to close trap's mouth.

As we became better fishers there was less need for the old feeding routes on land. Originally evolved to pluck fruit or pick up crawling grubs, eventually our hands could grab struggling fish and other animals closer in consistency to wet gelatin.

Thumb mobility counts because primates feed with their hands. Seals and whales could evolve forepaws into simple flippers because they don't feed with them, while their mouths show profound changes critical to their aquatic success. Our mouths remained much as they had been on land (with relatively stable dentition compared to, say, a walrus) because we didn't lunge after prey to clamp our jaws on it.

Hands, mouths and feeding behavior together make a key primate property. It is the central reason primates can quickly adapt to aquatic life when the opportunity arises, and forms a barrier to complete aquatic adaptation. Any primate can begin an aquatic transition the first time it grabs a crawfish from a muddy ditch. But a primate can never be quick in water because it's always thrashing around with its hands and arms, and can't afford to evolve them into flippers. When we returned to land after millions of years, it was because we never progressed beyond a feeding style which is, after all, not very efficient.

The Swimmer Appears

Our initial transformation to aquatic apes began when we first ate food on the beach. When it ended, we slept in the sea all night. Reduced canine teeth and opposed thumbs appeared early; the swimming ape came later.

Sucking lips and scissoring teeth measure ancient rivalries and nutritional advantage; competition to eat, not convenience. We left the beach for the water because it held more food, but we still had to compete for it. We evolved swimming skill because good swimmers ate better.

Children led the way. Primate infants start by exploring their mother's surroundings, then slowly move farther afield. Playing near mothers at the water's edge, infants splashed in shallow pools and flirted with waves breaking on the shore. Each generation was marginally more comfortable in water, knowing its rewards and confident in the face of its dangers.

Through successive generations, the farther out they waded, the more food they found. Adventurous ones ate better, and out-reproduced conservative people who stayed nearer the beach. Reluctant ones vanished by weight of numbers; the few who could survive on a narrow beach's fickle menu couldn't equal masses fed by the ocean.

Webbing was the evolving swimmer's first trace. Many vertebrates evolved webs of flesh connecting digits or limbs. Flying squirrels and bats fly on webbed limbs and fingers; penguins warm their young under skin flaps between their legs. With webbing so common we must assume its genetic recipe is deeply engraved, available in any species where the need arises.

Many primates have webbed toes. In the siamang and the bonobo, partial or complete webbing often joins some toes. Humans share this trait; a 1926 study found 9 percent of boys and 6 percent of girls had webbing between the second and third toes, and in some the webbing connects all the toes. Though Hardy and Morgan cite webbing as evidence for our aquatic past, it seems more likely a common primate feature of no value to our case.

Each species evolves webbing for its own needs (or accidents). Wading birds' webbed feet spread weight over muck or water, but don't grasp anything. A flying squirrel's webbed limbs spread its weight over more air but it catches no food in its flaps. Bats, on the other hand, use their webbed limbs to scoop up insects in flight, echoing their wings' origin.

The bat began by stalking bugs, grabbing them with paws instead of its mouth. Long hands and webbed fingers helped it grab erratic insects. Webbed hands then helped bats, as they leaped from branch to branch, and leaps evolved to glides. Gliding's value urged more webbing to evolve, extending down the body to the legs and tail, far beyond the need for grabbing with paws.

Our webbed hands might have evolved, much like the bat's, to help grab tiny fish, or to help us wade through waist-deep water. Apes have powerful upper bodies, and arms bear much of their weight when apes knucklewalk. Modern humans wade much as hairy partial-aquatics probably did, walking on the bottom, with hands scooping water to pull us along. When we came back out of the water webbing reduced again, leaving only vestiges. We still bear a trace of it; other primates lack the thin skin that connects a human's thumb and index finger.

Water had been no obstacle when we squatted on the shingle, scooping food from ankle-deep froth; we stepped high and walked over it. At waist-depth we moved *through* it, webbed hands helping us thrash along. We went farther out, found more food, so selective pressure favored better webbed hands. Soon we were at depths where webbed hands were not enough, and we added kicking.

Once we started kicking, we began evolving webbed toes, and our feet became less *hand*-like. Infants still clung to their mothers' fur with both hands and feet, because in the initial transformation we remained a fully-furred primate moving between water and beach. Also, a few would have groped for food with their feet (as we still do in some cultures). This was more common in our early transition, and faded as feet evolved webbing.

But these combined were obviously not enough to retain foot dexterity, so our toes became not only webbed but clumsy. Our ancestors could once grip branches between big toe and foot, as we grip between thumb and hand. With toe-mobility having little value in aquatics, toes now moved closer together to make a more-efficient paddle, a less-efficient grip. As toes became less mobile, fewer tried to grasp hidden prey with their feet. As feet became specialized as flippers, and legs became stronger, the swimming role of the arms and upper body began to decline.

With flippers for feet, we began to regularly move beyond neck-depth. So far our changes seem minor, but have profound implications for our subsequent development. From here on we were no longer adapting as terrestrial primates to a semi-aquatic life, but as aquatics in our own right. Though still tied to the land, at least for sleeping, our survival now depended on how well we swam.

Reshaping The Skeleton

Our biggest hindrance to swimming was not fur but legs, sticking out at angles as if we were glued to chairs. Fine for a quadruped, they were terrible for a swimmer.

Aquatic animals normally become streamlined, and a modern human's upright posture is our aquatic legacy. When our bodies adapted for swimming, our legs changed from akimbo obstructions to smooth extensions, eventually doing most of swimming's work. By the time we returned to land, our aquatic adaptation had crippled us. We now walk upright because we had no evolutionary route back to a quadrupedal pose.

The pelvis transfers strain between the spine and rear legs, and sets the angle of legs to spine. Most vertebrates have horizontal spines and vertical legs; the pelvis links them as a gusset holds post to beam. While the angle of legs-to-torso is relatively fixed for each species, all animals can vary that angle by moving their legs individually or by moving the pelvis itself.

Each leg movement has corresponding muscles, bringing thighs up to the body or pulling down and away. Muscles connect pelvis to torso and move the whole pelvis, other muscles connect legs to pelvis. No muscle connects one leg to another or a leg to the torso. To lift a leg forward, you tighten tummy muscles to pull the pelvis forward (or at least prevent it rocking back) and simultaneously contract leg muscles to raise the leg with respect to the pelvis.

Pelvic bones evolved projections to anchor muscles. The iliac ridges grew up and out to anchor flexor muscles that pull your legs up; the paired ischia (the bony lumps you sit on) anchor extensor muscles pulling your legs down. When climbing stairs, alternating ilial and ischial muscles are moving you. And paired pubic bones, along with part of your lower backbone, anchor muscles that pull your legs tight together. These seven bones, tightly bound, make your pelvis function as a solid ring, so that we often think of it as a single bone.

Any animal's leg mobility depends on hip-joint geometry, on pelvic bone projections anchoring muscles, on the muscles' strength and range. So how an animal moves determines its pelvic evolution, and we can deduce locomotion from pelvic evidence.

In a typical quadruped the pelvis lies parallel to the spine and the thigh is perpendicular to it. Flexors and extensors anchor at the pelvis' extreme ends (toward the head, or near the tail), giving mechanical advantage and a powerful running stride. Chickens, horses and kangaroos use this arrangement.

Primates have highly-mobile legs for climbing in trees, with muscle anchors placed to help swing legs far out to each side. But extensors and flexors move primate thighs the same as any other quadruped's, and primates can walk only when their legs are at roughly right angles to their pelves.

So an upright primate leans forward to maintain leg control. The spine slopes forward instead of being vertical, keeping the pelvis sloped forward too. Then the leg slopes down at right angles to the pelvis, and knees bend to keep the feet under the center of gravity. Though primates can do this (Goodall reports an injured chimp doing it consistently) the stance seems tiring.

Aquatic apes needed to align legs and spine (for streamlining) without losing mechanical advantage, with a pelvis evolved to handle right-angle stress. Evolution's solution was to bend our backs and tilt the whole pelvis. Individuals born with a more-arched spine had more-streamlined bodies, swam faster, and surpassed awkward contemporaries.

Our backs never would have become curved if we had needed *merely* to get our legs out of the way. A chimp can align its legs with its spine when lying on its back resting. But its legs are at their limit, with thigh-front muscles stretched taut. In the water, without gravity to help, relaxed legs made a permanent lap.

If we swam using webbed hands alone and trailed our legs behind us, over time thigh muscles would elongate, no spinal curve would occur, and our legs would dwindle as seals' have done. Our curved back's real benefit was in tilting the pelvis to retain mechanical advantage in the legs, and the only benefit in retaining mechanical advantage was to *use* the legs in that position. Our arched backs show that we swam with our legs.

Different Strokes

How we swam matters. Thrashing legs is a terrible way to swim, but being a clumsy swimmer kept our pelvis strong. Without a solid ring of pelvic bone we could not have returned to the land, so how we swam guaranteed we could come back when the opportunity arose.

An animal in water uses the same gaits for swimming as it uses for walking. Terrestrial mammals trot when forced to swim, forelegs and hindlegs alternating. Horses and deer trot crossing streams, a dog-paddle is a trot or walk, and small or big cats do the same. The first swimming stroke will always be a terrestrial gait; later aquatic adaptation grows from that.

Leg mobility determines the gait, and a fast trot takes very mobile rear legs. A bear, with relatively immobile rear legs, switches at low speed on land to a gallop — rear legs hitting the ground in unison. Galloping uses the animal's entire torso, bending and straightening to drive pelvis and legs as a unit. When an animal runs out of leg-mobility, yet needs more speed, it must gallop. Apes have no gallop; even at their fastest run, apes work their legs separately. We have, in short, leg mobility to burn.

Leg *im*mobility led to pinnipeds' aquatic success. As the sea lion's bear-like ancestor evolved into an aquatic, limited leg mobility made it swim in a gallop — torso muscles driving pelvis and rear legs in unison. This is energetically expensive on land but highly efficient in water. It also streamlines the body, as rear legs merge or become vestigial, while the spine may extend into a powerful tail. Whales, manatee, walrus and sea lions all began by swimming in a gallop, and show that by undulating their bodies up and down. Fish never had pelves, and undulate left and right.

We carried our primate heritage into the water, including highly mobile legs. While an ungulate might clumsily scratch an ear with a hind hoof, primates are nearly as adept with their legs as with their arms. This leg mobility gave us many swimming strokes.

We swam first using a walk, or dog-paddle, and wild primates dog-paddle across rivers. For quadrupedal mammals this is natural; an infant's crawl is a slow but perfect quadrupedal walk, on hands and knees. Our ancestral infants first experimented with this while holding Mom, learning to dog-paddle instead of learning to walk.

Our second stroke was a frog-kick or breast stroke. Among mammals the frog-kick is peculiar to primates, and has been reported in captive gorillas given access to water. Breast-strokes took advantage of our upper-body power and webbed hands; leg movements initially were weak. For a primate the stroke is so easy that we often use it as a resting stroke. We raise our legs up toward the body by bending the knees, extend them out to the sides and bring them straight together again.

Holding legs together is awkward for our relatives; ape legs naturally splay outward and appear bow-legged. Through generations of breast-strokes, our legs evolved to come tight together at the stroke's end. That gave a more powerful kick, and a longer glide at each stroke's end. The change is obvious in fossilized bones, showing close-set knees and parallel lower legs.

Our third stroke was the scissors kick. While the frog-style depended first upon hands and arms, the scissors marks our upper body's declining importance. It also implies our primordial feet had evolved into flippers, and our widely-spaced legs had become close-set. And scissors kicking requires a deeper pelvis.

Scissors kicking was not our goal, merely the natural result of the preceding two swimming styles. Pelvic rotation making us streamlined also gave our legs leverage for such a kick. Scissors kicks require parallel and powerful legs, and could evolve only *after* the frog-kick had made our legs strong and straight. Next time you go swimming, try scissoring your legs while keeping them a foot or two apart. Your body's rapid left-right oscillations will show you why the scissors kick appeared only after the frog-kick had given us close-set legs.

At the same time, we probably experimented with whole-body undulations, as in the butterfly stroke. Once we had evolved close-set legs and were kicking them forward and back, we could finally begin to move them together as other animals gallop.

As we evolved legs suited to scissors- or butterfly-kicks, we also evolved a hyper-extended lower leg. That is, your foot could kick out forward of the knee, as well as back. *Australopithecus* seems to show this in the fossil record, with a deep notch in the knee to give clearance for internal ligaments when the knee folds the "wrong" way. Our ability to lock our knees (something no other primate can do) is the only remnant of this hyper-extension.

Pelvic Adjustments

Swimming required an altered pelvis, able to move our legs in new directions. Flexibility with power was survival. But our new pelvis changed forever the way we copulate, and helped turn our path toward passion.

Flexibility came from pelvic shortening. Post-and-beam characterizes most animals' geometry, with the pelvis holding legs at a relatively stable angle. We needed a pelvis more like a giant wrist. Leg-muscle anchors are on the pelvis, not on the torso, so angling the pelvis while kicking resembles swiveling an outboard motor on a boat.

Chimps have a long and narrow pelvis tightly coupled to the spine, making them strong but inflexible, unable to hula convincingly. Over generations, small advantages in our swimming ability continually favored those with shorter pelves, thinner spines, and more-agile bodies. *Australopithecus gracilis* was named for this very litheness, with an extremely flexible connection of spine to pelvis.

Power came from widening and deepening the pelvis. The iliac ridges expanded toward both front and back, giving them a wheel-like appearance. The same bones flared out to the sides, giving *Australopithecus* a wider pelvis than our own, though he was much smaller than we. Each expansion of bone gave better leverage to the muscle it anchored.

The *gluteus maximus*, for example, had always attached (along with *g. medius* and *g. minimus*) to the iliac upper crest. Moving its attachment point to the rear gave a more-powerful kick. As the muscle grew and its anchoring bone deepened it became increasingly useful, and even more likely to further evolve in later generations. Now our largest single muscle, the *g. maximus* is unremarkable in other primates.

Leg mobility guided our pelvic evolution. A short, deep and agile pelvis was useful only because we used it to control the direction and power of kicks. We *can* swivel our hips now because we *needed* to swivel them then.

Rear leg mobility distinguishes aquatic primates from cetaceans and pinnipeds. Galloping animals who become aquatic use their legs together, in whole-body undulations. The sea lion's bearlike ancestor was a lousy trotter, but turned into an excellent aquatic. Once whole-body undulations take evolution's focus, tails can grow stronger and may eventually take over from legs.

Opposing leg movements distinguished our swimming style. Each leg helped us move only because the other leg balanced its kick, whether a dog-paddle, frog-kick, or scissors. Life in the trees gave us mobile legs before we entered the sea, and set us in an evolutionary channel. As our legs evolved to make them function more efficiently *in opposition*, we never had the need to evolve their *cooperative* use. Butterfly undulations never became the first stroke used by an infant. In the end, we returned from the water partly because this swimming style was flawed no matter how well we did it.

Since we consistently used our legs in opposition, we retained and deepened the basic primate pelvis. Whales galloped and evolved strong tails; rear legs dwindled to tiny remnants, and pelves disappeared. Pinnipeds swam with rear legs together for so long that they now can't separate them to walk. It is difficult to imagine either of these evolving back to a terrestrial life. But our clumsy legs required wide and strong muscle anchors, so we expanded our pelvis while other aquatics shed theirs. We went far from our birthplace, but we never lost our ticket home.

Basic Aquatic Sex

Shallow water sex was no problem. In our early transition, prehensile feet gave secure footing on the pebbly bottom. As we became more aquatic, we coupled while standing neck-deep. But our pelvis deepened to support swimming muscles, classic rear-entry sex grew awkward, finding a mate became difficult, and eventually we arrived at the Wench Revolution.

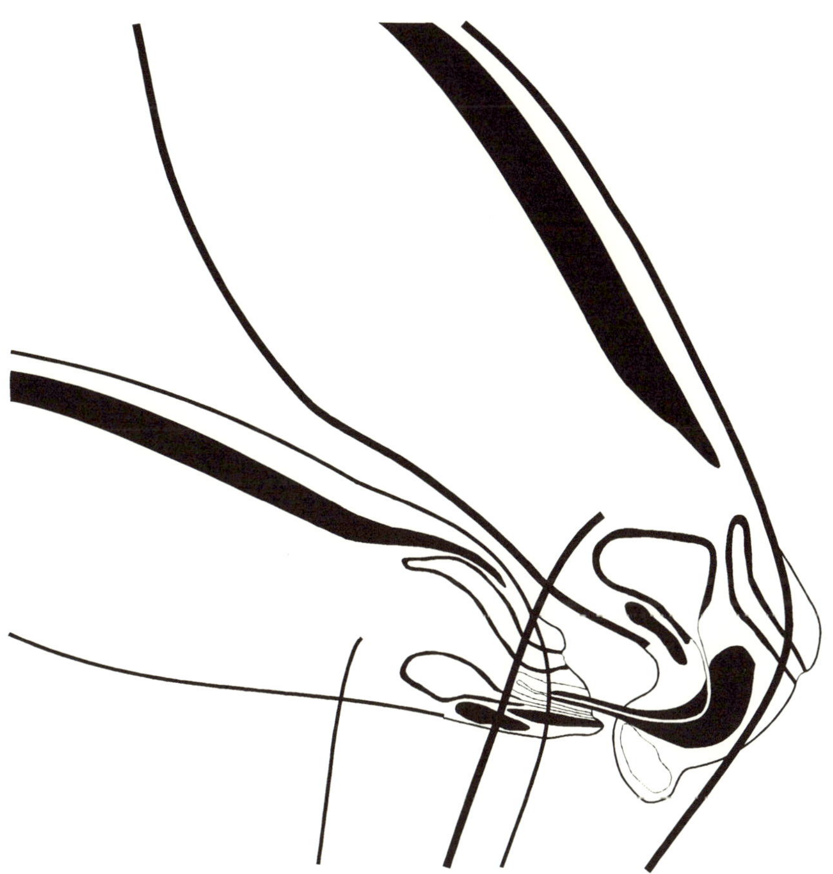

Figure 3: Primordial Rear-Entry Sex

During our early aquatic time, sex had the same casual air as on dry land. Near each available female, half a dozen males awaited their turn. Her fertile status brought them sloshing from a hundred yards away, clutching their sea-cucumbers. For several days they would remain near, enticed by her pink swelling and estrous odor, copulating in turn.

Throughout our initial aquatic transformation, we used the typical primate approach. The female with an itch would solicit as many males as she needed to scratch it. Facing away from each, she bent forward and presented; he could hold her with one or both hands and enter her from behind.

Bending over was a signal more than a physical requirement. Her early-aquatic pelvis was as shallow as our terrestrial ancestors'. As with modern chimp females, during coitus her back could easily touch the male's chest.

Over generations our bodies restructured to meet aquatic needs. The pelvis deepened and buttocks grew to give kicking leverage, carrying the vaginal opening farther forward, more distant from the penis. Rear-entry sex started becoming awkward long before our pelvis had fully deepened.

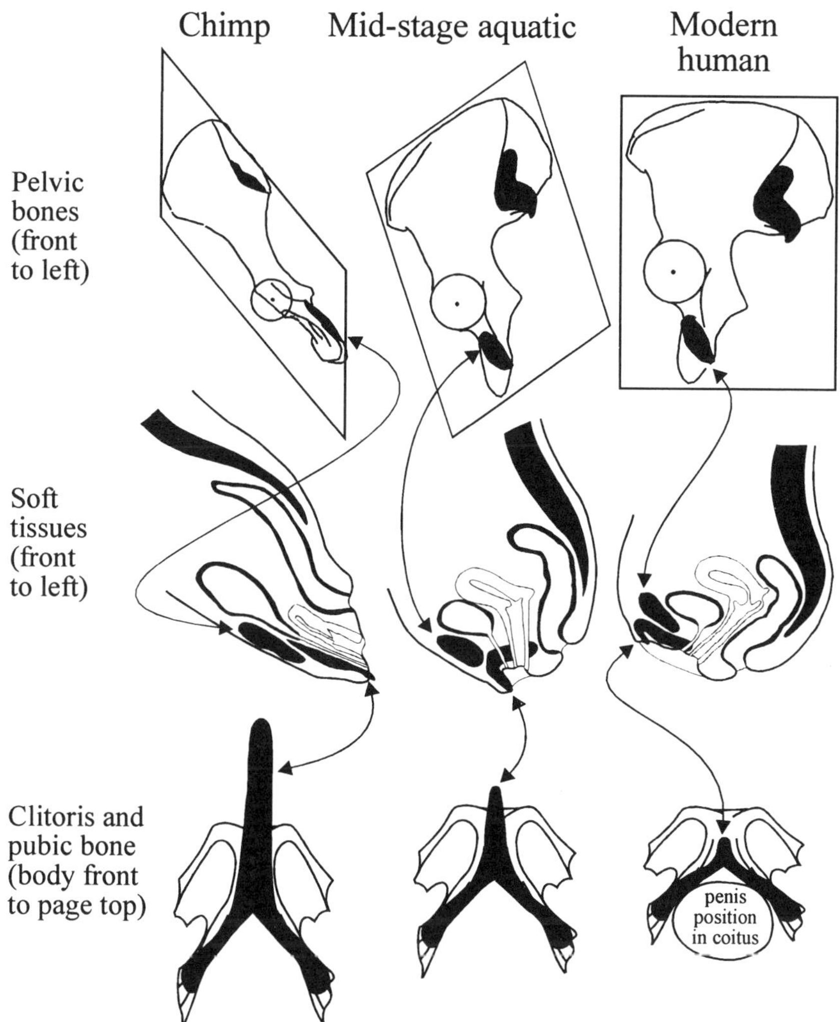

Figure 4: Deepening Female Pelvis

I'll often refer to a "mid-stage" aquatic in discussing our change to frontal sex. The figure illustrates this using female genitals, those being most relevant to my case. I show pelvic bones going through a simple linear distortion, along with corresponding soft tissues.

Notice the clitoris. Some reptiles have paired penises; in mammals these evolved into one Y-shaped organ, retaining separate anchors on each side of the pelvis, with only the joined portion visible. The clitoris grows from the same foundation and has the same Y-shape, straddling the vagina. As our pelvis deepened and swung forward it carried the clitoral legs along, while the clitoris' joined part shortened and continued to point backward. Even some sex-education texts dismiss the clitoris as merely a small patch of sensitive skin, but it is a mostly-hidden erectile organ of considerable size.

When its target retreated between female buttocks, the penis followed; men with shorter organs were less successful at impregnations, and the human penis evolved greater length. We evolved a large penis not because it gave coital orgasms to women, but because anything shorter was unable to cause pregnancy.

A deepening pelvis was only one of their problems. In knee-deep water the parties stood on the sandy bottom, as once they stood in the forest. Deep-water stability was trickier, because she couldn't crouch and brace a hand on the ground. Deeper water carried more of their weight, so footing was no longer secure. If they didn't pay attention, any little wave would knock them off their feet; whenever they got swept away ... they got swept away.

Males had to handle the situation. If a female (facing away) had nothing to hold, the male could at least hold *her*. But her hips were becoming smooth and rounded, larger and deeper to support swimming muscles, with subcutaneous fat replacing body fur. Males were losing their grip.

Copulating men are not famous for precision aiming. Penises are often completely withdrawn from the vagina by accident, perhaps reinserted in the wrong place, occasionally ruptured like water balloons by a misdirected thrust.

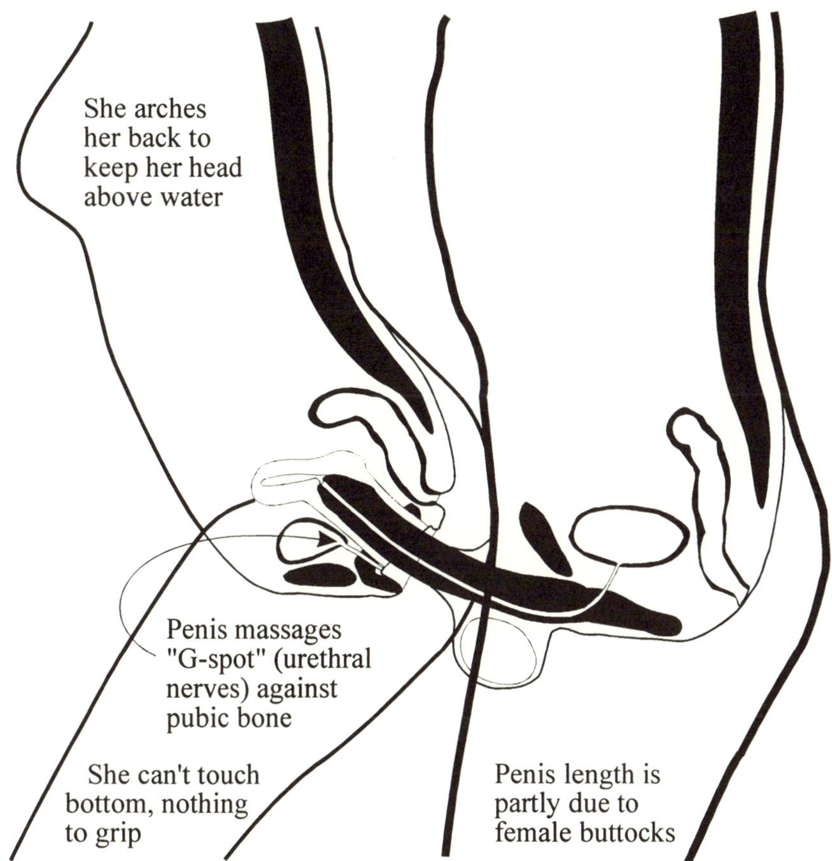

Figure 5: Rear-Entry Sex - Mid-Stage Aquatics

To corral unruly penises, the female evolved an extension of the labia minora. This apron-like flap (the "tablier egyptien") is still present in Bush females, drapes down from the front of the vaginal vestibule, and funnels the penis into the vagina during rear-entry sex. That is, when males had trouble aiming, females with a more-cooperative target reproduced better than their sisters. Bushpeople appear to illustrate human genital evolution by retaining traits of this transitional aquatic. (In a later chapter I will discuss the Bush male.)

Estrus Concealed

More than clumsy sex, uninterested males frustrated females. Over time, men evolved the ability to be aroused without estrus. We state it as "women are always sexy," but the change was in males. To see what the change meant, imagine a world where men weren't interested in sex until a bell rang. Scantily clad women could pose for men, kiss them, touch them, with no effect until the bell sounded. Once it rang (with or without naked women in view) males would be instantly aroused and ready.

Estrus was just such a bell for our ancestors, triggering male sexual interest by smell instead of sound. Immature or non-estrous female chimps may crouch and present, but males ignore them completely. Even a female with genital swellings evokes no male interest unless he verifies her estrous odor. Imagine a woman's outrage if, after she has invited a man to her bed, he demands such a test before agreeing.

For a hundred million years estrus guided reproduction, called the pairs to sex. More than female odors or genital swelling, estrus is an *interaction* between sexes. Female signals and male responses belong to the species; each animal carries both sets of genes. Her odors and behavioral signals evolved only because they evoked a male response; he evolved to notice both only because they led to fertile females.

Even heavy rain on our terrestrial parents left undersides dry and odors intact, but the aquatic move washed away estrous odors. Splashing shallow water weakened genital smells; deeper water drowned them. An aquatic ape male could reproduce only if he could get excited about a female without smelling her.

Arousal is *possible* for a male without a female crotch wrapped around his nose. Males (including primates) routinely masturbate without females. Our aquatic forebears had to be as indifferent to female odor when they copulated as they were indifferent when they masturbated. (Some would say men *are* masturbating when they copulate, near as matters.) Males could reproduce only if female odor was not a *requirement*, however much they might like it.

Estrus was not yet lost; males remained interested in female scents and females continued to generate them. In water estrus was only hidden, and remained deeply-embedded in our genes. Any estrous female who dried off would gain an immediate following. But men evolved to not *need* odor for coitus.

Dispensing with odor (a seemingly trivial change) was the first fracture in our sexual structure, where we now stand in the debris. To see a female as sexually available without estrus is to see her as an inherently sexual object, constantly enticing whether she wishes or not.

Perhaps more disruptive than masked odors, we also destroyed visual signals. Neck deep in water, no female could crouch and present her rear. Males couldn't recognize receptive females by the pink flash of swollen genitals or visual appeal of an offered backside.

The female rump's erotic effect is hardly news; the sight draws men instinctively, and women inherit the same instinct. After four million years of frontal sex, we still respond to the ancient signal. In other primates, a female's crouch-and-present instantly invites sex; females use this to deflect male attacks because the response is quick and certain. Females can be quite insistent; Goodall reports females not merely showing their behinds but pressing them against the male. Orang females occasionally bounce their genitals on male heads to get attention.

Water shrouded this basic signal. A soliciting female showed a male only the back of her head. Sexy this is not. Because the aquatic male couldn't recognize her instinctive pose, blatant insistence became the minimum requirement for a horny female.

An instinct was born here that we can call "an eye for beauty", with more discussion to come later. Any male could mate if he found a female sexually attractive by merely recognizing her as female. Primates are intensely face-focussed and easily read others' expressions. Our cousins can recognize another's sex from their face, but appear to not measure sexual attractiveness by facial features.

With scent drowned and visual cues missing, males began to find females attractive based on faces alone. Females didn't change, but males began enjoying the female face because it was all men saw. A male who was more attracted to a female face sought out females; males who found no facial lure waited for females to solicit sex as they always had. Beginning with a crude instinct, successful males passed this sharpening eye for beauty to their sons. Males started finding all females attractive by simply being female.

The female's lost odors and less-visible solicitation both meant that each act took negotiating. Males were oblivious to a female's horny condition, so she had to seek out partners to get things started. Her swelling and odor no longer collected males automatically. Even if males were hanging around, alerted by sexual activity or drawn by her femininity alone, she still had to solicit them in turn. She had to persuade each that she did indeed want her itch scratched right now, though she could supply none of the stimuli he'd evolved to expect. For the first time females wondered what it took to get males interested.

Dominance

Sex became elusive when we were aquatic. There were fewer copulations, longer intervals between swivings, and more males never got any. Our aquatic lives began to focus on sexual competition through personal dominance.

Longer intervals separated couplings because each took longer to negotiate. On land, surrounded by erect males, a female could get her itch scratched a hundred times a day. When simple bending-over was enough, each coupling took negligible effort for an easy orgasm. But when each aquatic male had to be explicitly notified, she had to spend at least a few minutes on him instead of a few seconds.

There were fewer copulations because females got tired of this. When she could have easy orgasms, she'd take them even when only a little horny. But when aquatic males needed clear invitations, a smaller part of her cycle found her horny enough to insist. An aquatic female managed perhaps a third the couplings her terrestrial mothers had earlier enjoyed. She might easily decide it was simpler to get one male's attention, and let him scratch her itch only a few times — to settle for what she could get without too much bother.

Decreased sexual activity gave a population of frustrated males, and this normally creates more-aggressive males in following generations. Frustrated males compete harder for sexual access; the more-aggressive pass on their genes. This basic mechanism created male aggression long before the first prosimians, in a pattern so prevalent that maleness and aggression are nearly always linked.

Dominance and aggression popped out of our behavioral toolbox to distribute suddenly scarce aquatic sex. Competition became *important* because sexual access was hard to attain, but water impeded movement and made competition *difficult*. Water buffered individuals, limiting the interactions that once built our loose hierarchy.

On land aggression is physical; the threat of force dissuades a rival. If a frustrated male sees another near an estrous female, he might shoulder the rival aside or try to monopolize her. If a male can intervene and recognizes intentions, he needs only seconds to chase away an interloper.

In water we could identify others but couldn't always judge intentions. A female exuded no scent and displayed no swollen genitals. The male she solicited might be oblivious until she pushed her bottom against him. Other males ten yards away might never catch on at all. Even if a second male recognized her intent and rushed to intervene, he might arrive too late. Chimp copulation takes only a few seconds and we were not yet excellent swimmers. By the time an interloper sloshed through ten yards of neck-deep water, coitus was over.

On land, body size had always backed up threats, automatically implying power. Adult males' greater size frightens smaller males and females. But water concealed body size by putting everyone's chin at sea level. Nobody looked up to anyone else. On land it was painful to have a large person jumping up and down on one. In water no one carried any weight at all. A larger body on land meant longer legs and a faster run. Litheness in water, not size, made a faster swimmer. Size didn't matter, if you could elude those you irritated.

The early-aquatic male was trapped: He could not evolve any combination of size and speed to make dominance work. A larger body might make him more imposing, if he could rise up to show how big it was. But a bulky body kept him from moving quickly to intervene. Some pinniped species (seals, sea lions) have the greatest dimorphism of any mammals, with huge males competing savagely for brief sexual seasons. But they fight only on land, where size helps.

Aquatic aggression had no benefit because intervention didn't work. The male attempting to monopolize females would quickly exhaust himself rushing back and forth between suspicious looking couples. Without a payoff for aggression there can be no evolution toward it, so water disconnected us from aggression and dominance as it had disconnected us from estrus.

Status vanished when dominance drowned, creating anarchy. Status comes when dominance is recognized without individual threats: If enough individuals fear me, the whole group will be polite. Male status depended largely on physical power (though female support was important in ape species), while female status depended on friendships and family. When combat became valueless in water, the threat of combat vanished too; as male *dominance* was disconnected, the *perception of dominance* went with it. In water no male had any status at all.

We lost our social form when we became aquatic; we saw no dominance, no status, no hierarchy, no leadership. We plunged into anarchy as we entered the sea, our already loose society dissolving on the way. We started down the beach as a troop, but by the end of our initial transformation we were just a mob.

Impediments to movement give freedom to residents. Greece spawned democracy partly because of geography. With hundreds of islands, no single ruler could control all activity, no one view could predominate. Islands have always sheltered pirates, insurgents, smugglers, religious cults and other independents. Dense jungle and mountainous lands give the same result, and guerrilla wars work best where movement is difficult.

In water, dominance slept. As aquatics we were free to ignore the once-alpha males; unable to interfere, they had no path to power. Anarchy left us confused, vaguely isolated. Relationships might be prickly in a group, but they were clear. You knew who outranked you, though being outranked was no fun. But after we became fully aquatic, generations might pass between fights, aquatics never knew who was stronger, and all lived in a leadership vacuum.

Early in our aquatic transition, dominance depended on our location. As we entered the water for a few hours, we dropped hierarchy like a towel, and picked it up again when climbing out. An alpha male could not enforce his will in the water, but all knew his status; an underling who transgressed while swimming paid for it back on the beach. It was like considering a minor sin, knowing you'll answer for it when your parents get home.

As aquatic hours slowly took larger parts of the day, land-based dominance faded. Near the end of our initial transformation we spent very little time on land. Male rivalries formed no coherent ranking, but degenerated to apparently random outbursts. Alpha-male aggressiveness was a small danger on land for a small part of the day, as safe to ignore as a substitute teacher.

Sexual Selection

Female mammals measure mates partly by status, but aquatic life stripped away their yardstick. Females were left with no good way to evaluate males — a predicament persisting today. Choosing sexual partners became highly subjective, often leaving friends gaping in disbelief.

Males play a minor role in mammalian sexual selection. A male may limit female choice by chasing other males away, holding a territory, or herding females into harems. His goal is to be her sole recourse when she needs her itch scratched. Modern men, when they limit women's freedom, are at least partly driven by a harem-keeping instinct.

Another male option is displaying dominance, often by combat where winners gain self-assurance. Each male's anger feels genuine, each fight is real, but it's a winner's self-assurance that gets female attention. Females see self-assurance as status, and are drawn to victorious males.

Male rivalry seldom injures females, though male harem-keeping often does. Hamadryas males savagely correct captive females, who occasionally die from injuries. Male chimps sometimes attack females, yet the females often remain (apparently voluntarily) with males who hurt them. When male gibbons are *too* dominant, however, females avoid them.

Whether holding harems or fighting rivals, males are sexually passive. Even if a female can choose only one male, he must still await her consent. A Hamadryas baboon male is the most brutal of harem-keepers, yet he cannot rape imprisoned females. He can only hold them captive and hope for some action.

Females prefer dominance. A dominant male's child carries genes of one who not only survived but prevailed, and genes of a mother with the instinct to make that choice. In any species where females can choose, both dominance and an instinct to select it can quickly evolve.

Female preference matters only because males are sexually passive. If males routinely raped, feminine preference would be meaningless. Rape's impossibility (for virtually all species) means male dominance and female choice evolved as complementary traits, each sex's innate behavior rewarding the opposite sex with superior offspring. Males fight each other because females can choose; female choice improves the species because males fight.

Male aggressiveness remained strong but ineffective in the sea; female preference remained strong but blind. With males unable to gain ascendancy, females could not detect superiority.

Individual preference had always played a minor role in our mate selection. Female baboons groom or copulate with a favorite, even at some risk, bypassing the hierarchy of sexual privilege. Male and female chimps often strongly prefer one or another partner. Not yet love, this is surely more than mere cooperation.

Absent other cues, friendship was an aquatic female's only yardstick for mates. In estrus she had an itch, and needed males to scratch it. With movement slow and soliciting awkward, she limited herself to fewer partners. Her personal preference limited those partners to friends.

Friendship always guides primate companionship, and if dominant males don't interfere, friendship also determines sexual activity. The water barred outside interference and freed each couple to follow their own interests. Fortunately, the common interests of males and females tend to create babies.

Next Generation

Food brought us to the sea and made us swimmers; childbearing cut us off from the land. Pregnancy's burden grew as mothers and infants evolved for our new life, and we abandoned the beaches when our burden grew too great.

Every early-aquatic mother suffered the same discomfort modern pregnancies cause. Her swollen belly pressed down on her bladder in a rotated and broadened pelvis, and dragged her forward when she stood.

Pregnancy was little problem for us as quadrupeds, but in newly aquatic bodies it became increasingly awkward.

At evening we straggled up the beach to sleep. Each night, increasingly streamlined aquatics again became terrestrial. With spines aligned to legs they tottered like penguins over the sand, and made nests for the night. Each night swollen females waddled with the group. Aside from their fur, they looked quite like modern women.

More than anyone else, pregnant females liked the water. It was a joy each morning to trudge down the beach and become suddenly weightless and free. Day's end, with its obligatory climb up the sand, was wearisome. A female surely was first to nap in the sea, late in her pregnancy, holding up her head as she dozed. Generations later another female slept there all night, and so began our fully aquatic life.

Childbirth itself posed small danger. Our brains expanded during our initial aquatic transformation, but not enough to cause much risk to the mother. Her own mother assisted her, as both human mothers and chimp females now help daughters in labor. A newborn wouldn't drown as long as someone lifted the infant into the air.

Sometimes assistants forgot, or mothers slept, and babies died. In our aquatic transition, infants began evolving three survival aids: A drown-resistant metabolism, self-sealing nostrils, and fat for insulation and buoyancy. Still present, they distinguish us from other primates.

Many diving animals turn down their metabolism underwater. In humans, children show the strongest response, and often survive immersions fatal for an adult. A University of Michigan study mentioned an 18-year-old who survived under very cold water for 38 minutes. For most, nothing in modern life exercises this metabolic switch, and by adulthood it fades. If a child dives daily from infancy the response should grow stronger, not weaker.

As aquatic infants rode their mothers' backs jockey-style, they unthinkingly stopped breathing when her movement submerged them, resumed when they surfaced. Holding its mother's upper body and shoulders, the infant had its head next to Mom's, in the air when hers was. Breath-holding and the metabolic switch continued during naps, as a modern baby sleeps on in spite of being moved.

Waterproof noses gave babies a second safeguard. Baby noses seal like eyedroppers, small and inflexible, too narrow to pass both air and water at once. No water can enter unless the baby breathes in, human babies instinctively breathe through *only* their noses, and they stop breathing when water is present. As an infant grew and needed more air than eyedroppers allowed, it developed the controllable nostrils of an adult.

Marine mammals typically have self-sealing nostrils. Elephant seals napping on the beach must wake every twenty minutes or so to breathe. Whales must intentionally open their nostrils when they come up. The only modern primate who routinely swims (the proboscis monkey) is named for its elaborately-covered nostrils. A relaxed nose, for any marine mammal, is closed.

Our protruding nose evolved as the simplest way to let nostrils seal, by elongating holes into slits. A sphincter is complex, but a single muscle pulling sideways can open a slit, and a closed slit is the nostril's relaxed state for aquatic humans. Typical nostrils of modern adult humans are elongated front-to-back, echoing their ancient form, though now permanently open. Look down your nose at your image in a mirror. Your nose sticks out not for beauty, but to lengthen and flatten your nostrils.

Our third protection was fat, for warmth and buoyancy. Aquatic mammals may grow vast amounts of blubber; ours needs no special explanation. But our brains (also made of fat) are unusual and crucial.

Human brains first evolved larger in water because they were buoyant. Half a billion years ago fat cells began evolving into nerve tissue and later into brains. These cells (like other fat) are lighter than bone or muscle, and infants with larger brains floated with their heads a bit higher in the water. They needed less maternal attention, spent less time in metabolic shut-down, died in that state less often, and passed the trait to their children. In aquatic humans, brains began growing because fat floats.

Our skulls also grew lighter, with thin and delicate bones. Heavy bones had no value, marine animals are always calcium-short, and light bones gave a buoyancy edge, starting in infancy and continuing through life. Light-skulled big-brained infants grew to light-skulled big-brained adults, with infant needs leading the way.

Brains expanded only minimally; we needed only enough fat to lift noses into the air. A modern baby floating on its back breathes easily because its nose is higher than necessary. Aquatic infants floated a little lower, but high enough. Brain size remained stable in aquatics for several million years because it had accomplished the necessary function and needed no larger size. There will be much more about brains, later on.

Complete Adaptation

Every evolving aquatic faces three hurdles: It must copulate, give birth, and sleep in water. (We assume it finds food and doesn't drown; dogs in a pond can do as much.) Primates have no trouble with the first two; modern humans couple in hot tubs, and women can give birth in water on a whim.

Our final requirement for an aquatic life was that we sleep there, and females led the way. With pregnancies' burdens and expanding body fat, our nightly waddle up the beach grew increasingly awkward. Pregnant females often dozed in the water, occasionally all night, though they had slept on the beach when young. New mothers, reluctant to abandon pregnant friends for the beach's lonelier comforts, began staying in the sea with their infants, first as seasons permitted, finally all the time. Infancy shapes adulthood, and soon no one slept on the beach.

So we became a new people, never walking on dry land. Children paddled into shallow water and stood there while parents fished. They mirrored their ancestors, who as children played with sea-foam while their parents combed the beach. Habit now confined them to water as once it confined them to land.

Once we no longer slept on land, we needed no fur for babies to clutch jockey-style. A baby primate clings to its mother, rocking to sleep as she walks, waking to suckle when she stops. When we plodded up the beach each night, infants rode their mothers' backs; returning to the sea in the morning, each still clutched Mom. Her hair also kept them warm in chilly nights.

Infants gripped no hair below the neck, once we stopped walking on the beach, and body hair had no value in the sea. But hair always hindered, holding water like a mop, slowing those whose survival depended on speed. When infants no longer needed Mom's body hair to hold, fur dwindled to let us swim better.

Head hair grew long in water because that's what babies held. An infant might grip fur on Mom's back when riding down the beach, but as she moved out to neck-depth, the baby moved up. Babies wanted their heads out of water, and grabbed Mom's scalp hair — the highest fur they could see. Her hair floated around her in the water, instantly available for her baby's clumsy grip. He might pull hard but usually didn't need to (he floated very well), and she pulled him along as a big boat tows a skiff.

Long scalp hair evolved early; we probably stood at neck-depth as soon as our fishing took more than a few hours a day. When we started sleeping in the water, it was already ancient. It was a handhold only in the water, for if the infant tried to hang by its mother's scalp hair on the beach, she'd shift it to more comfortable fur.

Beards were another of children's rest-stops. Common in primates, beards evolved fuller in us when aquatic life made them useful. Beards grew on men because they serve a specific sexual purpose, but that discussion can wait. Having evolved, they were handles for a tired child.

Breasts

When we became hairless, an infant couldn't maneuver across its mother's slick rib cage to find the nipple, so breasts evolved as flexible tubes to pass nutrients across the gap. Several primates have breasts, enlarged only during lactation, and normally lying flat against the rib cage.

Aquatic mammals often have perceptible breasts, and the dugong and manatee (both called mermaids) attracted sailors' attention with well-developed breasts. Both species hold their young in their flippers to suckle. Seals and whales have no breasts, and their young nurse while either swimming or lying next to them. The rule for aquatics seems to be: If mothers hold their infants, females evolve pendulous breasts. Human breasts followed the norm for aquatics, not the norm for primates.

Young breasts may be firm and round, but they evolved to be long and limp. If never restrained they may hang to the waist, and in some cultures women stretch them so they can feed a baby slung on their back.

I suspect that breasts in our aquatic ape were not pendent but buoyant — they floated up, not down. Like brains, breasts are largely fat, perhaps evolved partly for insulation. Fat also floats, and our aquatic mothers probably evolved fatty breasts (instead of merely long ones) to keep the business end near the water's surface, constantly available for a floating child.

(It's difficult to test *everything* I'm proposing. I considered putting an ad in the paper:

> **Wanted: women with pendulous breasts,**
> **willing to be photographed floating**
> **naked in salt water. No weirdos please.**

But some would misunderstand.)

With a very young baby the mother might stand in shallow water, cradling her infant's head as a modern mother holds a child. She could also float on her back or swim slowly backward to elevate her breasts a bit. Both the manatee and dugong suckle in this pose.

An older child could find the nipple on its own, and would do so for most of its nursing years. With slowed metabolism, a child could float face-down in front of its mother, one hand on her hair and one on her breast, nursing for minutes between breaths.

Our metabolic switch and breath-holding worked from birth. As an infant floated next to Mom, each time she ducked under for a fish she carried him along. With each dive he shut off as easily as you flick off the light when leaving a room; with each surfacing he breathed again.

In my mind I see them both; she finished nursing him and is ready for lunch herself. She grasps his forearm and swings him around half floating, her other hand gathering her hair, and she pushes the child's hand into it as girls clip ponytails. She swims with a living hairclip bobbing behind, supported by her back and cradled between her shoulders. He sleeps and wakes like a baby in a stroller, submerging and surfacing unaware, protected from danger by her body and his instincts. Mothers in Tierra del Fuego still swim this way, with children floating along, holding their hair.

Swimming For Two

Females always dragged infants along. Growing children paddled over to other adults to play, as young chimps do, and by age five could keep up with the group. But before it can go visiting, a child depends on Mom. Any adult *might* help a child, but only a mother *inevitably* finds herself in that position.

Swimming ability had no time to improve after childbirth, so females *all* needed to be excellent swimmers. Women now predominate in long-distance swimming. This amuses some anthropologists, who mention it parenthetically while busily proving male superiority on land, blind to female superiority's value in water.

By five months an infant chimp rides on its mother's back, and an aquatic baby started exploring in earnest. Since birth it had been carried on its mother's dives, seeing underwater, holding its breath. By five months it ate some solid snacks, watched her collect food from the bottom, started reaching out to do the same.

 Explorations wouldn't take it far. Mothers and other adults stood in neck-deep water, and that's where babies played. Security for an aquatic infant (just as in the trees) came from a firm grip on an adult. Since mothers with younger children would have less time for older ones, a two-year-old's anchor might be an older sibling or adult friend. Males might handle a couple hangers-on, and hominoid adult males often help children, particularly when trying to impress the mother.

A child had only two resting poses: Embracing an adult, or gripping their hair. Both came directly from terrestrial clinging, and infants embrace their mothers while clutching fur with hands and feet. Our fur dwindled, feet lost their grip, and embraces grew more common. Aquatic infants probably clasped with legs while clutching hair when asleep. Long before frontal sex evolved, aquatics spent most of their days in constant frontal contact.

Aquatic children lived in a world of limited aggression and many friends. Where baby chimps see adults of varying size and danger, aquatic children saw furry places to stop and rest — a world of hairy stepping-stones, where all adults were useful.

Weakness In Arms

Arms and upper bodies evolved for quickness in the water; our arms grew frail but longer as legs took over swimming's role. Unused power did not wither, as bedridden people become weak; instead we traded strength for speed.

Calcium is scarce in the ocean, so marine animals conserve bone. Whale pelves withered because they were unused, but also because other bones needed the calcium. Our body's ability to quickly extract calcium from idle bones may be a relic of our aquatic calcium shortage.

Gorillas, chimps, and orangs are the most-powerful primates, and our common ancestor shared that strength. Pre-aquatic hominoids could have cheerfully dismembered modern humans as easily as today's chimp or orangutan might do. Yet *Australopithecus* (likely our aquatic ancestor) was slim and light-boned (though with a pelvis so broad some call it pot-bellied), the mark of an ape caught in a calcium drought.

Strong arms *can* aid swimming, swinging wide in the crawl or backstroke, with the whole arm (not just the hand) oaring through the water. If wearing flippers, though, we often let arms trail and swim with legs alone, and we probably swam that way after our legs evolved for scissors kicking.

Our arms evolved for grasping prey at the expense of swimming. Our arms and hands adapted to strike like a snake from a coiled start. This gave us the distinctive human reach, extending the arm and hand straight to a goal. A human boxer jabs with a fist, a chimp cuffs with a sideways swing. The difference is so clear that we use it when humans mime chimps.

Speed came from a straight reach because water resists the sweeping arm. An ape flailing arms after fish misses every time. Our ancestors hunted by lying in wait, watching for a passing fish, holding their hands near the shoulders as a boxer holds fists cocked near his face.

Those born with faster arms ate better; over generations we evolved speed at the expense of power. Muscle attachment points moved in toward the elbow and shoulder, so a small muscular contraction would move the joint through a wider angle. This same geometry means exerting a given force takes more effort in a human than in a chimp, though each muscle fiber is as strong. We are weaker but quicker than our cousins.

Yet the design was badly flawed, because an elbow must move sideways through the water. (Put your wrist near your shoulder and try this, slowly reaching for something at arm's length.) At first the hand moves quickly, but halfway there the hand slows, while the elbow picks up speed. At the end the elbow moves much faster than the fingers. Clumsy, but the best we could do.

Our elbow limited arm length by limiting our speed. As arms evolved longer, elbows had to move farther sideways in the last stage of the strike. Sideways movement creates drag, so longer arms were slower. An eight-foot arm, say, would move so slowly in its final few inches that even the most lethargic fish could escape. So the Australopithecine arm length was a compromise between range and speed.

Fossils show that some australopithecines had arms relatively longer than ours today. This puzzles anthropologists because it suggests an evolutionary course-change, as the penguin now swims with organs that once evolved for flying. That is, *Australopithecus* as a hypothetical hunter-gatherer seemingly benefited from long arms in some way no modern hunter-gatherer does. But *Australopithecus* might make perfect sense as an aquatic; perhaps the optimal arm for an aquatic ape is longer than the optimal arm for a terrestrial one.

Speed And Legs

Our legs and feet reflect more compromises, evolving for both standing and swimming. Over generations, our knees grew close-set, legs gained power and finally surpassed our arms. While our arms evolved for quickness and dexterity, our feet evolved for propulsion and support at dexterity's expense.

We never became habitual deep-water aquatics, not fast enough to elude ocean predators; our safety lay in shallow bays and protected coasts. Our tree-living heritage made us uncomfortable when not holding on to something. Just as children held their parents, adults felt better standing on a firm bottom. Though we needed to swim we never stopped standing, just as in swimming pools people often stand on the bottom. Swimming with opposing leg-strokes kept the pelvis strong; standing preserved our weight-bearing physique well enough (as it happened) that we *could* return to land, when conditions permitted.

For swimming we evolved long webbed toes, like modern swim fins. To support that webbing, our big toe moved nearer the other four, instead of angling outward like a chimp's. Aquatic toes were longer than a modern human's, and probably longer than a chimp's as well. We have little evidence (small bones rarely fossilize) but long toes would match known leg lengths, and *Australopithecus* had relatively longer legs than chimps or humans.

Your toes and fingers each contain three bones (two in the big toe and in the thumb). Behind toes and fingers, one more bone (metatarsal in the foot, metacarpal in the hand) connects each digit to ankle or wrist. A chimp has fingers about as long as its metacarpals, toes as long as its metatarsals. You have fingers as long as your metacarpals too (i.e. the length of each finger is about the same as the distance from that finger's base to the wrist) but your toes are much shorter.

Many think our metatarsals evolved for speed; as horses rose up onto several toes to run, and finally a single toenail, so we rose onto our metatarsals, with the big toe slowly taking over. But our run more-closely resembles a seal walking on a hinged flipper, and being one of the slowest land mammals implies running was never our primary focus.

Speed suddenly became important when we returned to the land, stealing bird eggs and scraps from kills, and you can't run if your toes drag. As returned aquatics, we'd have been in the position I found myself at age eight or so, playing a Martian in a Boy Scout sketch. With a papier-mache' helmet, long underwear and swim fins, I was probably as authentic a Martian as you could want. Leaving the stage at the end of the sketch, I was nearly run down by stagehands carrying our flying saucer. If it had been a leopard instead of a prop I'd have been hominid pie.

Once back on land, with no need for long webbed toes, and powerful selection *against* them, our toes quickly became short to clear the ground. Running on the balls of our feet was faster than running flatfooted, so metatarsals stayed long. We didn't evolve long metatarsals for running, we evolved short toes to avoid stumbling.

So much for feet. In the next chapter we begin to talk about sex, then love, and treasures we carried unknowingly.

My sketch of aquatic ancestors is brief, adequate for a book concerned more with emotion than physique. Genitalia aside, we need little more to finish describing the aquatic male body; the remainder of our discussion focusses almost entirely on their mental state. Women's body changes were much more complex; their mental changes are the hinges where our evolution turned.

During our aquatic transition we changed as might any other mammal. We grew fat and lost hair, we evolved flippers and a streamlined shape, we gained salt-expelling tears and a metabolic switch. Eating with our hands sets us apart, more like sea otters than like seals. Our primate heritage condemned us to be only mediocre swimmers; our arboreal need for security tied us to shallow water.

Our fully-aquatic forebears were much like pudgy modern humans. Aside from shortened toes, we needed few body changes to convert back to terrestrial life. We could as well say a pudgy modern human is an adequate aquatic; our readiness to get back in the water confirms the place where we once evolved.

Chapter 3 - Wench Revolution

Face-to-face sex prevailed in the water, first as a practical solution, then as a habit, then as our sole approach. A true revolution, it brought a chain reaction of unforeseeable social upheaval. Frontal coitus began as merely convenient, not radical, and only a small aid to the participants.

Frontal sex is standard for aquatically copulating mammals. Whales and porpoises use it, and even beavers couple face-to-face, sitting in shallow water or floating on the surface. Manatee and dugong (elephants' cousins) copulate frontally; pinnipeds couple face-to-face in water, but always use rear-entry on land.

Apes use frontal sex more than do other primates. Young gorillas and chimps experiment with it, though adults prefer a rear approach. Orangs couple face-to-face routinely, because hanging from branches makes rear-entry sex awkward for them. When our rear-entry coupling became awkward in water, the frontal alternative was immediately available.

Converting to frontal sex marked not only the *habit* of face-to-face sex but a shift in instinctive preference. Women's erotic focus moved to their body front; embraces became sexual gestures for the first time, breasts became sexual objects. Most important, frontal sex changed our brains. In several stages, new instincts gave us confidence, religion, and the power to decide our fate.

Copulating face-to-face coincided with better reproduction, and we use it today because we are descended from thousands of generations conceived that way. We evolved the preference because of their success, not because frontal sex is automatically superior. My following descriptions are not about clear choices but about subtle degrees of fun and awkwardness, altering habits over millennia.

Awkward Genitalia

Rear-entry sex became awkward as we became aquatic. Muscular legs and a deeper pelvis made a female's hips larger. Her ancestor's vagina had pointed to the rear, meaning that her uterus was closer to her belly button than the same part of a modern woman. But as the *gluteus maximus* evolved larger, female buttocks grew. To keep them from blocking the male entirely, she had to lean forward when facing away, so her vagina eventually pointed downward instead of rearward, aligned along the body's central axis.

The deeper female pelvis gave the penis a more-distant target. No hard evidence survives, but penises probably became longer in response. As aiming grew difficult, the *tablier egyptien* evolved — extended labia majora, draped apron-like, guiding the wayward penis as it approached from behind.

Standing neck-deep, he had to hold her hips to maintain the rear-entry pose, but increasing fat and lack of fur turned hips into terrible handles. (At this point I call them "mid-stage" aquatics. This matches the middle illustration in the preceding figure on female pelvic evolution, and the figure on rear-entry coitus.)

It worked, but not well. She had to bend forward and might submerge during sex, but holding one's breath was routine. By this time she may have evolved enough fat to float herself, or he might hold her higher out of courtesy. I suspect she held her head above water, arching her back, and he angled his lower body to meet her.

Now imagine frontal sex instead, with a female embracing a male using her arms and legs, nuzzling, then coupling face-to-face. First, frontal sex took less work. Her legs (and largest muscles) held them together, in the same confident embrace she'd used since infancy. Instead of hands slipping on smooth hips, his hands and hers were entirely free. She could pull the vaginal funnel out of his way, he could reach around her thighs to ease his entry, or tickle her breasts floating in front of his face.

She no longer needed to either hold her breath or arch her back to keep her head out of water. Frontal sex let *both* lean backward, crotches pressed together and faces elevated. They could go on this way as long as they liked.

Most important, genital contact immediately improved, even though her vagina and his penis both had evolved for rear-entry sex. When her vagina had retreated between her buttocks, his penis had grown longer, with only part of it able to enter her. As her vagina aligned with her body center, and she leaned forward, he evolved an upward curve to meet her halfway. Unfortunately, his curve reduced her pleasure, because the penis no longer pressed directly on the vaginal front wall — the most-sensitive part.

When we turned around both these changes became advantages. With her legs around the male she could flatten the bottom of her pelvis against his crotch, with only tummies in the way. By facing him, she could let his penis enter down to its base, giving better vaginal access than when she faced away and bent over.

Her vagina's angle was a little higher than his normal erection. It's difficult to bend a penis down (with any luck at all) but easy to bend it higher. The penis tends to bend near its base, where the main erectile tissues diverge and attach to each side of the pelvis. Firm vaginal walls easily align an entering penis, pushing it in this case to a higher angle.

In rear-entry sex the penis' underside had massaged the G-spot, until the penis evolved an upward curve and lost contact. With the vagina now pushing the penis higher, the top side of the upward-curving penile tip pressed hard against the G-spot. Frontal sex gives a closer fit, G-spot pressure was greater and she enjoyed it more. She could stimulate an exquisitely sensitive area by tilting her pelvis, and he didn't mind a bit.

Finally, she probably gained direct stimulation of her clitoris. Clitoral shape varies widely among primates, and plays different roles. In modern women, penile movement stimulates the clitoris indirectly by labial traction. Frontal sex also directly compresses it between the female and male's pubic bones. This rhythmic compression closely mimics movements many women use in masturbation.

Though its size at any stage is guesswork, the clitoris had to evolve from the chimp female's large and protruding organ to the modern woman's hidden one. If her clitoris (like her pelvis and vagina) were at some mid-point, then it may have protruded enough for direct penile contact as we changed to frontal sex.

Clitoris aside, this was some darn good sex, and the advantages weren't lost on the bystanders. The farther we progressed as aquatics, the deeper our pelves grew, the more advantageous it became. By the time we were completely hairless, frontal sex was probably the only pose we used.

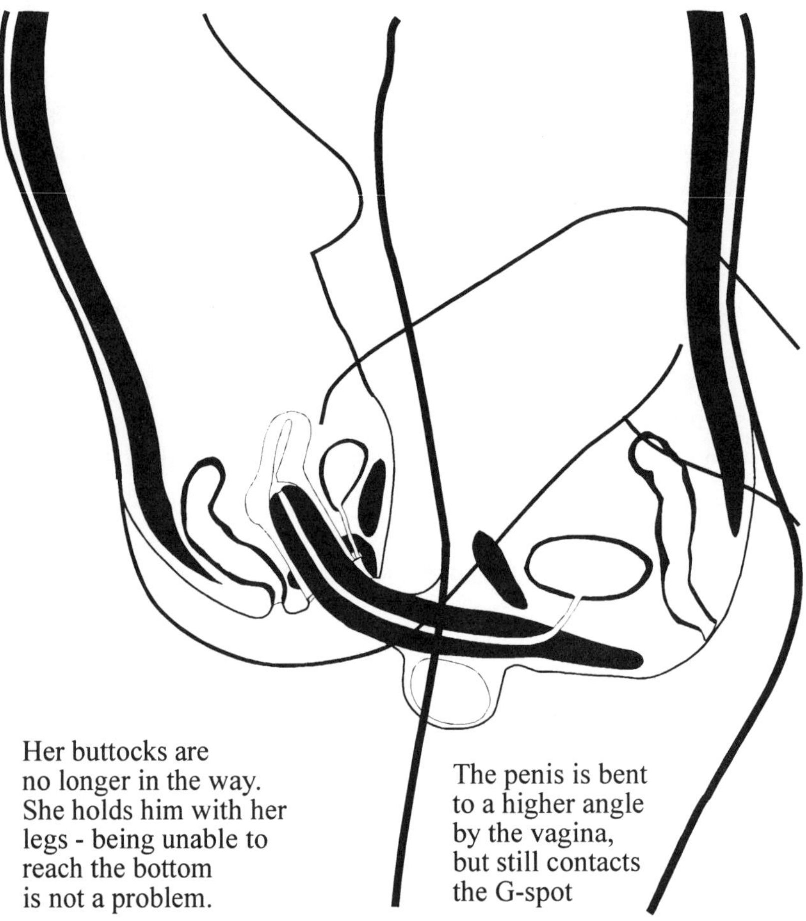

Figure 6: Frontal Sex - Mid-Stage Aquatics

The Process

But the frontal revolution didn't happen instantly, because frontal sex is scary. Facing an adult male invites combat; looking him in the eye is the most hazardous thing a primate can do. We recognize the threat instinctively; in conversation, men turn slightly away to reduce tension. For females, any contact with a male can be stressful, so female chimps often avoid adult males and during sex don't *normally* face them. We took this instinct into the sea; not all mid-stage females could tolerate face-to-face sex.

Frontal sex prevailed because rear-entry sex was flawed, and its problems left evidence in our bodies. If the labia minora evolved into a funnel, then females without that feature missed pregnancies. If penises grew longer, short males left fewer children; if they curved up, males with no curve lost intromission. Evolution happens when life culls failures; if our genitals evolved, we must have had problems.

In each generation a few aquatics played with frontal sex, as all apes will. At some point conception grew more certain face-to-face, as rear-entry sex became increasingly awkward. From that point on, any mind or body feature aiding frontal sex had value too, even if entirely trivial before. We evolved two themes at once: Our increasingly aquatic body urged frontal sex on us, and we simultaneously evolved features and mindsets to help frontal sex work better.

Beards

Beards encouraged frontal contact between males and young females, creating frontally-copulating adult females who otherwise would avoid that pose. Eventually females evolved a *preference* for frontal contact. Beards were an evolutionary bridge, spanning the time between advantage and preference, and they could do that only in a primate.

Other primates have beards, we didn't evolve them afresh; if we had no beards, frontal sex would have arrived by some other route. We did not first evolve beards in order to later evolve frontal sex, for evolution has no goals and doesn't plan ahead. The practical value of frontal sex *encouraged* beard growth in a species that already had rudimentary ones.

Relatively few youngsters held on to beards. Children past puberty stood unaided on the bottom; nursing infants stayed with their mothers. Young aquatics held on to beards when old enough to swim and young enough to need an anchor — between two and six years old, say.

Only females gained a reproductive effect. Each child grew up facing adult males for several hours a day while eating, resting or sleeping, desensitizing to the instinctively-felt danger of facing a male. As a cat raised with dogs may never learn to fear them, a female raised clinging to a man's front doesn't fear it as an adult.

Individual bearded males might benefit, because adult females would recognize specific ones as the helpful friends from her childhood. A female could generalize her childhood experience to favor all bearded males.

More important, breeding pools containing bearded males had more adult females who could tolerate frontal sex. By adding moderately brave (and now desensitized) females to the brave few who might copulate frontally anyway, a larger number of females could copulate face to face. Frontal sex gave more reliable impregnations, and beards increased the ratio of females who could tolerate frontal sex.   By out-reproducing smooth-faced groups, bearded aquatics prevailed.

As a desensitizer, beards marked a boundary between nature and nurture. Males enhanced a pre-existing *physical* feature because in young females it caused *behavioral* changes that persisted into adulthood. In each generation beards started over, assuaging young females' tensions. Nineteen generations of cats may live with dogs, but the twentieth can learn fear as easily as the first.

Beards had value only because female fears were instinctive. Beards did not cause fear of confrontation to disappear; we show the same fear today, and turn aside to defuse it. Once the pelvis deepened enough to make frontal sex work better, any attribute helping frontal sex would inevitably be favored. Beards were something evolution could work with.

Over uncounted generations, beards lured young females into face-to-face sex. During that period females evolved to prefer that pose, their preference to be discussed later. But it takes time for instincts to evolve, just as it took thousands of generations for our innate love of water to appear. So facial hair was an evolutionary *bridge*, a simple body feature carrying the behavioral "responsibility" for frontal sex until more complex emotional evolution could happen.

Now set beards aside for a moment.

Female Masturbation

Masturbation gives sex's first lesson. Elephants, porcupines, dolphins, cattle and rodents all masturbate. Those who don't masturbate when young often fail to reproduce when mature; auto-eroticism teaches sexuality as playing teaches coordination.

We often miss other species' masturbation because human observers look for human-like acts. A red deer masturbates by stroking its antlers through leaves; if you didn't know it, you might not see it. A porcupine female straddles a stick like a girl with a stick horse, pressing it into her crotch.

Primates need masturbation, and females need it the most. Chimp females acquiesce good-naturedly to sexual demands by virtually any male, even infants; male apes are rarely frustrated. But males satisfy females only during estrus, so female apes probably masturbate more frequently than do males.

A female ape can masturbate with no visible sign; women masturbate by leaning forward on a seat or sitting on a firm edge. Polynesian women masturbate by squatting with a heel pressed into the crotch. The *adductor longus* muscle connects to the pubic bone near the clitoris, and women (often unaware) massage their clitoris by tensing the muscle, thinking they attain orgasm by imagination alone. Ape females masturbate in front of an unseeing crowd, while male masturbation leaves no doubt.

Some see the clitoris as embryonic debris left over in the fetus, because sex hormones first appear when the genitals are partly formed. In this view, the clitoris is merely an aborted penis, and female auto-eroticism is an accidental function of accidental tissue. But female primate genitals evolved specifically to make auto-eroticism easy and safe. Natural selection ensured that primate females learned sex's first lesson, preserving masturbation at all costs.

Far from withering when female hormones appear, the clitoral complex grows more erectile tissue than the penis. The clitoris attracts feminine attention from infancy (baby girls reach inside their diapers to masturbate), and remains the auto-erotic focus for life. But the clitoris alone doesn't *prove* either masturbation-as-accident or -as-goal.

The vagina sends a clearer message, because its lower portion *discourages* masturbation. A membrane blocks it during years of infant exploration, and the lower vagina remains insensitive for life. An infant cannot penetrate her vagina; an adult gets no pleasure from it. The hymen and vaginal numbness together limit female behavior to prevent injury, and only masturbation provides a convincing reason.

Some argue the hymen evolved to help emotion-based pair bonds. Morris (1967) suggests copulation's initial pain slows youthful sexual experimentation, ensuring that the woman will not copulate until she has developed a relationship worth more to her than the discomfort. His argument fails because all primates have a hymen or its equivalent. In many primates, tissue fills the vaginal core, sloughing out just before the first heat. In some, the vagina re-occludes after each estrus or each pregnancy, and the female becomes a "virgin" again. If the hymen evolved to slow human first pairings, then the same structure evolved for completely different reasons in our near relatives.

Vaginal insensitivity is well recognized. The lower vagina has so few nerves that minor surgery there requires no anesthetic. Vaginal walls thicken in estrus, further numbing them. All primate females have virtually identical and numb lower vaginas, while external genitals show broad diversification. Sherfey (1966) takes this to mean that the lower vaginal structure is trivial, evolved without special selective pressure. But a design must be *extremely* important to remain consistent across hundreds of primate species, while adjacent tissue varies widely.

The hymen and lower vagina both limit a young primate female. Exploring her body, she soon finds the clitoris, and she understands from infancy that penetration gives pleasure to adults around her. When she attempts to mimic the adult act, the hymen stops her. If not a virgin, she finds numbness instead.

Numbness puts deep nerves safely out of reach. Some primate females solicit 50 copulations a day; Masters and Johnson report 20 to 50 orgasms when women use electric vibrators. A rough stick and this much enthusiasm could kill. But the hymen and numb vagina bar a female, because she explores tentatively. She stops when the next inch doesn't feel any better, and so limits her exploration to tougher tissues at the vaginal opening. Numb vaginas are valuable for primates because females with sensitive lower vaginas penetrate them for pleasure and injure themselves. Insensitivity's uniformity shows its value.

Coitus renders numbness trivial, because the male isn't numb. The penis passes the lower vagina, hits the G-spot, and she has a reliable orgasm. Male confidence leaps over a barrier evolved to block only females. And if she later tries to recreate the penetration, she again encounters an unrewarding zone that discourages her.

Our numb zone and G-spot complement each other. Only an animal with vaginal numbness needs to evolve some other sensory reward for sex. So bladder-nerve eroticism in primate females results directly from female readiness to masturbate. Females evolved the deep-touch response because the lower third was off-limits, the clitoris does not give reliable coital orgasm, and sexual enjoyment is vital to reproduction. Absence of sensation is not absence of function. Vaginal numbness has value only because primate females can handle tools.

Why didn't primate masturbation simply disappear? Self-repression can easily evolve, and my thesis will later portray humans as the ultimate self-repressed ape. But evolution arrived at masturbation instead of self-repression in over 200 primate species. Masturbation is so vital to adult sexuality that the genitals of *all* female primates evolved to help *safe* masturbation, rather than evolve for *less* masturbation. This is a remarkable testament to its importance.

Unwilling Seductresses

Beards, then, helped us convert to frontal sex because females masturbate. Beards encouraged young females to dangle at the one location where both sexes could benefit. Females were, in effect, led by beards to play seductresses, and it changed us all.

I have not yet used the terms "men" or "women", "boys" or "girls". These aquatics have come only a short way on the path from chimp to human, and I don't want you to misunderstand who I'm speaking of. Our aquatic ancestors were far less inhibited than modern humans. Any infant female might solicit an adult male, as Goodall says of wild chimps. My aquatic description may seem beyond belief, so much have we changed.

Masturbation remained a young female's only release, but her ancient options had vanished. Her world was hairy stepping stones spread across the water, with no rocks to press one's crotch against, no sticks to play horse with. The only solid objects were bodies of adults and other children around her. As captive dolphins masturbate against the bodies of human visitors, aquatic females worked with what they had.

Converting to frontal sex *used* beards and our cling instinct, but did not *demand* them. Whales, porpoises, manatees and beaver copulate belly-to-belly, and none have beards. We took a common (not universal) primate feature and used it in a new way. When rear-entry sex became awkward, and a few played at frontal coitus, the primate instinct to cling gave beards new importance.

By shifting clinging children to males' fronts, beards led young females to masturbate against penises. Female chimps often masturbate against the rump of another female, because (having no fleshy buttocks) their rumps are hard surfaces to rub on. After direct clitoral stimulation, women still prefer this method most. Penises, lacking a layer of subcutaneous fat, were the closest thing to a firm surface that females could find.

Dangling from beards, a young female could rub along the penis as girls ride stick horses, and mimic the adult females she saw around her. So beards and masturbation meant that young females were repeatedly enjoying the sensations available at a male's front. By puberty she was not merely *accustomed* to facing a male, she *sought* that pose. This ensured she would accept frontal coitus, and reliable impregnation from frontal sex paid off in population growth.

And males learned their own lesson from masturbating females. Males show little interest in non-estrous females, and the water masked estrus, yet males easily masturbate when thinking about sex. Now females were giving the same direct stimulation the male might give himself. Males were massaged to climax by young females, and for both this was a repeatable encounter, a lifelong commonplace experience.

The lesson repeated in each generation. Though a male innately sought estrous odors, recurring female masturbation showed him other means to erections. As masturbation desensitized females to the danger of confronting a male, it hyper-sensitized males to the sexual possibilities of non-estrous females.

The lesson was passed not culturally, but in the combination of beards and female masturbation. Once beards appeared, the lesson remained stable over perhaps millions of years, giving time for supporting instincts to evolve. We lived in the water so long that we now innately like to live near it; we were taught frontal sex so well we now prefer it instinctively.

In gratifying orgasmically insatiable young females, and ensuring their future frontal copulation, the penis had value long past the point where a male would go limp. Since the male who entertained young females gained an increased likelihood of their favors when they matured, a reliable erection lured future partners. So males evolved long ago what many now dream of — a permanent erection. The male who possesses it will have more female attention and children. The gene pool with such a trait has a better avenue for desensitizing females, converting them to frontal sex.

East African Bushmen still display permanent erections from birth to death. Both male and female (with her *tablier egyptien*) appear to have genitals frozen in time, living fossils from our aquatic days. The *tablier egyptien* suggests clumsy rear-entry sex, while a permanent erection helped the frontal persuasion. Their genitalia, male and female, may be a snapshot of a single moment in our past.

The aquatic penis also became thicker, giving humans the thickest penis of any primate. Many think it evolved to deeply massage the pubococcygeal (PC) muscle, or to indirectly stimulate the clitoris through labial traction. These both imply that women sought that effect, and selectively mated with well-endowed males, leading to more children by those men.

I suggest thickness simply gives strength. Thick penises evolve where strain is great; seals and sea lions who couple in water have thicker penises than those who couple on land. They (and some other species) also have a baculum — a reinforcing bone in the penis. Masters and Johnson noted that a broken penis rarely straightens, and usually precludes normal coitus. Under the dual stresses of aquatic adult sex and childhood masturbation, our penis evolved thicker because thinner organs suffered damage.

Greater size also signaled partners more clearly. Children respond better to oversized signals, and Signe Hammer noted how huge a father's penis looks to a little girl. Perhaps well-endowed men did attract more visitors, but young ones.

The Preferences

We like frontal sex because those who used it out-reproduced those who did not. Each pair coupled face to face only if they preferred to, and we inherited their liking. If frontal sex hadn't brought reproductive advantage, we wouldn't now care for it.

Face-to-face sex is never inevitable. In any evolving aquatic, when bodies become slick and rear-entry coitus gets tricky, frontal sex grows increasingly likely. When frontal sex begins giving some advantage in each species, those coupling that way pass on features that helped them do it. The inheritance makes coitus even easier for the children, who bequeath it again. Features of body or mind do not *lead* a species into frontal sex. They merely exist, indifferent to their own form as an antler or claw is indifferent, available for selection.

Once frontal sex brought benefit, natural selection focussed on it. Preferences sharpened over generations as more-enthusiastic couples used it, out-reproducing those who liked it less. The ability to *find* rewards in frontal contact became a competitive arena, but only because face-to-face sex was bringing practical advantage.

Palm Of Her Body

Frontal sex erotically awakened the skin of our body front, and skin contact there became a sexual thrill. No longer purely genital, sex for women began requiring embraces, too.

Long before our aquatic time, skin eroticism evolved in primates to support grooming and social contact. Primates groom to court, to soothe, to make peace. Animals who enjoy being touched will tolerate others' presence to get it, and groom to prime the pump. Goodall says that grooming's importance cannot be overestimated as a force in primate society. Humans share the need: In a controlled study (a century ago) 37 percent of human infants died when deprived of simple skin touching. Of those who were touched, none died.

Grooming over millions of years has shaped primate parasites, giving lice that freeze immobile when they sense fur being parted. Primates are so hungry for touch that even seriously injured animals (in one case, an infant with bone protruding from a fractured arm) become calm when groomed.

Female aquatic apes evolved body-front eroticism; their facing skin became focussed, sexual, electric, growing by degrees from classic primate skin hunger. Frontal skin eroticism grew *stronger* in women, rather than *began* with them, because women who liked frontal contact made more babies. We all inherit this forward-focus: When chimps groom, they start at the back; when humans groom they start at the front.

Frontal eroticism evolved from once-trivial variations. Infant primates clutch their mothers by splaying arms and legs across her body in a frontal embrace. When we became hairless in the sea, embraces lost their value, except when they helped desensitize females to frontal contact. Females who enjoyed frontal contact spent more time in the pose, became better-desensitized and out-reproduced the rest. Minor preference grew to a hunger by being selectively favored over generations, and the embrace itself became sexual for women.

Women frequently masturbate lying face-down on a bed, with a mound of cloth or a teddy bear to grind their clitoral/vulval area against. Not all women do this; only 5.5% of a sample of 1844 was reported by Hite (1976). Kinsey (1953) reports the same technique (including one three-year-old girl using it) but gives no statistics. Skimpy data suggests perhaps one in twenty women uses this pose, while no other primate does.

Frontal-contact masturbation does not specifically argue for *aquatic* frontal sex, but is likely in *any* evolutionary path to that pose. If frontal sex implies stress, then any pleasure found in frontal contact will help overcome a female's hesitations. Body-front eroticism evolved by subtraction, when more genitally-focussed women failed to endure a frontal embrace. Women enjoy body-front contact because our mothers coupled that way, and passed their liking on to their children.

Sex became a body-front event. Women thrilled to a full embrace, face-to-face, each with her chosen mate. Sex remained more than embracing; her automatic orgasm still worked, and other emotional rewards soon evolved, but body-front contact quickly gained importance. A female no longer coupled with only genitals — she coupled with everything that could be touched.

We had long been highly sexual, and our Miocene ancestor (the hairy promiscuous picnicker) lived in a riot of sensuous freedom. Close contact with its mother and mutual pleasures of nursing started each infant on a life of embracing, grooming, playing and group sex. Genitals were simply part of one's body, used in their place and time like hands, teeth or fingernails. An estrous female presents her genitals because they itch. She holds out to the male that which *she* perceives as sexual, not which *he* shows an interest in.

A woman's front became like the palm of a body-sized hand, wrapping her in a sexual organ. Many think the female front was chosen by men as her most appealing side; men wanted frontal contact for arousal, decided breasts were buttock-substitutes. If women resent wearing bras, men are easy to blame, but women unknowingly shaped their own ends.

As a woman's sexuality became less genitally focussed, she did what any soliciting female would: She presented to the male the parts that needed his attention. Her diffused eroticism made each woman see herself as a wholly sexual object enclosed in electric skin. Not a *more-erotic* female, she had merely changed focus. Women offer bodies for sex, while chimps offer rumps, because women are wearing what needs attention.

A chimp female invites sex by facing away and bending over; a woman invites it by facing her lover and baring her skin. More than a hundred million years of mammalian evolution tells females to face away and look over their shoulder when interested in a male. Seventy million years of primate evolution tells them facing a male is a dangerous pose. But for at least five million years a woman's electric skin has urged her to face her intended, to bare the palm of her body to be touched. We are all a little confused.

Frontal Enticements

By soliciting sex with skin instead of odor, women competed for men's eyes instead of noses. Women unknowingly blurred the boundary between their selves and their sexuality; aquatic men started seeing women as sexual possessions. In the ocean we started on the road to beauty contests and the fear of aging.

Eros, slowly diffusing, changed women's self-image accidentally. Skin hunger augmented orgasms in early frontal sex, helping hesitant females endure the pose. It didn't matter if they thought of themselves as wrapped in sexual skin. It mattered only that they couple face to face, and evolve appetites helping them to do so, because frontal sex brought more efficient reproduction. You need to eat when hungry because eating keeps you alive; you need not understand the biology behind hunger's discomfort.

Whole-body eroticism became vital in the sea. It was skin-hunger's function rather than skin's appearance, her wanting to embrace rather than thinking herself embraceable, that aided each female's reproduction. As frontal sex continued, the automatic orgasm faded (discussed below). Skin eroticism became one of several features slowly replacing orgasms.

When a woman wants sex with a man she faces him; frontal eroticism shapes her response. She holds out the palm of her body, electric with the wanting to be touched. Starting in the water, women sought sex for their whole bodies. Each woman held out the parts needing stroking, saying in effect, "Here, please touch this, but not too quickly."

They did this to ask, not to compete, as a hungry child asks for food by pointing to his stomach.

For millions of years women asked this way, guided by skin hunger to give men this single cue. Men didn't understand electric skin, saw not skin eroticism but whole women, and courted the women they saw. Women competed for attention, and mated most-readily with those men who could see the sexual prospects. When women solicited sex with skin, they bred men who saw what women presented as sexual.

Women display skin and shape to compete against what other women offer; men value women by criteria women selected. A woman has frontal enticements because *she thinks* she does, not because men tell her so. But she competes with those enticements because other women also want attention. An orphan in a crowd, like it or not, must compete against his neighbors with more blatant pathos. Men's modern economic and physical dominance permits them to push female competition often to brutal extremes, but men judge the features women unknowingly choose to present.

While a chimp female attracts with estrous odors, a human female solicits with appearance. No matter how snaggle-toothed or malformed, each chimp female takes her turn as the center of sexual attention, knowing she will shortly be ignored again. Human females have no time of clear availability, no balancing time alone. So all women must continually display what feels to them like a sexual organ, encircled by men constantly looking.

A female chimp owns her genitals; she uses any penis for her satisfaction without ever becoming a possession. A woman seeks not only a man's genitals but also his embrace to scratch her local and diffused itches. She may not intend to be a possession, but neither she nor the male can easily distinguish her skin from her self. When men and women couple, men often think they possess the person beneath her skin.

Women's view of themselves had minimal effect in water. When estrous odors were masked, beards and young masturbating females helped males evolve constant sexual interest. Until we returned to land and men became worth competing for, it mattered little that a woman *thought herself* pretty.   But in the water we set this new course, where appearance counted.

By the time we returned to land, female automatic orgasms were long-dead, replaced by love and skin eroticism.   In this new world, a woman needed a man for emotional survival; attracting him was vital; female competition was essential.   Like any ape she presented what needed scratching, and measured herself against her competitors.

Competition by appearance doesn't imply any one standard of beauty.   For women to compete by displaying skin does not mean smoother or younger skin will invariably be chosen.   Our instinctive male preference for younger women, though real enough, does not inevitably follow competition by appearance.   It arrives much later by another route, but when men begin choosing younger women it will be by appearance.

Beauty started in water in response to skin eroticism; it continues because the hunger continues, and we have no way to put it down. Women as a whole created a new sexual competition where each individual must eventually lose.   Each is perpetually on display, competing against her sisters similarly condemned.   Though women score other contests in different ways, on this facet of their lives women must find mates to win.   Whatever the exact face or figure in vogue, there will be some times of her life when any given woman can attract no mate, there will be some women who can *never* attract a mate.   No aquatic female wanted that result; they wanted only to be touched.

<u>Vaginal Rotation</u>

Frontal sex caused the vagina to change direction and destroyed automatic orgasms for women.   The vagina had already changed in the sea, withdrawn behind expanding buttocks, centered in a deepening pelvis.   In early aquatic frontal sex, the penis had to bend upward and pass under her pubic bone, to line up with the vagina paralleling her torso. Now women evolved a lower vaginal angle to make the bend unnecessary.

A thrusting pelvis pivots at the lower spine; the pelvic top remains almost stationary, the bottom moves forward and back.   The penis, pointing forward at the lower end of the pelvis, moves fore-and-aft as well.

Still pointing down for rear-entry sex, the mid-stage vagina forced women (in frontal sex) to lean backward to match the man's angle of entry. If she leaned forward she could feel body-front contact, but her vagina forced his penis to a higher angle and he risked flipping out. If she leaned back she obtained some penile pressure on her G-spot but lost the embrace. Her skin hunger asked for closeness; her vaginal angle needed some distance.

His job grew awkward too; instead of thrusting forward he had to lift and lower his pelvis on each stroke, or ejaculate in the open ocean. Though better than rear-entry, frontal sex often caused men to accidentally pull out with climaxes too close to stop, or too-soon-triggered by friction against the excited female's crotch. It may not matter to the participants (an orgasm feels about the same either way) but the pair passed on their genes only when they successfully *completed* the act.

Minor variations in vaginal alignment affected women's ability to reproduce, because in some it aligned better to the forward-entering penis. Over ages of frontal sex, those females with a more-sloped vagina lost contact with males less often, giving a reproductive advantage to women whose genitals were better aligned to penises.

Our deepening pelvis had caused rear-entry problems, culling women with a too-deep pelvis, slowing the pelvic changes we needed for swimming. Those problems eventually made frontal sex, though not perfect, a better route. Once it appeared, frontal sex freed our pelvis to evolve any depth it needed, to power our kicking legs. And in this deeper pelvis the vagina had all the room it needed for tilting over.

After converting to frontal sex, each change in vaginal angle reflected increasing success; fewer females felt like shifting position, fewer males accidentally pulled out. Females with a more-rotated vagina left the most progeny. Their children inherited the angle and set the stage for the next small adjustment, and slowly our vagina angled to meet a front-entering penis.

For our Miocene and early-aquatic ancestors (as for most species) coitus meant male action and rear-entry sex. Female movements rarely helped, because they required whole-body motions, while the male thrusted by simply rotating his pelvis. Most females could do nothing more cooperative than hold still.

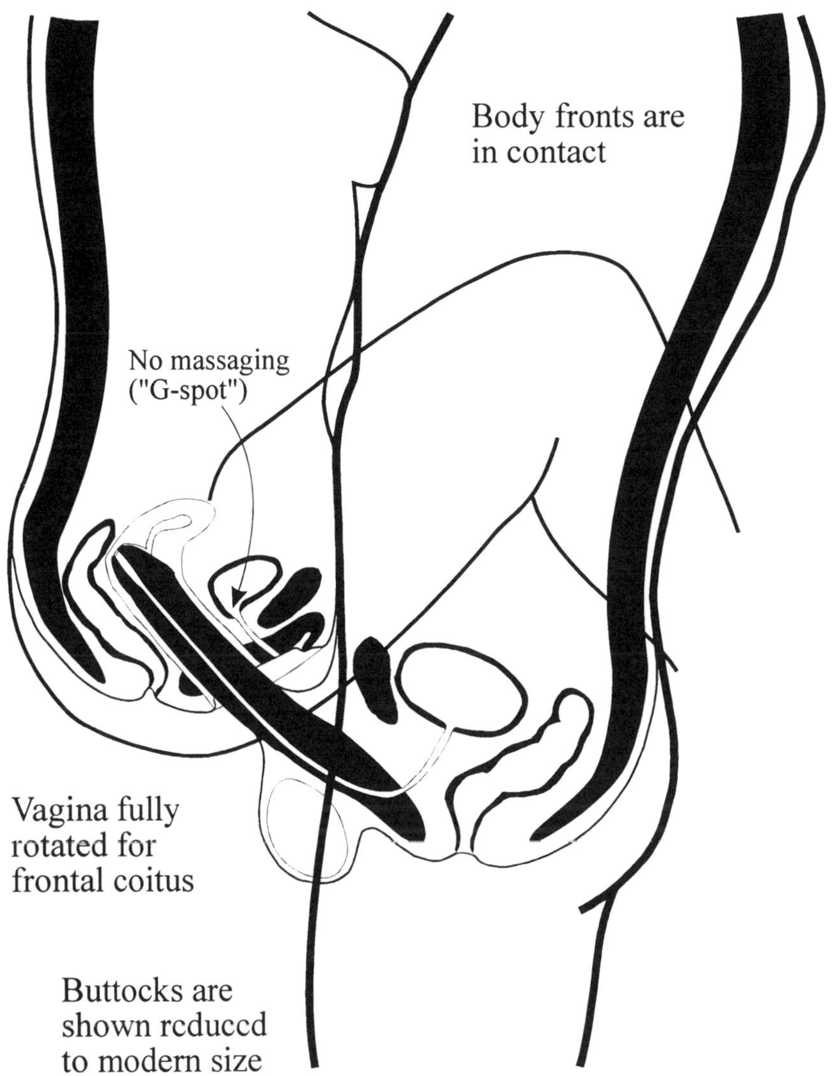

Figure 7: Frontal Sex - Late Aquatic and Modern

With a vagina tilted for frontal sex, women now became active participants. Our newly-tilted vagina lined up with male thrusts, but it also lined up with *female* thrusts for the first time, and her movements now paid off. She could guide penetration depth and angle, controlling labial friction against the penis, applying pressure to the clitoral area.

She learned her movements through her introduction to sex. Pelvic movement along the penis shaft had been her masturbation technique when little; her attempts to duplicate that movement in coitus may have caused males to lose intromission. Now her vaginal orientation allowed her to participate enthusiastically as an adult. Other little primate females masturbate by clitoral stimulation, but in adult coitus there is no clitoral stroking. Only our vaginal angle gives us clitoral massaging, leading to its importance in human sex when more direct sources of pleasure were withdrawn.

So a modern woman's vaginal alignment reflects the *sum* of both male and female pelvic thrusts; active and passive roles can change at will. If the male's permanent erection existed (as the Bushman suggests) then females were perhaps the more enthusiastic sex. With access to a permanent erection, she might have had the twenty or more orgasms a modern woman achieves with a dildo, the male being mostly along for the ride.

Vanishing Orgasm

Direct G-spot stimulation ceased as the vagina's inner end moved from the belly button toward the spine. When deep-pressure orgasms slowly dwindled, females at first achieved orgasms through slower coitus, but finally lost reliable climaxes entirely. That we *could* lose orgasms, and still reproduce, shows how strong compensating emotions had grown.

Clitoral and deep-pressure nerves are dual pleasure sources for all primates, their importance varying by species. Females must masturbate to learn what sex is; they must get something extra from a male or masturbating is all they'll ever do. Deep-pressure eroticism (i.e., vaginal numbness) evolved in primates because female masturbation was a necessary hazard. An insensitive vagina means only males stroke deeper nerves, which became eroticized to reward females for coitus.

The clitoris' placement limited it to light stroking as the penis passed near, en route to deeper nerves. With couplings only a few seconds long, clitoral stroking couldn't compare to intense G-spot contact. (But repeated copulations might add up to climactic clitoral impact, and may cause female orgasms in short-penised species.)

Bladder and urethral nerves define the G-spot, pressed against her pubic bone by his rear-entering penis. Frontal sex caused the vagina to tilt toward the woman's spine, matching our new thrusts, moving away from the pubic bone and the bladder. Our pelvis continued to deepen, giving leg leverage to a swimmer and more room in the pelvic bowl.

Women started losing deep-touch orgasms a thousand generations after vaginal rotation began. A woman might tilt her pelvis to bring the penis back into more-forceful G-spot contact, or tilt the other way. Women varied in vaginal alignment; each knew how her body responded to various partners.

Each woman had a stable vaginal angle; her ability to orgasm was fixed for life. Before deep-touch orgasms vanished we had a language (next chapter). Females compared notes; older women told young ones how to achieve deep-touch climaxes. Within any generation all had nearly the same anatomy, tiny variations on the same orgasmic ability.

They could make no comparisons across generations. They likely had no statistical tool to measure orgasmic frequency or communicate it. Mothers today rarely tell their daughters *whether* the earth moves, much less give *percentages*. Yet to track the orgasm's loss, accurate reports had to last for thousands of years.

As vaginas continued to rotate, the deep-pressure orgasm slowly became a rare thing; a woman might have only a few deep-touch climaxes in her life. She had no reason to think they were more common a thousand generations earlier. Her sexual pleasure was just something she was born with, like her nose. When the deep-pressure orgasm went extinct, like dinosaurs, no one noticed the last one.

Rear-entry sex no longer offered an orgasmic solution. Vaginas did not independently change angle, with our coital pose changing to match. Vaginal rotation *recorded* frontal sex, showing we used no other pose. When G-spot orgasms faded, we had not used rear-entry sex for perhaps a quarter million years. No one had ever seen it, no oral tradition hinted of it. Any inventive pair trying it would find slipping hips, a misaligned vagina, the woman's head submerged.

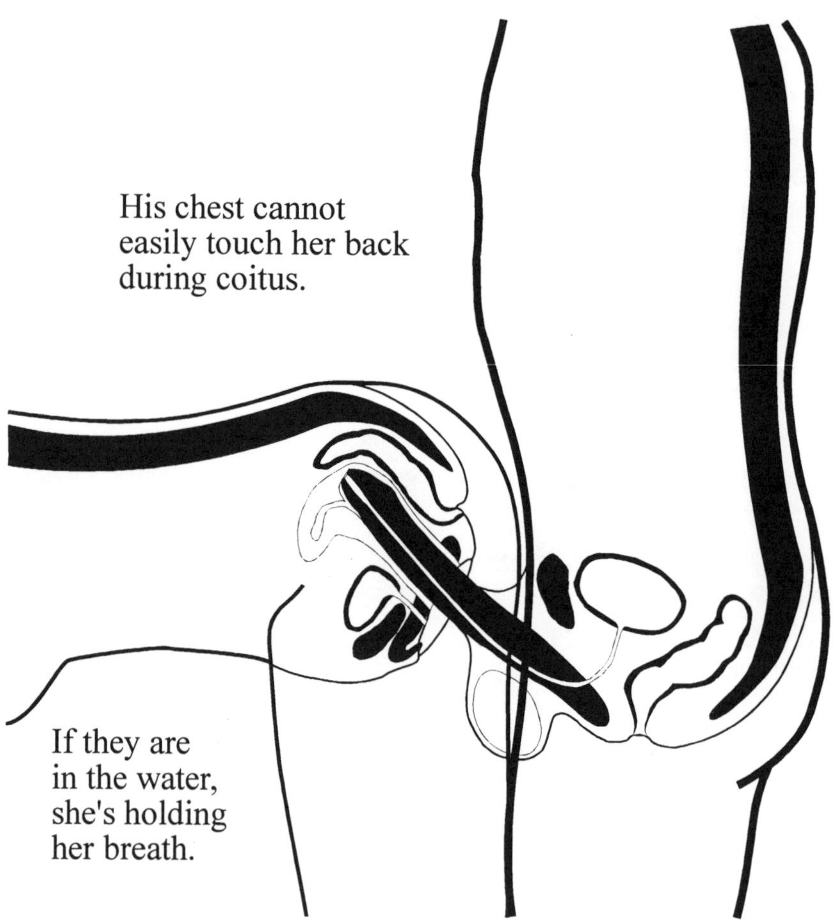

Figure 8: Rear-Entry Sex - Late Aquatic and Modern

Our instinctive preferences for frontal sex continued to grow. Beards, permanent erections, and girlish masturbation lured us into frontal sex for a million years. Women evolved body-front eroticism and passion; the next chapter explores love's birth. By the time deep-touch orgasms died, passion and electric skin were basic to each woman's sexuality, and women were deep in an evolutionary channel that would remove nearly *all* their climaxes.

Women could still achieve slower orgasms via clitoral-labial traction. Masters and Johnson suggest that any woman can climax this way if stimulation lasts long enough. Frontal coitus improved labial traction, the man had a permanent erection and the thicker penis that aquatic sex demanded. So the deep-touch loss may have led us to *extend* coitus, replacing quick orgasms with the cumulative climaxes successive matings can give.

And still our preferences grew. Each small loss of orgasmic ability culled some women who cared only about orgasms. Each generation had more women who appreciated other rewards — embraces, affection for a particular man. It will take the rest of the book to trace the roots of what we value in each other. The more elusive the climax, the larger emotion's role became. Women who didn't climax at all had only emotional ties and skin hunger to lure them.

Our emotional preferences alone finally compelled face-to-face sex. Born with an instinct for frontal sex, no girl needed desensitizing. Capable of arousal without estrous odors, no male needed it either. Beards and permanent erections no longer had to urge frontal sex, and the bridge was obsolete because we had evolved to want it.

As instincts grew, the permanent erection disappeared. Forces had always opposed vulnerable and cumbersome permanent erections, but no danger outweighs failure to reproduce. As soon as erections weren't being selected *for* by women's hesitating at frontal sex, they were selected *against*. Men became flaccid as soon as it was safe; transient impotence lowered its head when women found other reasons to like men.

Without a man's permanent erection, women quickly lost slow labial-traction orgasms along with fast deep-touch ones. Women were too far down an evolutionary channel to turn back, instinctively seeking the pose that robbed them of climaxes, already thinking love was more important than sex.

When orgasms became rare, they became shattering. A modern woman's climax may leave her gasping, in sharp contrast to the placid reaction of females who have them all the time. The contrast leads some to think that only women have orgasms, but it is the difficulty, not the climax, which is uniquely ours.

Women might have selectively mated with men who were more erect, causing us to retain permanent erections. Fisher theorizes that a similar choice caused the human penis' large relative size. But fear seems the more powerful force; once it is overcome, most women settle for what their chosen mate can supply. Permanent erections (and female orgasms) vanished because women coupled without them; women's emotions gave permission for orgasms to fade.

Women evolved to *find* rewards in frontal sex where once few rewards existed. By the end of our frontal-sex conversion, women had evolved instinctive hungers of awesome power. A woman's attraction to a man, her desire to be held by him, was strong enough to induce her to couple though she knew she would not climax. Still is.

Chapter 4 - Dating Games

In courting, we calm fears to get close enough for sex. Courting gained importance for aquatic humans because our fears had grown. Life in water destroyed old mating cues, sex became confrontation, and women lost simpler rewards. Men and women were separated by growing confusion and hesitation, and had to evolve courting behaviors powerful enough to close the gap.

This is the eye of the storm. We might plot various lines of cause and effect linking each evolutionary turn in my essay, yet it all rotates here; how we courted in water sends out waves in all directions. Up to this point we have adapted to water; from here on we adapt to frontal sex.

We must discard myths of superior human bonding, and see the promiscuous primate at our root. I outlined early changes in aquatic sex, as arousal became elusive and fear grew. These reduced sexual chances, so a female would often choose a single male to scratch all her itches. This began as a convenience, not a bond. It is convenient to take a single taxi all the way to my destination, but I don't bond with the cabbie.

Early tepid pairings became competition's focus, with female emotion as the key. Sex became less rewarding for a female but she had to reproduce, no matter what mental gymnastics were required. From an animal who picked sexual partners for convenience, we evolved into one for whom a partner is life's most important choice. We were once able to rate mates on a clear and unambiguous scale, but in water we could scarcely measure them at all. Aquatic courting was a new game for our still-promiscuous (though sodden) picnickers, an unremarkable ape living in the ocean, who played by the old rules until old rules failed.

Status And Mate Selection

Choosing mates became a woman's most difficult task in water. She instinctively sought status and dominance in males, but aquatic life obscured status and replaced once-clear evidence with subtle hints and new competitions. What females wanted was hard to find; what males wanted didn't matter.

Dominance comes from winning; status comes from an audience. Two may fight and one dominate, but status comes when another doesn't bother to fight, knowing the outcome in advance. Status is *perception* of dominance rather than dominance itself, varying by how much the viewer knows. When dominance drowned in the water, each viewer had few clues to status.

Status guides mate selection in many mammal species, because females who choose strong males have healthy young. Males battle to dominate, to swagger before female eyes, because swaggering males are chosen. Once females began to recognize dominant males and prefer them, long before the first dinosaurs appeared, males began obediently butting heads. (But males also fight to hold harems or territory by force, ignoring female desires.)

When dominance brings great gains, male aggression and body size can reach extremes, as in sea lions. When dominance gives small advantage, male and female body sizes match more closely. In some primate species, only dominant males have regular sexual contact; in others, individuals pair for life, and (after accepting a mate) dominance probably means little.

Chimps recognize status in others and also in themselves. A bull seeks cows and fights other bulls but does not see himself as a player in a larger game. Bulls probably don't know what cows think of them. By contrast, a chimp knows that swaggering counts.

Status guided but did not control our promiscuous ancestors. Promiscuity partly disarmed male dominance; being boss means little when sex is casual. Even so, females innately preferred dominant males, and males competed for female attention. Though less vital in our Miocene parents than in many species, hunger for supremacy kept nudging males onward, and females kept seeking it.

In the water, female sexual hesitation started to change mating's consequences. Choosing a dominant male makes good sense if you don't have to live with him. Males and females meet only briefly, in most species, and then go separate ways. When aquatic women began hesitating at frontal sex, they had fewer partners and longer pairings. Suddenly a women's choice was something she'd have to put up with for a while. (This led to a disastrous misunderstanding. Women don't want to *be* dominated by a male — they want to know they *have* a dominant male. It's not always easy to find one dominant enough to be exciting, docile enough to share the chores, subtle enough to understand a shade of difference.)

With sex awkward and harder to find, sexual competition became necessary and useful again, but the aquatic environment making competition important also made it difficult. Males couldn't interrupt coupling pairs. They didn't see coitus until it was underway, couldn't detect estrus (and know whom to watch). Unable to move quickly in water, they couldn't enforce rights of rank or punish the disrespectful. Unable to rise to full height while treading water, they couldn't impress by body size or power.

Dominance disintegrated in the sea. The male power structure (alpha and omega) vanished when physical threats lost meaning. When dominance collapsed, status disappeared. A woman didn't know which man to prefer because men weren't swaggering.

Elusive status did not stop women's seeking it, any more than living on the ground stops us throwing out our arms when we fall back. With dominance tough to display yet instinctively preferred, new forms of competition arose in men. In effect, females forced male inventiveness because women were confused about which man was better. A male's status depended on the opinion females had of him.

Traditional male behavior became irrelevant. It never really mattered whether one rushed around waving branches or thumping logs. All apes compete by threat and bluff, choosing acts that had the greatest effect on other males regardless what females thought. You used whatever threat made another male whimper and groom you.

Females didn't *need* the tree thumped or the branch waved; Becky Thatcher didn't need a gymnastic display by Tom Sawyer. Stags could play tiddly-winks instead of clash antlers, so long as victors displayed the swagger females instinctively sought. Males had always chosen the *form* of competition while females recognized the *winner*.

On land male physical dominance had some practical advantages; in water the only thing that mattered was the female view of it. Anything she saw as status made her a more willing partner. Any male qualities that made her receptive were valuable for the man, regardless whether they had ever before been used in dominance displays. For the first time, women could choose the weapons for men's duels.

Lost Identity Labels

We lost more than status in water — we lost some identity. A primate is not merely a rank, but an individual with ties and alliances. Family and group memberships peeled away in the sea like layers of clothing, leaving us out of uniform.

We use personality, family, and group to weigh each other. Alone, we infer another's status from personal bearing and the deference others show. With acquaintances we have shared experiences that help us predict their acts. We judge also by physique — size and power tell us whom we can safely pester.

More broadly, we see others as members of a family, taking their share of the lineage's privilege. A politically powerful dynasty grants high status to a fool we would otherwise scorn. Or we understand that we know nothing of their relatives, and hold a little of our judgment in reserve. Before we marry, we like to meet the family.

Finally, status comes from group memberships. Modern communication permits nations of great size, but we evolved to focus on tens and twenties. We still focus on our own small groups today despite millions of neighbors, moving through crowds as if through shrubbery. Members of our group are automatic allies; outsiders are nearly transparent.

Individuals readily adopt goals of their group, often nothing more than boundary patrols. Each howler monkey troop has its woodland plot; they tour boundaries each day at regular times. There they yell at neighboring troops on similar missions at appointed hours, everyone seemingly gratified by the hollering.

When a group has goals, accepting those goals claims membership in the group. If a certain politician seems useful to me, supporting his cause makes me his teammate. When territorial primates enter a group, it is their help in home defense that shows their full acceptance. Struhsaker reported a vervet male joining a troop: At first he was chased away, but two months later he was helping to defend the territory. Carpenter indicates that howler monkeys take longer. Membership in a group means regular meetings with other members. Repeated contact and chummy feelings from chasing strangers are basic lubricants of primate relationships. This gives the football team's camaraderie, the elite military unit's elan.

Personality, family, and group are the three layers of primate identification, essential in recognizing and evaluating others. In the sea, aquatic family allegiances weakened, lineage's power withered, groups dissolved into mobs. Only personality remained in women's view.

Physical Boundaries

Group membership dissolved in water because boundaries vanished and territory lost meaning. Never strongly territorial, as Miocene terrestrials we still preferred known companions in familiar places, and we ganged up on strangers. The sea washed away our tepid territorial interest, and group identity disappeared.

As early aquatics we still focussed on land. Water was a foreign place, safe only a few feet from shore, bordering our coastal feeding range. Over millennia we evolved to chase shallow-water prey, to swim, and to hold our breath under water. As we became aquatic and ceased going inland, beaches became mere nesting sites.

We moved first as terrestrials, expanding along the coast to find marine foods. When wading was our limit, a broad river would stop us. Too timid to swim across, we could pass only by moving far inland up one bank and down the other side, fording in the shallows. A wave-pounded cape (or a coastal lava field too broad to see across) was equally impassable.

We filled the coast by small degrees, shifting nest sites over centuries, our population limited to the number who could live where feet could walk. Drought confined us to our strip of beach, deep water (still dangerous) confined us to the surf, so any obstacle on that beach could bar our path. Our freedom was defined by *land*marks.

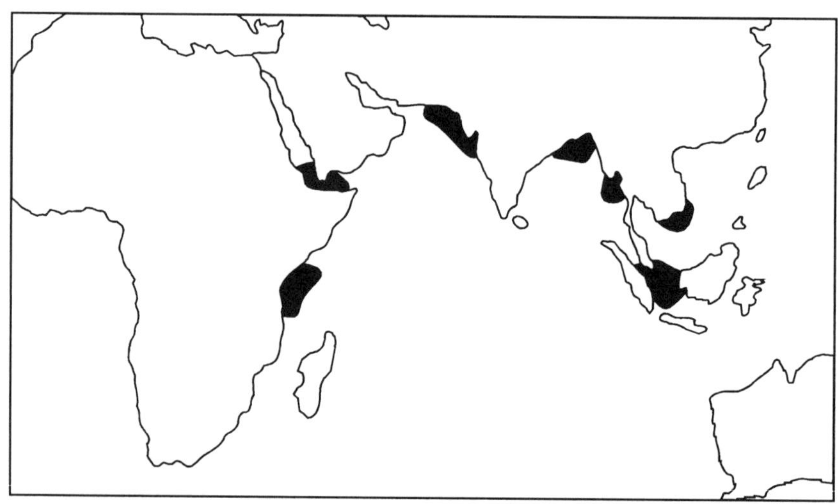

Hypothetical early aquatic ape populations, using a modern map.
These maps show only the concept, not the geology of that time.

Figure 9: Early Aquatic Distribution

As we neared fully-aquatic lives we gathered no food on the beaches, and they became *only* nesting sites. The water's richness alone limited our numbers. Like shore birds tightly packed, or seals cheek-by-jowl, we slept as close as we liked. Water was our focus, and only our habit of sleeping on land brought us back to the beach at all.

Like sea lions and birds we slept on islands and isolated shores where predators couldn't reach us, slumbering apes in colonies of thousands. In the mornings we woke by dozens and hundreds, coughing and scratching. Down to the water then, for breakfast and safety, carrying infants while children scampered ahead.

In evenings we straggled back, but we could straggle to a *different* beach. Comfortable aquatics, we had thickening fat, webbed feet and hands. Spine-to-leg alignment made us adequate swimmers, and we could easily cover several miles a day. A river mouth meant only a different feel to the water; a rocky cape made no impression on us at all. Land had value only if you could safely sleep on it.

Once a barrier, deeper water became a broad highway with sleeping beaches as roadside stops. Once a thin line dotted on segments of shore, we became a wide brush-stroke painted along the coasts. Yet until we slept in water, the land still defined location.

Finally, when fully aquatic, we formed this coastal brush-stroke of humanity by night and by day. Sleeping while standing neck-deep in bays, with children holding on or treading water as they slept nearby, we cared nothing for the land though we rarely swam in open ocean. Dry ground had no value; only water mattered to us. We measured our world by depths and currents, by oyster-beds and kelp.

When fully aquatic, we might have coated the coasts. This could explain some known locations of apparently-bipedal hominids.

Figure 10: Fully Aquatic Distribution

Polynesians have 30 words for coconut and we had as many for water. We could name a point in an apparently featureless bay by describing its water and bottom alone, such that a friend could agree to meet us there and be accurate within fifty feet, without ever mentioning a *land*mark at all.

Groups And Mobs

As our beach-front wading pool developed into an aquatic highway, the primate troop degenerated into a mob, and any sense of group membership vanished. When men shed uniforms of status and group, they left women confused.

A society's leaders form groups but also limit group sizes; we can define an alpha's leadership role by the number who follow. Baboon troops never have over 200 members, and are probably the most disciplined primates. Chimp and gorilla groups may contain only a dozen amiable companions and have a far looser structure.

Leading males lost status along with everyone else when we became aquatic. An alpha male woke in the morning on the beach, thumped an underling perhaps, and then his power vanished as his troop scrambled down to the water. He wouldn't thump another until evening. We were herd-like because we were many, and because none could lead or discipline the throng.

Primate troops split when they grow too large to lead. Lineages break away to form new troops, sometimes waging bloody war with the root population. These breaks demand leaders, both to hold the old group and to lead the new, but the sea washed away dominance and status and leaders. Other than mothers and their daughters running to several generations, there was virtually no aquatic social structure, though there was intense social interaction.

When we lived in the ocean, no group split off because of its size. A thousand slept on a beach as easily as a hundred, when all our food came from the sea. Some would move on if they chose, tired of crowds or seeking fresh bays or just curious. But it happened by individual choice, by a family's whim, by accompanying a friend, not by following a leader.

Our mobs grew out at the edges year by year. In a few centuries, a group that started as a hundred became a thousand, covering miles of shore. Members at one end might never see the other, for they had all they needed where they were. None knew he had a thousand cousins on the coast; each went wherever he cared to swim. Groups grew in numbers and softened in structure as we became aquatic, flowing out from their first beach like butter in a warm pan.

No boundary patrols watched the edges. Having lost landmarks and leaders, they had no boundary to monitor and no one to lead the mission.

With too many neighbors to count, there was no point yelling at strangers. The ape you hollered at might indeed be a foreigner, or just a cousin you hadn't met yet. Without a clear "us" there could be no "them."

Aquatic ape populations flowed together without ever knowing it. We began as thousands of breeding pools along Indian Ocean coasts, each evolving aquatic life independently during the Pliocene drought. We became many mobs and finally one mob, footloose strangers on an aquatic highway.

Nucleus And Isolation

We can begin to imagine the world an aquatic woman saw. Her view defined our courting requirements, because her doubts set the male's barrier. Frontal sex eroded our instinctive promiscuity, so fewer men found partners, though pairings lasted longer. Anonymous in an aquatic mob, a male had to make her notice him or she would choose another.

We each felt vaguely isolated, as if adrift in a distant city. We hadn't evolved for crowds of strangers, but we surely evolved for connections. Relationships had aided primate success for perhaps forty million years, making us socially adept and gregarious, and the sea suddenly disabled some of our social aspects.

Sea lions are often alone but never lonely. At sea they have no society, on land no family group larger than a mother and her current pup. Cetacean family groups are sometimes more like ours. Porpoises maintain tight relationships over long terms, and killer whales remain in their natal group for life. In combination, seals and whales tell us that social structures are possible but not required in aquatic life.

If an aquatic ape gained some advantage in solitude, we would have evolved to like being alone. Orangs evolved to dislike company, possibly because wide spacing helped their foraging, just as territorial birds' combativeness guarantees they will spread apart. If being alone were good for us, we would never feel lonely.

We felt isolated in water because we had evolved in a social matrix that lost part of its structure more quickly than we could lose interest in it. The aquatic ape's world contained no hierarchy, no personal dominance outside the family, though our minds looked for the order. Our stomachs led us to the ocean beaches, where the sea seduced us and made us a mob, and our brains had to cope. Any gregarious ape will likely feel vaguely isolated if its population explodes into an unbounded habitat. We all feel a little lost in a city.

So we focussed on the society that remained, clinging tighter to our families like refugees. We didn't know we missed leaders and rankings, and couldn't have created them if we'd known. We put our energy into what we *could* do, as a woman tied to her home may clean it obsessively; we created a society from what we had, as a lonely child talks to stuffed toys.

Families themselves remained much as they had been on land. Ape families center on mother/daughter bonds and sisterships. Daughter apes commonly remain near their mothers for life, and sisterhoods can lead to adult alliances. Since the mother has multiple mother/daughter bonds, she anchors a circle of siblings whose alliances will outlive her.

Families spanned several generations. With first births at about 12 years and a lifespan of 30 to 45, as aquatics we had living grandmothers and perhaps great-grandmothers. Each female had daughters who respected her and a mother she respected. More distant relationships came from habit and familiarity; a child learns how to deal with cousins and uncles by watching how her parents treat them.

Boys and girls grew as they had on land and still do, paralleling chimp families. Young children stayed close to their mothers, and girls tended to remain closer. Girls were also being selected for a masturbatory style that kept them closer to their mother's adult male friends. As boys grew they went exploring, but frequently returned at least into near-adulthood.

Seawater washed away outer layers, but our aquatic society shared its core with Miocene precursors. Terrestrial male combat was blatant and ephemeral, while female bonds and competition guided the group. The sea disabled male dominance, removed the old hierarchy of force. Female matrilineal relationships were left foremost by default. Always the strongest social group and usually the smallest, the nuclear family now became the only one we had.

Strangers Among Them

Mothers anchored family clans, with males moving between them. Each young female looked out from a stable feminine island, at a constant flow of males who wanted her attention. Near each woman-cluster floated males unrelated by blood, who had successfully courted a group female or hoped to do so soon. Males ranged from adolescents looking for their first mate to old men who sought only company.

Currying favor, men made themselves useful where they could. Primate males often join new groups by ingratiating themselves with youngsters or females. With no territory to help defend, men had little but babysitting to show their good intentions. Male primates often help infants, particularly when paternity is unclear or when males are strangers and the mother is watching. Aquatic males offered mooring-posts for children and masturbatory relief for girls — not unpleasant for either.

Men might have remained with female-clusters for five years or so, typical for human pairings with or without marriage vows. It matches male-residence times in other primates, and likely evolved to prevent inbreeding with daughters. Men *may* have stayed as little as a few months. That's a long time for a highly promiscuous ape, and longer than some modern humans can manage.

A male's stay directly depended on his courting skill. The more work it took to get sex, the less he listened to promiscuous instincts, and the less-quickly he abandoned a female. Men who found women more receptive were more sexually mobile, just as today. Each man evaluated his own chances and decided when to go, and our courting evolved by endless repetition of these choices by males in the sea.

If courting were no problem for any male, courting would not have evolved at all. If it were extremely difficult, key skeletal changes would announce the problem — a thesis I will try to prove when we return to land. As in other apes, aquatic courting occupied a middle range where not all females accepted all males, but persistence could pay.

When a male was near a family who accepted him, he also tested responses from nearby females — typical behavior for many married men. If he found a warm reception he would likely shift his lodgings. Some males set out on mate searches with no plans to return; others leapfrogged continuously, certain of the next female before quitting the last.

Female-centered families anchored points in a web of fluid relationships. Women remained together for life, raising children in a continuous flow. Males occasionally arrived or departed, each missed briefly and then forgotten. Sisterly bonds and current male friends provided allies in squabbles and things to squabble about. Family-islands, by default, gave the foundation and target for our primate social energies.

Men flowed slowly past female clusters as water moves through a marsh. Males passed by every week if not every day, trolling for sex, inspecting and being inspected by resident females. For the first time women had to evaluate strangers as sexual prospects.

Primate evolution hadn't prepared females for this. A mate had never before been a stranger, but had always been first accepted by the troop, helping defend the territory and working out his rank against other males. In species where alpha males controlled sexual activity, he had few chances for perhaps many months. For females, evaluating males had always relied on long acquaintance, on her awareness of his status and power. She also understood his individual nature, and her accepting him often depended only on her personal preference. In apes, his status might come mostly from female acceptance, and not too quickly.

Now aquatic, however, there was no group larger than the family. No hierarchy of males gave rank to the new male, and helped a female evaluate him. This defines the aquatic distance between the sexes: Males had to develop courting behaviors powerful enough to persuade females to mate with strangers. At the peak of aquatic courting, it was *inorgasmic* females he had to persuade. A male had to win the confidence of a skeptical female who was adrift in a distant city, unable to meet his family.

If all this sounds suspiciously modern, it is no accident. Behaviors we consider cultured and sophisticated are, I believe, very old. As I outline aquatic courting, you should start seeing these people as essentially modern. Far from rising meteorically since the Stone Age, we have changed little in the last few million years.

<u>Closing The Gap</u>

We measure distance between men and women by what we endure to bridge it. Sometimes parents arrange marriages, or matchmakers locate potential spouses. But most must leave the trenches and wade through no-man's land, seeking a mate from the other side. We do this repeatedly, risking heartbreak, disease, and rare but violent death. And after we find one, we may drop them and go find another. This is the species calling itself the peak of evolution, mind you.

Courting is a time for careful approach, a time to relax our guard and allay another's fear. Naturally armed species court slowly; bears may keep company for several weeks before mating, while grazing animals often need only a few seconds. Significantly, we court more slowly than any other animal, though we lack natural weapons.

A female aquatic ape had to choose from a succession of hopeful males. She had her mother and young siblings nearby, and perhaps children of her own, but she lived in a mob with no clear hierarchy beyond her family. She missed society's clarifying web of rank and privilege and leaders, but she didn't realize it, lacking anthropologists to tell her so. Yet her primate mind evolved its contours in a troop matrix, and now the matrix was gone.

Into her amorphous world came strange males drifting along the coast. Men moved from family to family, importuned any woman they saw, and were willing to pause for a week or month if they met a warm welcome. Our cautious female needed to couple with one of these males face-to-face, in the most dangerous pose for a primate. She had grown up masturbating against her mother's friends, and was used to the frontal position. Yet now she had to embrace a stranger. Males had much persuading to do.

Emotionally she was on her own. Though primate youngsters enjoy pestering coupling adults, other adults don't interfere. Goodall reported a female chimp screaming in terror during the entire time of her first coitus (though she maintained the correct position) and no one calmed her. In the intimate company of an adult male, her mother and other females would not aid her. With status drowned, other males could not chase him away, or warn him that hurting her would be a bad idea.

Yet she was in control. Surrounded today by blatant male competition, we often forget that males are sexually passive and must await female acceptance. Male sexual patience, no longer overshadowed by terrestrial dominance battles, became aquatic man's defining feature. There were no unwilling sexual partners in the sea; women could outswim any man they disliked. Unable to compel her (and lacking an evolutionary tendency toward rape), a man's only solution was to persuade her.

Each male vied to be chosen, without understanding feminine instincts guiding the choice, without recourse to classic male competition. Unable to dominate, men couldn't *display* dominance — a basic cue for female sexual interest. Too much bluster in lieu of real dominance could irritate or dismay their one-woman audience. So men trod a narrow path, trying to gain attention without provoking fear or distaste.

With classic dominance pulled out from under them, males fell back on simple attention-getting. Naturally hedonic, apes gain status by extravagant display. Eager to learn and show off, chimps are born clowns and easy to train. Baboons more often gain status agonically — by threat and fear. But agonic status depends on thumping the disrespectful, not easy to do in the wet. Worse, the agonic troop shows respect by *avoiding* the powerful; what men needed was something to bring females *closer*.

We did not *invent* attention-getting for aquatic courting. Already part of our nature, it was all that remained when dominance drowned. Without a man needing to understand women's fears, he left more children if he was fun to be with, if he had interesting talents. Over thousands of lifetimes women selected men for those traits. This left room for male behaviors not only to be invented but also to become innate — the birth of instinct given need and time.

Baby Talk

Conversation began with baby-sitting. On land or in the sea, a newly arrived male ape must introduce himself to the troop, which might present threats from resident males. Once they have ceased to chase away a stranger, he often starts his climb through the ranks by forming alliances with females. And winning female friendships starts by playing with their infants. Ape males frequently play with children, but it is never more important than when they join a group.

Aquatic males looking for adult females approached family groups as they would approach any other troop. Our aquatic society had no borders and none to guard them; few would chase a new male away from a family. But strangers still had to find a path from hanger-on to member, from awkward to comfortable, from sexual prospect to sexual partner.

When a new male played with children, he followed standard ape strategy. When a mother saw him, she grew to approve of and trust him; when a woman saw him, he evoked memories of her own childhood. And when he played with a girl child, they became partners in the sexual pastimes we evolved in the water. An adult female seeing this could hardly miss the point.

As an infant she had met such strangers before. Other males, seeking the favors of her mother, aunts or older sisters, played with *her* in her childhood as they joined her family years earlier. She was too young to care that they were strangers; she used them as she used any male, and they were part of her introduction to sex. An endless loop of erotic childhoods became increasingly important as aquatic society disintegrated and we converted to frontal sex.

Pleasing a girl-child would help her later to mate, but the male hoped to win an adult female. Men did not obsessively focus on young girls at this point, though they evolved to do so after our terrestrial return. Indeed, permanent erections suggest that male interest was relatively trivial. It didn't matter much whether he liked children, but it mattered that women *think* he did, and saw him as a safe choice. So males in all primate species gain favor by *displaying* affection to infants, as politicians kiss babies.

When sex became more frightening, men gained mates by kissing babies more loudly. We had complex vocalizations before we became aquatic; many primates use grunts, cries, screams and chuckles; gorillas, cats and elephants purr when relaxed. Aquatic men competed for sex by being more blatantly peaceful, chuckling and murmuring to the child, less for the child's benefit than for nearby females to hear. The more ostentatiously a man played with children, the more a child clearly liked it, the safer he looked to a young woman evaluating him as a mate.

Adult voices rumbled in her own childhood. Any adult voice sounded lower than a child's, with larger males' lower still. A newly-arrived male's voice, deeper than her own, evoked a memory of the men she had clung to some years earlier. Voices became a sexual cue; a man who made safe noises felt safe.

Deeper male voices not only *sounded* like she remembered, she could *feel* them. His chest vibrations carried over short distances; she felt them through the water almost as if they were already embracing. With rising skin eroticism, this feathery touch was at the very threshold of awareness. Though men did not understand female fear, males rumbled and purred to infants because it calmed nervous females nearby. Through countless generations women chose these safer men, and caused males to evolve constant calming sounds.

Any courting animal must say in some way "Here I am — I will not hurt you." Scent can say "Here I am" even at a distance. When animals get close the message "I will not hurt you" becomes vital. Though never *inevitable*, communicating by sound is within the primate context and common to virtually all aquatics.

Aquatic men had little to compete with besides finding a way to announce they were safe, and sound was the simplest way to do it. And when men started competing that way, their children carried the genes of vocalizing men and of women who liked vocalizing men. So female preference became innate as well; when murmuring men won the race, they carried along women's liking for them.

In terrestrial life, maleness meant dominance, and a woman would have coupled with all while preferring the strong. But in water, males held children and fought little, and girls saw men as embraceable. Only the destruction of standard dominance allowed women to focus so narrowly on their own preferences.

Singing In The Wilderness

Rumbling and purring led to serenades and duets. Primates use sound to maintain contact in dense forest, and forest-dwellers tend to be noisy. Combat between rival primate bands often means yelling contests. Many monogamous primates use long-distance territorial calls; often only the male sings, or he sings first. As in territorial birds, primate females are often voiceless when paired with a male. Gibbons are exceptions, with the female leading and the male responding.

Duetting announces monogamy, and tropical birds often sing extremely complex duets. Ardrey (1966) mentions a Kenyan shrike reported to sing as many as seventeen different melodies in a day, each melody used by no other pair. Gibbons (our nearest relative, after gorillas, chimps and orangs) pair on a territory for life, and duet to proclaim their state, with subtle differences distinguishing individual's songs. Sarah Hrdy (1981) says that if either partner fails to respond, unmated strangers hear the survivor singing alone, and presumably take it as an invitation to come calling.

For aquatics, singing was a natural next step. A man simply raised the volume as he murmured to children, knowing that females could hear him. A male's gentle noises can calm at three feet or greet at thirty. Song may have begun as a broadcast chuckle, saying he intended no harm, asking permission to come closer, and the female answering in kind.

Women began singing when men did. Women's interest in vocal men had set the genetic snowball rolling; daughters inherited their fathers' vocal talent, and women, too, could gain from noise. With other sexual cues hidden in water, a distinctively-female voice could gather hopeful males who would then have to confront each other, and she could pick the one she liked.

Aquatics duetted to claim not territory but mates. Lovers "make beautiful music together" to declare mutual interest; an answer assured each that the other sang to them alone. By telling each other they tell the world, making a sound greater than the sum of two individuals. Each of us can name paired singers who might be hard to think of as individuals.

Mutual recognition liberated couples. The scarier sex got, the harder it became to fit courting and mating between breakfast and lunch. As boundaries dissolved in water, the harder it was to know where your new friend might have got to. By giving couples a way to connect over distances, singing freed them to go about their daily business while their voices danced together.

New possibilities arose because singing worked best when everyone did it differently. We (and our relatives) were already the smartest and most inventive of primates. Now we had singing (a great way to get attention) in a world where the old attention-getters had failed. With no rules and no meaning, praise and renown came to anyone who could invent.

We had reached our first cultural explosion. Families evolved individual sounds like the tartans of highland clans, like family-songs of killer whales. In each family, individuals had their choral place and solo parts. An approaching male might sing the family theme while trying to strike up a duet with a female. His voice said "stranger", his theme said "friendly", his copying her personal variation told a female he wanted to know her better.

The family sound held many duets, each recognized alone yet fitting into the whole, as an orchestra conductor recognizes a single instrument while hearing the full group. In fact, I suspect that the conductor's skill marks our evolution for precisely that ability. If we evolved to hear only noises of the jungle, how could we pick out one misplayed violin from a roomful? No such acuity should evolve without powerful selective pressure, and the pressure seems missing unless song somehow helped us survive. I suggest we evolved to pick out our lover's voice from a chorus of singers.

Male competition guided song's evolution. With simpler dominance disabled, vocal innovation became the male's best way to gain attention. Needing a mate, competing against other males, men honed vocal skills endlessly in an intense exploration of song's possibilities. On a new instinctive base we created varied singing styles, solo and in concert, serenading for our mates. Our aquatic highway on the East African coast must have sounded like a Welsh village.

Women changed too, evolving to be swayed, competing to be courtable. Women who liked *some* male charms had a better chance to reproduce than women who had only the fading orgasm as a sexual lure. Perhaps not something the world needed done, singing was something men could do, like gymnastics at Becky Thatcher's fence. So women evolved sensitivity to voices, just as men evolved the interest in using them.

In singing we first developed vocal rhythms that now mark our speech, evolving part and counterpart. Two people conversing show matching EEG traces as their phrases synchronize — a clear but nearly hidden trace of our aquatic vocal dance.

Song still carries potent sexual messages, for women more than for men. Most popular singers are male, and no human spectacle compares to frenzied female fans. Women report orgasms from listening to music, and music is the primary sexual stimulant for girls in all age groups and economic states. When men had nothing but music for luring women, women's interest inexorably evolved.

Men sang for one woman; women sang to collect men. I think we often sing the same today. Culture has so many layers one can't be sure, but perhaps this sounds right to you as well: A woman sings *for* an audience; they hear her as a group. A man sings *to* an audience; each woman in the group thinks he sings *to her alone.*

The Spoken Word

From competitive singing came language — nouns and verbs five million years ago. Speech evolved by repeating and condensing, the result of having the right equipment at the right time. When we invented language we didn't *need* it, any more than we now need mimes or tapdancers.

The first songs had been only modulated murmurs, the wandering male's "hello" to new families, joining the chorus, seeking a duet. When we evolved mouths with small canine teeth and soft, everted lips, their flexibility gave us the power to babble. Chimps and gorillas lack flexible lips, and compete by finding new ways to clack their teeth.

As instinctive show-offs we strive for attention, and as aquatics we had nothing but sound to show off. Any new noise could gather admirers; one can only imagine the excitement greeting the first whistler. Having no practical use, babbling evolved freely, like avant-garde music.

Babbling quickly became an art, sounding at first like "scat" (a jazz style of singing meaningless syllables) or a baby's prattle. Gibberish carried no message, but *complexity* won female attention when nothing else worked, and males had to compete in winning doubtful female hearts.

Complexity also let us condense and accelerate the family choruses and solo parts. Songs identified group and individual just as in other species. A few notes of the family song requested an answer — "Sing back if this is your chorus, so I can find you." A phrase from the solo part could *be* the answer — "I don't know about the others, but *I'm* over here!" A gibbon or shrike can do as much.

Bob Hope had his "Thanks For The Memories"; when you saw Bob, someone would be playing it. My stepfather had such a signature song at his supper club, and when he walked in they played a bit of it. His friends knew he'd arrived when they heard his song. Eventually one recognized it in the first second, as quiz-show contestants can identify a tune with only three or four notes.

Identifying a tune in a couple notes is such an unlikely skill, one must wonder how it evolved. It seems trivial in a Mighty Hunter planning a stalk. I recall a contest where radio listeners were able to identify half a dozen songs in a fragment of noise a second or so long — an unbelievable feat until you have heard it done. Such an ability implies that recognizing songs quickly, or some skill very like that, had value.

Babbling made songs complex, as more information squeezed into less time. Each sound became more than a note on the scale; it was also the lips' shape at its start and end, and how the tongue moved. A half-second of consonants and vowels replaced five seconds of sustained cooing to announce identity. Personal names evolved from signature songs through repetition and intelligence.

Once vocalizing became an instinct, we sang and babbled even when not courting. Before babbling, it might take several seconds to vent one's anger over a missed meal. After babbling, a single word might do. One babble might (for example) announce that the little fish were mobbing on the beach. With repetition and condensation this became something like "Grunion!". The same sound, babbled wistfully when they were out of season, might be "Grunion?"

Attaching a sound to an event (or a person) was possible only because the sounds were meaningless to start with. Complexity as competition gave us a world of new sounds to try. All it took to start *intentionally* building a language was to recognize that one *had* attached a sound to an event. Someone noticed a habitual noise, and repeated it when thinking of the situation where it was used. Some suggest the first word was "mama", uttered by nursing babies who could babble. Makes sense to me.

Meaningful babble quickly overtook singing. Children learned name-songs through daily use. Patterns arose through habit: One's name followed by the "fishing" song might mean he had gone fishing, one's name *after* the "fishing" song *asked* if he were fishing. Chimps today can handle sentences tougher than that, given the tools to express them or training in sign language.

Over thousands of generations, men's verbal skills drove brain evolution. Well-spoken men easily sustained courting conversations, allaying female fears. Men slower off the mark left fewer progeny. Our brain began to specialize for speech, evolving syntax wiring making language easier to learn. Brain enhancements were first diffused, but with time they largely coalesced into Broca's area (where speech is controlled) and Wernicke's area (where it is understood). These, which Holloway found in two-million-year-old skull casts, appeared perhaps millions of years after language itself.

All this began in simple competition for attention, not from a need to communicate meaning, but meaning became what we used it for. As courting competition heated up, men were compelled to handle complex sentences. When meaning replaced babble, what he said and words he chose became the female's measure of his worth. A chimp can sign a sentence of half a dozen words; we evolved to work out sonnets.

Most animals have "closed" call systems, where each sound has only one meaning. "Look out for the snake" and "Hawk overhead" cannot be mistaken. Most theories of human speech try to explain how we evolved this into an "open" system, combining individual sounds in complex patterns to have complex meanings. The old closed calls were supposedly too limiting.

Yet we still have closed calls, immediately recognizable. I cannot write them down, but I don't have to. You know them instantly when you hear them, from scream of fear to shout of rage, spanning all cultures. You hear them and create them with a different part of your brain than you use for language.

Language did not evolve out of closed calls, to communicate more-complex meanings, but grew from a foundation which (at first) had no message at all. Language came from singing, not from a call for help or a cry of anger. Growing outside old mental channels carved by cries and shouts, language evolved freely, as dancing words.

All this was accidental. Mobile lips and fat narrow tongues were a neotenic response to diet, not evolved for language. Yet they gave us the power to babble. Language didn't appear because we had anything important to say, but because complexity grabbed female attention when males had little to compete with.

Ancient Conversation

Conversation sparked our second cultural explosion, close behind the first. Beyond mood or identity, language became a tool for exploration. Now gone forever, we once had a verbal communion nearly beyond imagining. Our loss is incalculable.

At first, soothing sounds only helped males gain female attention and allay fears; expressive murmurs communicated only peacefulness. With the birth of true language, pushed by nothing more than male competition, the *message* took on as much value as the tone. When all men could speak well, women's hearts went to men with something to say. Men's competition then pushed human intelligence to a peak.

Conversations became magical. For the first time on this planet, one animal could look into another's mind — an awesome journey for a social ape. Understanding others dominates our thoughts; life without personality is unthinkable. It gives power, bonds friendships, lubricates love. Many spend hours a day focussed on the feelings of strangers in soap operas or novels. But it happened to us once for the very first time.

Most couples have intense conversations early in their courting, marathon talks reaching deep into the other's soul. To accept another's secret joy and pain, and wrap them in our hearts, is the most erotic thing we can do without touching skin. A lover's mind presents a breathtaking vista not quite of this world, not quite out of it.

It was *this* conversation for which language next evolved — not for planning a hunt or dispersing spoils but for building trust. Many species announce submission by ritual surrender, where the defeated animal bares his neck or his belly to the victor. Courting conversation does the same: In words the male bares vulnerable places, shows the female how to hurt him. When a woman sees a man's muscles she may be impressed, but when she sees his fears she gains potent knowledge, and fears him less.

A male bird may offer the female a beetle; we evolved to offer ourselves. By handing the woman his secrets, a man gave her weapons of knowledge, the power to hurt, the confidence to face him. No ape ever held such a weapon before. Her armament rewarded the male who gave it to her, let him approach her, but penalized males who could not confide. Aquatic courting conversation focussed on mutual exploration, intimate verbal touching, urged on by the simple need to compete.

No grand plan made aquatic evolution give assurance to women, who so badly needed some then. Not every aquatic species evolved talking; where females were more easily impressed, attention-getting surely stopped at yodeling. If males had never competed, some females would still copulate. It was entirely accidental that male competition's ultimate result was to *communicate* more than impress.

Aquatic men invented language and talked better than women (I hear women gritting their teeth), yet many men now think that sports statistics suffice for conversation. Aquatic evolution focussed on male ability to hold female attention; speech was his performance and intelligence was its foundation. Men had only a small advantage over women, for each woman inherited half her genes from a well-spoken father, half her mother's genes came from such a one, and so on.

Eventually all this changed. Like canine teeth when we became aquatic, intelligence got in our way when we returned to land, and was discarded. Men became a bit stupid because duller men reproduced better; male courting skills collapsed along with intelligence. But women gained nothing from stupidity and did not regress as far. Their modern verbal advantage is part of the evidence for our intellectual collapse, discussed at length later. During that collapse women took the conversational high ground, and have held it ever since. We have far to go.

Though suppressing male intellect (once back on land) aided our survival, it also created problems. Stripped of verbal skill but retaining the instinct for verbal competition, men now look for "the line" that will win a female heart. Women, having evolved for intense conversation, are frustrated by taciturn men.

Just as male competition gave verbal skills to women by inheritance, modern male dullness reduced women's intelligence. When men got dumb, we all got dumb. Half her genes are from a male; a woman is half-built from one for whom suppressed intelligence was an adaptive response. Though a woman may gain by being smarter than men, she is also now inevitably dumber than she might have been.

Yet we cannot measure our intellectual decline, we have only evidence that we regressed, and my evidence must wait awhile. If I am right, if men evolved dullness, then verbal skills of women and men today mark the difference between being *selected* for a duller mind and inheriting one from your father as a little girl. Men would have had as much relative *advantage* in the water — the difference between being selected for verbal courting skills and inheriting them as a female child.

Come a bit further out on this limb with me and imagine the world we left behind. If we lost intelligence when we returned to land, then late-stage aquatics mark a peak of human intellect. Both sexes had powerful verbal and intuitive skills to smooth the courting ways. Men evolved to suit female likes; we acquired mates not by dominance but by affection.

Until three to four million years ago, when Pleistocene rains ended the Pliocene drought, we had a brilliant aquatic culture. Predators took a few of us, but within our species there was virtually no physical danger. We were intensely competitive, obsessed by personality; we never tired of talking, probing, comparing. As a result we had a well-developed sense of ourselves and others. We were introspective, thoughtful, philosophical. Cogito, ergo swum.

Language was our entertainment, our contact, our lives. We talked so long and so hard that we feel deprived without dialogue, and can stroke each other with words. Our oldest oral relics are epic poems, thousands of lines long. Yet Beowulf may be a regressed ape's *trivial* effort compared to aquatic forebears', who might win a lady by dashing off a hundred sonnets composed on the fly and remembered for life.

I believe that four to six million years ago we were a species of singers and poets. Need I say that this is radical? Yet if language came from song and was the focus of evolution, and if we became more stupid later, then such a picture fits exactly.

What did we talk about, when we could only talk? Perhaps we discussed higher powers. Later I will suggest other apes have some sense of gods, as we did before we became aquatic. Floating naked in the water, we may have asked who made us. Perhaps we traced philosophical paths never since rediscovered. A million years too late, with the female orgasm long dead, we may have debated how to couple. We probably were naked-eye astronomers, marking the movements of planets.

Many cultures include in their creation story a time when water covered the whole world. Perhaps this alone remains, passed down vaguely over millions of years, remembered well because so close to our hearts. This verbal relic may be the last faint trace of a story once told in the first person by barely imagined ancestors.

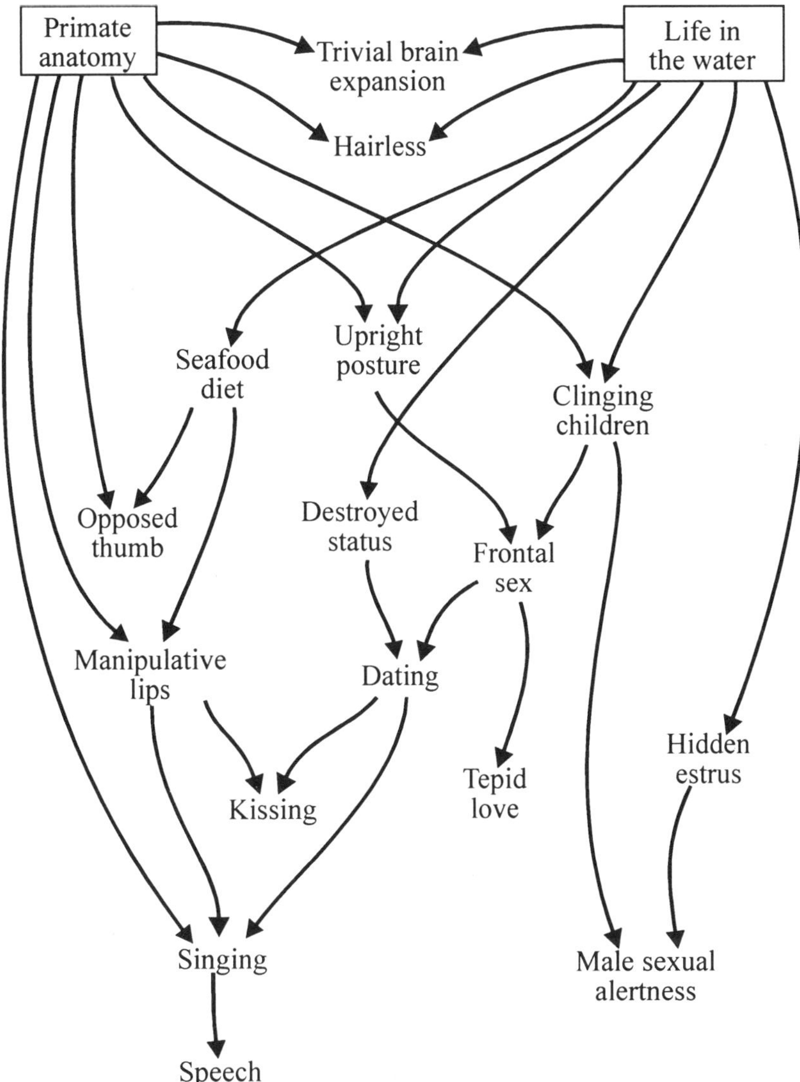

Figure 11: Human Evolution in Water

<u>With Fear And Obsession</u>

A nation of poets sounds utopian, but our subject is women's fear. Aquatic courting evolved to overcome female fear and indifference when sex got scary and dominance drowned. Speech came from predictable competition under unusual circumstances, and sets the stage for the modern human's entrance.

Utopian as it may sound, it was not. Hesitation, not desire for simple excellence, drove our evolution; men competed to overcome female fears, and many failed. If men were discussing the Meaning of Life, it was because women were seeking that kind of man, not because the Meaning of Life was waiting to be discovered. Utopian as it may sound, we cannot go back. For better or worse we returned to the land, and our brains no longer work as aquatic brains did.

The aquatic female was torn between her desire for the male and her fear of having him too close. She had no checklist of approved male behavior, which once completed would magically extract her consent. Rather she balanced her affection against fear, sensing whether she wanted to hold him in the palm of her body. That the male *intended* her no harm counts for nothing; her *perception* of peril built the male's barrier. Predators aside, the most dangerous being in her world was an adult male. Aquatic sex meant she had to approach him face to face, and take him into her full embrace.

Like ourselves, beavers mate face to face, sitting upright in shallow water or drifting on the surface. The smaller female will not couple with the male until she defeats him in actual combat, though she may lose a few rounds on the way. Once she has subdued him they are a pair, and will live in perfect amity from that day forward, sitting for hours close together, murmuring to each other.

Beavers conquer fear by combat; as aquatics we conquered it by persuasion. Ours was a more common style, for aquatic mammals don't usually fight in the water. (Beavers fight on land.) It's also normal for apes to compete by show rather than force. This let female preferences guide male evolution; males became elaborately persuasive because females could refuse.

And males were up to the task. On our return to land male skills became a tragicomic handicap and we discarded intelligence. Courting collapsed. Groping in the dark replaced aquatic competence, and dating started to make us throw up. Many today use drugs or alcohol to dull dating's stress; by anesthetizing fear we can mate in spite of clumsiness. It can lead to discomposure in the morning, when the medication is gone but the stranger isn't.

Aquatic women couldn't avoid changing along with men; while guiding men's evolution, women's preferences were themselves under selective pressure. If natural selection pulled men in from one side, it pulled women in from the other, as even the smallest orbiting moon slightly perturbs the largest planet. While men evolved to allay female fear, females evolved to both *feel* less fear and to *enjoy* the fear they felt.

Women had to tolerate and then embrace strangers chest-to-chest. Body-front eroticism, childhood desensitizing, conversations and surrendering weapons all were needed because her sense of danger persisted. So women who *felt* less fear, or learned to *enjoy* the thrill of that fear, had a reproductive advantage. Women evolved emotional responses to fear, creating what we now call "love", to make sex tolerable.

Nothing condemns skin-hunger or male courting as inevitably inadequate. If love had not appeared, these would have grown stronger and would now suffice. But once love evolved, it helped reduce women's fears, and allowed skin hunger and courting to remain insufficient on their own. Love gives men permission to be less than perfect.

In water we adapted to fear as if sliding down a gentle slope. Each girl child became used to close contact; adult females had only minor fear of men. But fear grew explosively when we returned to land, and suppressing fear became crucial. Desensitizing ceased when men stopped cuddling female children; courting collapsed and conversation died when men became dull; dominance by force returned when men could again fight, and women saw their peril. We came up out of water only to fall into a fearsome abyss.

Parents Of Passion

Love is a human life's peak experience, delicious terror, beloved imprisonment. Love is fear, focussed and compressed, as diamond is carbon. Women evolved love by their competition, as men evolved language, to fill the aquatic courting gap.

In most work on evolution, love has little place. Many think love proves human superiority, and only a sophisticated brain grows this mental flower. We began, some say, with a pair-bond that tied together unfeeling pre-human adults for children's sake. A hunter had to know he fed his own kids; his mate needed an instinct for faithfulness. With time and the human brain's growth, a pair-bond became affection and then love. Some call love a transcendent emotion of a superior being. This mindset cannot consider love in australopithecines. If small-brained *Australopithecus* didn't show human intellect, it couldn't have passion.

I disagree. Not civilized or genteel, love is primal and raw. Not soft and tepid, love is rough, clutching and brutal. Not a surface emotion, not gravy on the rice of our lives, love grips our innards. Not a sculptor's mallet to carve a flower, love is a sledgehammer to knock us off our feet.

We do not *learn* love; we all feel its power too clearly and too alike. We pursue love insanely. Losing it leads some to suicide or murder; rejecting it leaves one shriveled and hollow. It may take us unaware with breathtaking impact; we instantly dismiss lifelong plans, if love appears.

As powerful as it is, love remains curiously hard to identify. Magazines carry quizzes to help the reader recognize love, but never a quiz to identify anger. Though love's power betrays its instinctive quality, we somehow remain unsure of it as of no other emotion.

Love's feeling announces its lineage. The adrenaline rush, pounding heart and stomach impact all feel the same in love and fear. The weak knees and vertigo of seeing a loved one are like nothing so much as the weak-kneed dizziness of being narrowly missed by a car. We did not receive passion as a Mighty Hunter Big-Brain graduation gift, but evolved it from older instincts.

Love's first parent was fear, one of many fears we can feel. This hazard evolved a pleasure reward, just as jumping from airplanes gives pleasure to some. We evolved an innate hunger for the pleasure this fear brings; we seek risks for pleasure's sake. We have become so sensitive to love's joys that we never see its fear-thrill origin.

Love's second parent was obsession, focussing our attention to a point. When attention became elusive in a species born to show off, indifferent individuals failed to reproduce. When only personal preference could lure women to sex, preference evolved into obsession hard enough to scratch any indifference.

Passion is a uniquely human hunger for a specific focussed fear, evolved when tolerating fear was crucial to our survival. What we think of as a superior emotion is no final flower of chivalric tradition; love was working smoothly four to six million years ago. Lucy (*Australopithecus*), tiny brain and all, fell in love as we do today.

Love From Fear

Aquatic women evolved love to enjoy danger they could not avoid. Passionate women reproduced better; it had nothing to do with faithful wives and returning hunters.

Love began as tepid stuff, for aquatic fears were mild. True, a female had to face males in a pose made threatening by forty million years of primate confrontation. But water cushioned her approach, it let males desensitize girls, it subdued dominance and force. Still fears remained.

We began face-to-face sex when rear-entry coitus grew awkward, not because frontal sex inherently pleases more. Facing a man bothered some women less than others; when frontal sex began to give advantage, those women turned around. When females using frontal sex out-reproduced those who did not, their children by sheer numbers converted the whole population.

Women who best-responded to desensitizing became our frontal-sex pioneers, more sensitive to body-front eroticism, able to handle the fear this pose evoked. Not *oblivious* to fear, they sought it as a parachute jumper relishes the leap. They liked it because it was a little scary; by chance but not by plan, manageable fear became a sexual thrill.

Animals regularly seek stressful situations. John Calhoun set up rat cages linked by passages, giving some parts uncontrolled traffic while other parts could be guarded by dominant males. Uncontrolled sections degenerated into a riot of cannibalism, rape and infanticide. Yet rats from protected spaces regularly visited the chaos, apparently seeking nothing but the thrill, since they already had all they needed at home.

Primates, too, seek danger. Youngsters love to pester adults and each other, seeing how far they can go before being punished. We enjoy sports that offer risks; we probe boundaries of permitted behavior and survivable games. Human adults living in cities show the same interest (and the same violence) as overcrowded rats.

Women evolved an instinct for excitement into a sexual lure. Without a fear-thrill, only the meekest of men might leave progeny; the more dominant a male, the more danger of confronting him. Not a fatal evolutionary course, but not the course we took. Women enjoying fear's titillation gave permission for male dominance to continue. When such a female thought a male dominant he became *more* interesting, not less.

Many species use female fear as a sexual lure. Lorenz noted it in fish (cichlids), where a male's fear immediately extinguishes his sexuality, while female sexual interest is impossible *without* some fear of the male. Fisher suggests men choose small women because such women were less scary; I suggest women choose larger men because they are *more* scary. Not by accident do women speak of finding men they can look up to.

Male dominance vanished in water, and a woman attributed status according to whatever competition men could manage. Her ability to refuse led to male song and speech — the feminine ideal of aquatic masculinity. But she couldn't ignore a forty-million-year-instinct that males were dangerous *simply as males*, and fear (or, more mildly, respect) is the innate root of dominance.

Because women lacked an objective yardstick for men, love's fear-thrill grew as a subjective loop. A female alone attributed status to the male. *(He was exactly as wonderful as she thought he was.)* His increased status in her eyes made frontal sex more exciting. *(She could always imagine she had the alpha male.)* More-intense females reproduced faster, giving the same advantage to their daughters. *(The more easily she imagined dominance in any male, the more babies she made.)* Women evolved into parachute jumpers, enjoying a danger men never saw.

When we grew fat and slippery, frontal sex became practical but not automatically warm and fuzzy. Over thousands of generations the vagina tilted backward, helping maintain intromission and costing us the deep-touch orgasm. Without some other sexual reward we'd have slowly gone extinct as orgasms faded.

But love was stepping in, weak and tepid and growing. Women who found confrontation exciting reproduced better. As an emotional reward for *any* contact, love gave permission for orgasms to depart. Sex became more exciting as the fear-thrill strengthened, and when deep-touch orgasms died our fear-thrill had grown enough to help bear the evolutionary burden. The worse sex gets for them, the more love women will evolve to feel.

So enjoying a certain danger became a sexual reward for women, as for the female cichlid. Unfortunately, love hasn't had time to acquire a unique character. We still confuse it with fear, and use self-help quizzes to decide what we've got.   Confusion often leads to heartbreak when people think they have love and it turns out to be merely excitement.

We cannot separate suppressing fear from female sexuality. Danger is possible in any relationship, but habitually denying danger can also blind a woman to true peril. The thrill of being with a strange and exciting man can quickly turn to the terror of rape or death.

Willingness to endure fear came from a thousand generations of forgotten braveries, as women who could not face men were culled. Earlier I noted that females may have gained sexual rewards by clinging to males. While boys explored, girls showed a different courage which too had value, bearded the lion in his den, faced the most dangerous of animals.

Often women and men display courage in different ways. My wife once showed me astonishing bravery, confronting adversaries who never recognized her fear. We praise a man who passes rites of manhood, using only a spear to kill a lion. A woman shows equal courage when she confronts dangers that threaten her alone and brings home trophies giving *her* life meaning.

Obsession From Focussing

Obsession became love's second parent, evolved like our fear-thrill from earlier feelings. Obsession is compressed and focussed attention, as diamond is compressed carbon. It became perhaps our most important mental asset.

Apes consistently care about personality. While both male and female chimps will usually couple with any available partner, some prefer certain individuals. This sometimes conflicts with hard-won sexual rights, and illicit lovers must hide. Males may spend weeks pursuing a single female; females may take pains to mate with a favored male. When the sea muted normal dominance, personal preference was all we had.

Courting slowed when frontal sex made mating scary, and a woman narrowed her choices. Instead of a number of men, she sought one or two. Women who *already* tended to prefer certain men gained a small advantage over women who resisted monogamy to the end. Focussed women wasted less time soliciting.

Focussed females had a better chance of being impressed. As instinctive showoffs we gain confidence from attention, and confidence announces dominance. Women who listened better *created* confident men from clumsy stutterers, cutting alpha males from whole cloth. Since his confidence is her innate mating cue, the female who pays attention has an increased chance of finding a mate.

A woman who focusses on one man creates isolation in a crowd. Isolation enhances courting in humans; love stories frequently feature pairs who ignore each other only until chance leaves them alone together. Humans often contrive isolation to aid courting; chimps, gorillas and baboons also arrange honeymoons. With people all around, the female who concentrates creates a mental island for herself and the man.

Aquatic females with a gift for focussing lifted the male's status by paying attention, and removed his competition by ignoring them. The more quickly a woman could go from first handshake to obsessive interest, the more likely his showing off would meet with applause. In a world where dominance lost its cues, where not every female could have the alpha male, such a female could imagine value in even a klutz, make him seem an alpha, launch her progeny.

Women competed against other women through their ability to focus. Easily distracted women created fewer alphas from klutzes, remained unimpressed and unpregnant. Females whose attention was more easily caught were more easily courted, and passed their susceptibility to their children. It had little to do with the alpha-ness of males who courted them.

None of this could have evolved on land. Terrestrial estrus guaranteed a female would get her itch scratched as often as she liked. She could easily solicit any male and knew who was alpha. Personal preference guided her choice occasionally, trivially, harmlessly. But aquatic sex created new competitions, one of which weighed different women's success at overcoming their fears. Impressibility was merely one property that helped conquer fear.

Once focussing appeared, like our fear-thrill it grew stronger. In an uninterested troop, attentive women make more babies. When all are attentive, obsessive women win. Focussing led by degrees to obsession, desiring one person to the exclusion of others. The object of our obsession automatically becomes the world's most valuable person, the most-desired partner. Reacting in this way evolved to become a mental earthquake, powerful enough to shake any stable objection, because the susceptibility overcomes doubt when mating is subjective. We oppose this force when we try to dissuade a friend from dating a creep. Mere logic rarely prevails.

Stop for a moment and imagine a massive bowl, a magical container, filled with an enchanted liquor of wisdom. Its capacity has no limit; the more you pour in the larger it becomes, the richer the mix. And with each addition to the bowl new miracles can be drawn out of it — wonderful devices, curative potions, answers.

Many see the human brain in these terms. In the human-supremacy view of evolution, our brain is an intellectual bowl, a repository of information. We grew this brain to hold this data, to give us a competitive edge and conquer the world. For many, our brain is the Holy Grail of evolution.

Yet we barely outscore chimps on intelligence tests. Up to the age of a year and a half, chimps easily outpace human children. Darwin noted that the human differences may merely *seem* to be peculiar to us, and may simply be due to our use of language. I will later show why I think brain size has little to do with intelligence, and how brain growth came from other forces.

Our brains-for-smarts myth clouds the brain's most remarkable function. Turn over the bowl, dump the knowledge soup and see it as a bell. The human brain's key feature is its susceptibility to being rung bell-like by the presence of the beloved.

Our most important mental asset is an ability to pay attention, to focus on a problem. It doesn't matter if other animals are as smart, or if it isn't the smartest human who labors on a problem. Even an omega anthropologist might find something of value if he chews on a puzzle long enough. More than intelligence, focussing sets us apart from other animals.

We evolved a knack for focussing as a courting aid in the water, but we can focus on anything. Obsession is a mental trick, not a sexual response. Humans invented airplanes because we worked over generations on the problems of flight. We cure diseases because at the end of an autopsy we still remember why we cut the body open. Chimps never get beyond the attention-span of a child; gorillas are often too apathetic to realize a problem has come up.

Humans never needed great intelligence (and I will later describe an excellent reason to be a little dull) but it *was* necessary that we tremble at the presence of another, to ring like a gong when struck. Women evolved hyper-impressibility when men had little to impress with, giving us a mental trick we later turned to our advantage. In the Great Scheme of Things it is much more important for you to fall in love than to invent a spear. For now my evidence is this: Nearly all fall in love but few invent spears.

Love In Men

Women evolved love (fear-thrill and obsession) to keep sex tolerable and men desirable. Men were beyond their depth; they didn't need love and couldn't instinctively balance its parts. When women evolved love to ease fear, love in their sons started to get scary.

Not inherently sexual, fear-thrill and obsession are weak spots in women's emotional armor, helpful because they reduce hesitation, evolved to clear hurdles raised by women's fears. Fears were constant but mild in the water; women chose their mates and none were raped. Once love overcame women's doubts and frontal sex worked, passion was strong enough. Women's passion-competition worked unseen, weighing female against female, favoring those willing to copulate, culling those who avoided frontal sex. So love in women automatically rose to a level that ensured procreation, and little more.

Men inherited love from their mothers, but didn't need passion to overcome masculine hesitation. Male focussing certainly eased courting; men who gave their attention to just one woman were more likely to get hers in return. So their mothers' *necessary* instinct was at least *useful* in men, but men had no *need* for new feelings. Like a squirrel hunter with a buffalo rifle, men had too big a gun for the job.

Love in men is unruly and confusing; men fall in love more quickly than women, and are probably hurt more often. In part this came from our intellectual collapse when we returned to land. Men shed specific mental skills when those skills became a hindrance, and in the process became oblivious to subtle signals which otherwise might avoid much grief. Our sexual collapse, so distant now, obscures love's process in aquatic men. Modern man, for better or worse, is not the man women loved in the water.

Fear-thrill and obsession seldom balance in men. Both men and women inherit a taste for danger, but danger itself we learn from birth. Women know that men can hurt, and our innate fear-thrill evolved to overcome women's learned fear.   But men rarely feel threatened *physically* by women, don't have acquired fear to overcome, and men's sexual encounters lack the threat of a powerful partner. Not exactly instinctive according to sex, the need for fear-tolerance is inherent in body sizes of men and women, and so works *as if it were* instinctive. While we all learn fear from life's dangers, obsession is purely inherited and therefore inevitable in both men and women. Without fear as a counter-balance, obsession often leads men to acts that terrify women.

An unruly instinct does not excuse, but it may cloud motives. Men often shrug helplessly when asked why they stalk or rape, for instincts resist rational discussion.   There will be more to say about male misbehavior; I will later suggest that most rapists are normal men, though badly trained. When women protest "He had no right to do that to me," I can only agree he hadn't the right, but perhaps he had the instinct.

In women the separate forces of fear-thrill and obsession work in tandem so frequently that we can hardly distinguish them. We call this the correct recipe for love, and put quizzes in magazines to validate it. These two evolved to a balanced and common goal; the recipe is correct when women mate. But men have no such congruent function, and are clumsy with the ingredients.

Love In The Water

We are fast approaching our return to land. Though I described love as evolving while we were aquatic, in fact it only began there and set our evolutionary course. We can now separate it into two stages — mild incentives acting in the water, and powerful guides acting on land.

Love evolved to help women overcome fear. Love in the sea was a weak and tepid thing, because danger was small and it took little to persuade a woman to couple. On land sex became terrifying, and only powerful passion could overcome the sense of danger.

Love evolved to counter fear: It adjusts generation by generation to stress, and to the need to make babies. In the sea fears were mild, since males desensitized females to frontal contact and swayed them with sophisticated courting we have now lost. So in the water love never evolved far beyond powerfully liking someone.   But when we lost automatic orgasms, this tepid love alone sufficed to lead aquatic women to couple.

Fear exploded when we returned to land. In water individual size was immaterial; on land the female had to face a larger male at close range. In water she swam faster and could elude unwanted men; on land she ran slower. Aquatic girls clung to men's beards and masturbated against their bodies. On land no male tolerated a child supporting its full weight from his face. Finally, after our move to land, men evolved a less sophisticated style of thought and male courting skills collapsed.

This presented a daunting challenge to the pubescent female. Without childhood desensitization, with no personal experience of man-as-pleasure, she still required the desire to copulate with him. He lost the ability to sway her, and could do little more than announce his availability and interest. She had long since lost automatic orgasms, and could get more reliable satisfaction by masturbating.

Easy alternatives and meager rewards quickly weeded out all but the most passionate of lovers. Women for centuries have asked why they expend so much passion on men who return so little. Change the question to a statement and it makes more sense: Women expend so much passion *because* men return so little. When male skills collapsed, passion was what it took to bridge the gap. If men were as perfect as women-in-love think men are, women could afford to love men less.

From humble beginnings as simple personal preference, love exploded onto land as an overwhelming force. Our brain evolved acute susceptibility to this new force, hunger for the gong-like reverberations the beloved could cause. Love evolved first to overcome doubts in water, and grew to counterbalance terror on land.

Chapter 5 - Water Shed

We returned to land over the mind's horizon. Beyond lived barely conceivable poets in water; on the near side appear hunter-gatherers we easily recognize. Returning to land was accidental and crucial; all we have done since, we could do precisely because this return was so difficult.

Sexual fear hindered our return. Though far from promiscuous in the sea, our coupling remained fairly casual. Back on land, terror limited our sexuality. Sexual play between adult males and young girls was not coercive or scary in the water, but on land could become abusive and traumatic.

Sexual conflicts guided our evolution. Males and females have separate dominance styles in most mammals, but complete agreement on sex. Once a male wins female attention (by courting, ritual combat or territorial possession), they have simple coitus, mutual consent, and equal pleasure. But humans constantly fight over sex's meaning, their own intention, and permitted behavior. Women live in fear of unwanted sex, with rape a constant danger. Some men delight in women's submission and sexual humiliation. Though other species do have rape, in no other species do males use it as a threat or warning to females, nor did men before we returned to land.

Sexual disagreements owe much to miscommunication between men and women. Our emotional makeup changed radically as we returned to land, as men and women moved on different courses. Women evolved passion to breathtaking intensity; men shed mental skills when they became a hindrance. Conflicts multiplied.

In water women had the initiative, but men took control as we returned to land. No longer could a female *always* refuse sex; no longer did males *always* compete for her attention. The result was brutal, terrifying, and extremely successful. For promiscuous picnickers sex was entertainment, for aquatic courters a career, and for returned terrestrials a necessary evil.

Terrestrial Transition

We were committed to the sea; no more-aquatic primate ever returned to land. We spent more of each day in the water than seals do, though seals look more marine than we did. We ate, slept, mated, and died there. But we were a snippet of a trifle; few Miocene apes entered the water and fewer left it. Terrestrial returnees came from fringe groups, and represent all aquatics no more than whales represent all mammals.

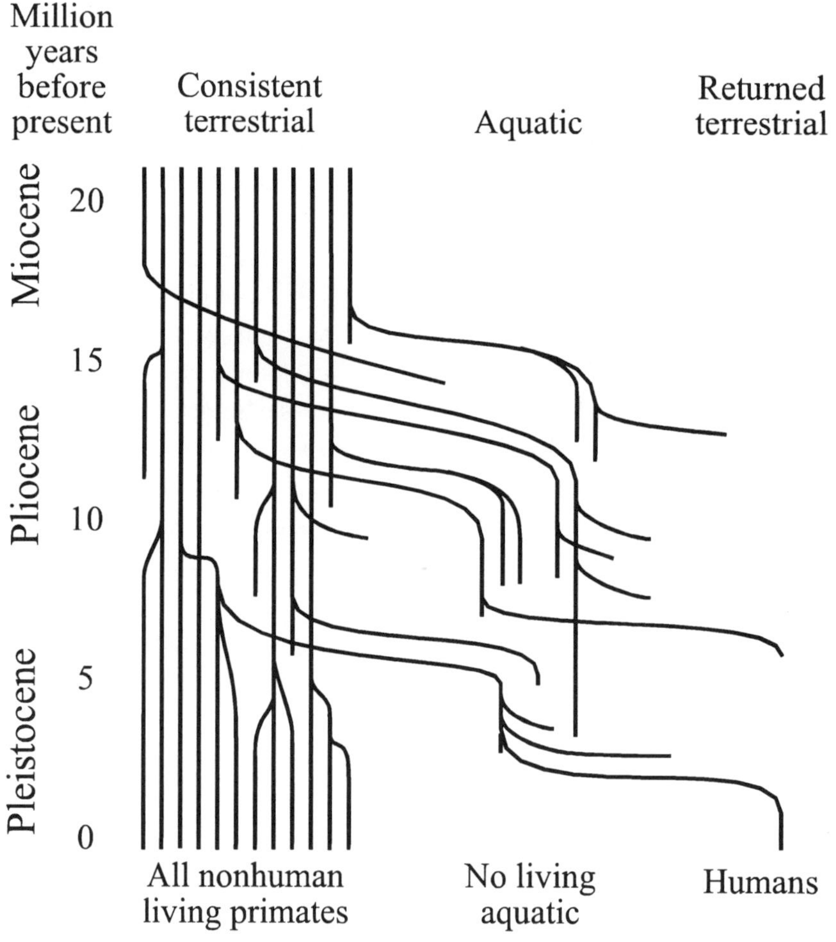

Figure 12: Aquatic Population Flow

Diet led us into the water and out again. More plentiful food on land brought us back as the Pliocene drought ended, tribe by tribe evolving out of the water. We slept on desolate beaches in early aquatic days, when ground was merely for sleeping on, but we came back only where the land was green.

Our aquatic path had a dead end, though we didn't know it, for our feeding style barred further evolution. Primates always use their hands for feeding, and so do a few others, but most animals grab food by biting it. Like a raccoon's, our feeding style shifted nicely to shallow-water foods, and quickly let us become aquatic. One can grab a clam as easily as a fruit.

Evolving to reach ever faster in water, we never reached a point where we had to catch food in our mouths. Our canine teeth grew smaller (to help suck mussel innards) making teeth useless for grabbing prey. Increased brain size (for flotation) rotated our faces forward instead of lining up with the spine, making it impossible to get jaws out in front to grab a fish.

Our evolutionary path was narrow and short; our primate heritage separated us from other aquatics. Feeding with hands, we couldn't evolve fore flippers or much speed. Primates can easily *become* aquatic, but can *exist* as aquatics only under favorable conditions.

Time and luck and our limitations gave a few aquatics the chance to return. We were good enough in water to live there and move around, poor enough to benefit by returning to land. Given time and chance, some of us wandered into places where the move back could happen. Very few began the return and fewer completed it. Most lived in water until they went extinct, but fringe groups occasionally found local forces urging them up on land.

Our limitations pressed us close to beaches, though we did not walk on land. Slow and vulnerable in deep water, we stayed near land to avoid predators. Every ocean beach had some of us nearby, slowly flowing on a broad coastal highway. Population growth forced us into marshes, up rivers, oozing into every cranny where an aquatic ape could find food.

Pushed against beaches and up rivers, we occupied a broad range of habitats. Since we caught prey with our hands (and needed to see it), we liked clear water in ocean bays and rivers. Unlike others whose dental equipment commits them to perhaps only one or two foods, our grasping hands and versatile mouths let us change diet at will. Since we don't breathe or drink through gills or skin, we could move from salt- to fresh-water whenever we chose.

Along Indian Ocean coasts from the Cape of Good Hope to Java lay thousands of sites ideal for an aquatic's return. Just as some of us specialized in certain fish or mussels or shallow-water plants, a few specialized in life on the beach and then the plains. Our return was never inevitable, but given enough apes and time it might *appear* inevitable; it was not a single event but a flood of groups evolving back to land.

Fossil remains of apparently bipedal hominids have been found all along these coasts. While East Africa and the Rift Valley hold modern attention, our missing links' first relics came from the far east. In 1890 through 1894, Eugene Dubois found Java Man on the Solo River in central Java. Fossil hominid remains in China and elsewhere show a broad distribution of similar primates over a long time. More hominids have turned up in Africa, but African remains may reflect where we've looked more than where we evolved.

We moved into coastal water to escape the drought, and spread along the Indian Ocean coasts to reach Java, Sumatra and the South China Sea. Many places offered more food than the drought-stricken beaches where we entered the sea. Rivers made rich deltas; trade winds kept some shores lush with rain. In those greener places some aquatic apes quickly evolved back to living on land. If they left the sea as early aquatics (with fur intact and legs not yet aligned to spine), they probably thrived, bearing no visible hint of an aquatic sojourn.

But mid-stage returnees were crippled. I defined "mid-stage aquatics" as those who would not revert to quadrupedal walking if they returned to land. Leg-spine alignment committed them to an upright posture; hairless bodies meant they always carried their infants. Females with young needed help, or easy pickings.

Mid-stage aquatics still mated for orgasms, not love; the pelvis had started rotating but had not yet fully deepened. Too early for poets and singers, they hadn't evolved emotion-based pairing. Independent as a Miocene chimp, they lacked the cooperation needed to help females survive. Barely able to manage in fruitful places, they had no resilience and died off quickly when greenery faded.

In our early aquatic time, the coastal "highway" was *only* a highway — a path from drought-stricken lands to greener places. Thousands of groups moved into it, troop loyalties collapsed, and a mob slowly flowed along, thinning and mixing while bodies adapted. Where large rivers formed marshes or trade winds watered islands, our primate limits urged us onto the land. Populations of mid-stage aquatics evolved back, and died when local lushness again turned brown.

But our aquatic highway was also a training camp. Wherever apes entered the sea, wherever they flowed, time alone made them more adept. Collapsing dominance made them court with language; deepening pelves made emotions replace female orgasms. Oblivious to the grand scheme, they evolved bonds of affection and mutual need to replace independence and promiscuity.

Over millions of years, our chances of returning to land improved. Love and language spread along the coast, evolved and interbred. As sex got worse our chances improved; apes who could mate without orgasms could do anything. Returning groups became more closely paired, more cooperative, resilient in hard times. Once winking out after only fifty generations perhaps, populations of returning aquatics now wavered like candles for a thousand generations or more.

Finally some groups became successful terrestrials. With powerful languages and emotional bonds, they survived though crippled by bipedalism. Supporting weaker members, adapting to what they found, they could live on less-promising land. No longer did a handicapped ex-aquatic need lush greenery; when humans cooperate they can thrive in the desert.

The Road Back

The better a primate adapts to water, the less likely it will return to land, and the more predictable its returning-site becomes. Paradoxically, completing a difficult return increases its chance of long-term survival. Returning to land resembles crossing a ridge: After the top it's all downhill, but you need to pick the right pass.

For millions of years, on coasts spanning a third of the globe, countless aquatic apes became terrestrial. Early returnees cleared low obstacles in easy transitions, and went extinct. We need to measure the hurdles facing late-stage aquatics, including our ancestors. Understanding what makes a successful move can help show why easy moves lead to long-term failure.

Like any specialization, moving back to land requires breeding isolation. Any individual's liking for shore foods would be lost in a gene pool of deep-water feeders. No matter what one rebel prefers, only breeding isolation of a small population lets his preference survive long enough for natural selection to weigh it.

Some aquatic groups moved to land on isolated islands' lee sides, where greenery crept down near the water. Perhaps many groups moved to land on many neighboring islands. Even so, these must have been segregated from the larger population to permit specialization. Aquatic apes' movements determine the minimal isolation distance. If they routinely migrated thousands of miles, then it would take a lonely island indeed (and an unusually stable local population) to give breeding isolation. But apes don't migrate, as mediocre swimmers they avoided deep water, and movements were moderate.

Many islands offered enough isolation. Island chains splay across the seas, with some chains' ends too far apart to let shallow-water swimmers cross over. Though we continually moved from island to island, each archipelago's extreme end was almost-completely isolated. It exchanged only a few individuals with the next island, which in turn passed only a few to the next, and so on.

Many islands offered enough food, too, dangled near sheltered beaches where an aquatic might see it. Once their diet included land-based foods, they could spend time on and near the beach looking for them, being weaned from the sea. Even productive oceans have leaner places; a generous island in miserly seas could draw many populations from the water. But island transition led to oblivion because islands small enough to be safe for us also limited us to small populations, constantly facing extinction through climate shifts or disease. Islands gave no room to spread the risks.

Rivers made better paths. Like islands they offer breeding isolation, but rivers also allow subtle behavior changes, gentler on-the-job training. Though we evolved in the oceans, we moved upriver before returning to land. River channels reached away from the ocean like a watery bush, trunk to branch to twig, each stream bearing populations like fruits. Headwaters of separate streams a mile or so apart were as biologically remote as continents.

Difficulty of moving upriver completed the isolation. Cataracts severely limit upstream movement, and even slow water is hard to swim against. Going downstream was easy, but isolation comes from limiting *upward* movement. Any aquatic ape *could* swim up rivers; we had language to discuss it, obsessiveness to take the challenge. Lone adventurers matter less than what the average tribe can manage, with women and babies, unlikely to move far or quickly.

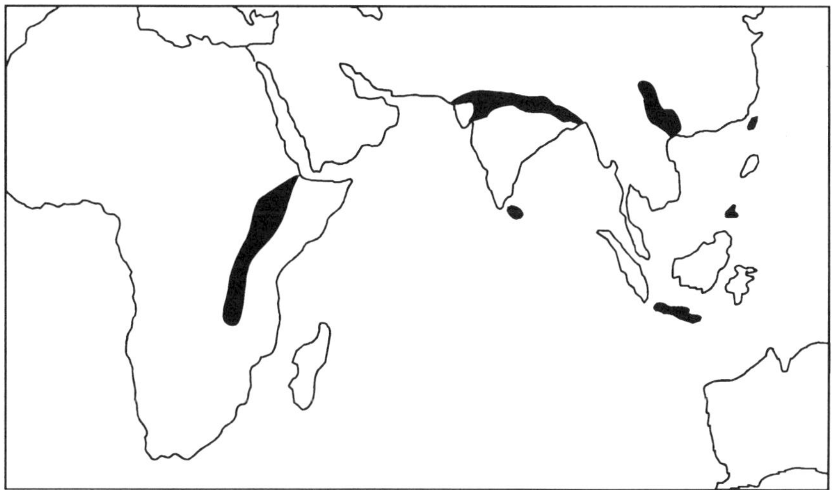

Aquatic apes colonized river systems, and foraged in floodplains.
This accidentally put them in ideal starting points
for populating the continents.

Figure 13: Rivers Seed the Continents

Amazon porpoises illustrate riverine isolation, living far from the sea in streams and flooded jungle. The most-primitive of porpoises, they either evolved there or arrived early in their evolution. They have no breeding contact with salt-water porpoises nor with the evolutionary forces the seas present. With little need for speed, they are agile but not streamlined, resembling pink bags of rocks with fins. Like porpoises we went upstream, detached from our parent population and evolved in new ways.

We seeded the land; thousands of populations slowly filled thousands of tributaries. One lone tribe might have sufficed to create us, but we grew from hundreds of returning populations. On a coast spanning a third of the globe we reached out countless genetic tentacles, each unaware of the rest. Up the Rift Valley, the Ganges, the Indus and the Yangtse, swimming hominoids went where they found food.

We gradually found new dangers. We knew about aquatic predators, and watched for crocodiles, but now terrestrial predators began fishing for riverine aquatic apes as a bear takes salmon. So we became more cautious, listening to the land before we saw it. Rivers were quieter than beaches, and let us hear alarm calls of animals and birds, and notice when sounds stopped. As a blind man navigates by sound, with our ears we saw life beyond the riverbanks, and knew well the jungle before we stepped out.

We began a land-based diet long before we walked there. No windswept beach barred us; while wading we gathered berries, and eggs of birds, crocodiles and turtles. Amazonian porpoises swim through the jungle during floods, returning to the main channel as water recedes. By doing the same we trained for life beyond the bank, and learned what could be found out there in the trees.

On some river stretches we got more food from the banks than we could by chasing fish. Such lush zones weaned us from the water. Knowing the sounds (or silences) which meant "danger," we cautiously collected what we could and scuttled back to safety, webbed feet flapping. We probably stole now and then from a predator's kill. Weak, fearful, awkward, and slow, we were fish out of water.

Terrestrial Evolution

Ideally, we would have shed aquatic frills when we returned to land, becoming a terrestrial quadruped, chimp-like and perfectly adequate, but evolution has no plan and rarely reverses its course. We came out of the water with most of our aquatic baggage, and carry it still. It doesn't all help us, but hasn't yet killed us.

Our ancestors needed only minor physical changes to cope on land, and we evolved them by first adapting to shallow rivers. Because we frequently stood on the bottom and always swam by opposing our legs, we retained a functioning pelvis. Any aquatic ape could stand on the beach, if it wanted to and was not too fat.

For an aquatic in shallow water, as for modern humans, walking is often faster than swimming. We chased trapped fish in foot-deep pools, lifting our feet high; webbing and long toes slowed the runner, and rapidly vanished. Our swimmer's long metatarsals remained long, plunging down spear-like through water to the bottom, where stubby, web-less toes splayed out to launch our next stride. You can see the change in your own hands and feet. Your soles are twice the length of your palms, partly due to the metatarsals' extreme length, yet your shortest finger is longer than your longest toe. Our hands show the same bone-size ratios as those of chimps, but our feet are virtually toe-less.

The same running style worked as we moved up the beaches. We chased nothing faster than fish in shallow-water cul-de-sacs, and probably needed speed most as we ran *back* to the water for safety. We slowly changed from running in shallow water to running on sandbars and mudflats. Unlike any other ape we run on the balls of our feet, as a seal moves on the tips of flippers.

Slow as we ran, we were always faster on two legs than on four. When we became aquatic, steady selective pressure gave us a streamlined body and an effective swimming kick, with each tiny straightening giving some benefit. We returned to land with a short broad pelvis and long swimming legs, and could run on all fours no better than a modern human.

Moving quadrupedally might have *become* faster, if we had gained some incremental benefit from it and evolved to suit. Too awkward a pose for anything but play, we never used it and therefore couldn't evolve toward it. So we remained bipedal and one of the slowest land mammals, not by superiority but by lacking an evolutionary path to anything faster. We carry things because a biped's hands are free to hold with; we didn't evolve bipedalism to free our hands for carrying things.

As we walked, our knees bore our full weight for the first time in perhaps a million years. Increased strain made bones grow thicker all through our skeletons, but the greatest danger was in our knees. I suggested earlier that our knees could hyper-extend, and now a swimmer's asset became a runner's liability.

Bending your leg back at the knee presents no problem; when absorbing a shock it can bend until your heel hits your butt. But knee hyper-extension was a limited *reverse* bend, enhancing a scissors kick. Only knee tendons' tensile strength kept feet from swinging forward to hit bellies, and a life of swimming strengthened them for the role.

Great for a weightless swimmer, this was terrible on land, for tendons are strong but not elastic. A stride too heavy or a jump too far could rip apart the knee like a football player's, so the reverse knee-bend quickly vanished, commemorating generations of injuries. We show only a trace of this hyper-extension now when we lock our knees, as no other primate can.

Moderate changes to feet and legs could make an aquatic ape barely competent in the shallows and on river banks, able to find food and survive. But aquatic grace gave terrestrial awkwardness; competent walking hindered swimming. Short toes and webless feet made running faster but swimming slower. Limited knee hyper-extension prevented crippling injury but degraded the vital scissors kick.

We teetered briefly here on an evolutionary "ridge"; we could roll down either side but could not balance for long. If we evolved terrestrial adaptations, we became slow and clumsy in the water; if we adapted to water, we could not move quickly on land. Every returning aquatic encountered this ridge, with its height determined by how long the animal had been in the water. Those whose bodies better-suited the water needed lusher conditions to lure them out.

"Choosing" between a terrestrial or aquatic life reflects net survival; the value of food weighed against the danger of pursuing it, for instance. We became aquatic because a drought lasted long enough and we found food in the shallows. We returned because the pickings in some spots on land were very good or in water very poor. Our return shows a net advantage in finding land-based foods (for a tiny fraction of aquatic apes) despite our obvious limits.

But moving to land held no attractions for women, and discussing their reluctance will consume the next two chapters. We were breeding populations, not adventure clubs, and until all could manage on land, all were bound to stay near the river. We took this gamble unaware: Not wonderful in either environment, we had to all adapt to terrestrial life before the transient local lushness faded. Then we could learn to survive in leaner places, slowly colonizing the continents.

Spread along Indian Ocean coasts and up the rivers, each green pocket with its aquatic ape population marked an evolutionary race. A time limit was built in, though none saw it, and the time varied for each locale. Perhaps many began evolving for land, then returned to purely aquatic life a few thousand years later as land-based foods grew scanty. If they had too little time to re-evolve webbed feet, they may have starved and died as demi-aquatics.

An aquatic's return to land was no sure thing. We can see how the most adventurous might explore river banks. Our survival lay in persuading the *least*-enthusiastic (a woman carrying her infant) to do the same.

Children Went First

Children left the water first, men next, and women last. It was never inevitable that we return to land; of three suspected aquatic primates, only we clearly did. But given that we did return, it *is* inevitable that some individuals do so more readily than others.

Children led as in our earlier move to water, yard by yard, with each generation playing farther up the beach. As adults they remained comfortable where they played when young, and their children in turn went a bit farther. By accident children found new foods which became treats, then habits, then staples. We supplemented our diet with vegetables, berries, and or slow and helpless prey. By degrees we moved into shallows and up the bank because at each stage we found more to eat. We were a *product* of local lushness, not an *instigator* of change. We never chose to leave the water, and most aquatic apes had no such path.

As dietary success lured us farther up riverbanks, we adapted for it. We collected eggs, fruits and vegetables, and like most primates we rarely shared. The farther we explored, the more we needed speed for running back to water. Toes continued reducing in size, webbing faded, but we remained tied to the river, for women were not yet on the land. We were stealing bird eggs and running home, not evolving for long-distance chases on the plains.

Children pushed the boundaries of their world. Their forebears once dabbled in shallows as mothers scavenged on beaches; now they scampered through the shrubbery near mothers still in rivers. They invented new games, stopped to snack or nap on the bank. Once we slept on ocean beaches and hurried to the water each morning; now we slept in the water and (some of us) in the mornings clambered up the bank. The greenery gave places to hide, gang up, pounce on; who could resist?

We moved to land not because we *intended* to, but because individual likes can give long-term reproductive advantages, and lush land gave success to those who liked being there. It took a generous bit of jungle to support a half-crippled ex-aquatic ape. All children played in the shrubbery; as boys grew to men, they found more to like on land, and women found less.

Adult male chimps stalk prey, including pigs and monkeys; male baboons kill young grazers. Male chimps go on hunting binges, rarely sharing meat with females or young. They don't need this for nutrition (grubs give equivalent food value), so chimps apparently hunt for sport. While we were returning to land, males resumed stalking prey and scavenging kills of other predators.

Killing small prey never required weapons; chimps rip captured animals apart by brute force, peeling and fracturing the skull to get the brain. But skeletal changes giving us quick grabs in water also made us weaker. A chimp can tear apart a human, but a human cannot dismember a chimp.

With weakened jaws and reduced canine teeth, we needed rock hammers for even monkey skulls, as we used rocks to crack shellfish. Each aquatic mob, slowly evolving up the riverbanks, received the same lessons. Over hundreds of generations they smashed bones left behind by predators, or cut off meat with sharp stone edges, killed injured animals with them, then began stalking.

This created the bipedal hunting ape. No ape would give up quadrupedal walking for a slower two-legged stride; no ape would trust a rock or stick when it had canine teeth. Only with bipedalism forced upon it, after evolution had disarmed and weakened it, would an ape regularly use weapons in place of teeth. We armed ourselves only because we were otherwise too crippled to survive.

We became upright hunters in an evolutionary kindergarten. An arboreal ape forced out of the trees by drought hasn't the natural weapons to defend itself nor the skill to make and use them. As Morgan notes, such an ape is doomed. But a lush pocket of riverbank, with enough food to make scavenging merely interesting instead of necessary, gave our ancestors freedom to make mistakes along the way.

Men Went Home

Peace of mind kept men on the beach, though food sustained them and drove evolution. Men went half in a daze, as adults might return to a beloved home last seen in childhood. Water had always frustrated men because they couldn't display dominance, and if males couldn't display it females couldn't see it. So singing and language evolved as attention-getters for courting, helping a female to find *something* to cheer about in her mate.

In water, male aggressiveness had its hands tied; men had no way to assert their will. Each male saw this as we became aquatic, and hated it. Each morning the mob raced down the beach, his power vanished and he couldn't stop it. But now a young male playing on the riverbank could race across the sandbar, thump someone, and run away! During our aquatic time, males always had the wistful feeling that it would be awfully nice to pound a victim now and then. Suddenly they could.

A sleeping giant, male dominance woke and shook itself. On the beach men could thump or be thumped, establish dominance, recognize status. On the beach men swaggered again. As children they formed gangs and picked foes; as adults they knew who they could displace or to whom they must yield. Male dominance could be *selected for* again as well as *satisfied*. With land-based foods newly important, land-based dominance parceled it out. As in other species, dominant individuals live better, grow bigger, and leave more children to inherit the dominance and muscles.

Females never stopped looking for dominance, they just lost clues for measuring it while in the sea. Now that men again confronted each other, women again saw the victor's strut and loser's slink. When women instinctively responded to old dominance cues, male status began replacing verbal skills as a courting aid. Women don't pursue football players for their poetry.

For the first time in millions of years a hierarchy appeared; a male knew his place in line. Not everyone could be the boss, but everyone knew where he fit. For forty million years we evolved within a complex social structure, until the sea yanked it away and left us with only personal preference. On the beach, for the first time since the Pliocene drought began, men had peace of mind.

As much as food, this lured men to land; hierarchy gave a crystalline view of the world, and fed a different hunger. More than women, men will accept less if they can live where hierarchy is clear, will be a policeman or soldier though underpaid. It's easier to persuade a boy than a girl that soldiering is fun. Men stayed on land simply because they *liked* it there.

We have arrived at our first generation of football widows. Like guys clustered for a televised game, our forefathers spent idle hours on land, comfortable where victory and defeat left no doubt. Unable to explain it, they agreed that women just didn't understand. Real men sit on the beach.

Female Limits

Women and men lived in different worlds. Women's fears meant little to men; thumping victims meant little to women. The beach made women nervous and uncomfortable. Aquatic women's sexual hesitations, though mild, had been enough to foster men's elaborate courting skills. For women, being on dry land was so unpleasant that aquatic doubts pale by comparison; only a revolution from within could persuade women to go.

Women and men largely ignored the others' competitions. Pre-aquatic males had rolled like teenage toughs through the playground, frightening mothers and their young. Females never needed to *see* male dominance contests, though the male hierarchy always gave females a valuable mating cue. Women noted the swagger without having to see the fight. Noisy male contests and brief supremacy bore little resemblance to female lineages, holding power over generations.

As infants, males and females started out roughly matched, with young males slightly more aggressive. All played on the river bank, first under a mother's close supervision and then more independently. So each young female and male had an equal chance to run and tumble.

With advancing age, males grew rougher. Young men thumped each other, ganged up, found their place. Aggression brought them satisfaction not felt for millions of years. Each also saw combat's rewards displayed by older males. Alphas gained power through judicious threats along with useful alliances; every day young males saw power's value. So their instincts, role models, and personal experience told them that aggression was the key.

Little females didn't see the point. Though they understand physical dominance, females usually compete more carefully. Naturally smaller and less-aggressive than males, they gain more by cleverness than by muscle. Though females *will* fight, subtlety and wise alliances give more certain success. Lessons of push and shove, which taught each young male how to win, meant little to a female, and being on dry land meant even less. Being able to run and thump someone doesn't help much if you're smaller, if you have no inclination to thump, if your competitive style doesn't demand it.

Women found rewards in the water where older women and role models stayed, where alliances took shape. Competing lineage by lineage, females had long evolved an ability to function as a family member. Aquatic life never weakened female alliances, and when social structures dissolved in water the female's family focussed her life. Any intelligent woman stayed in the river, for women found discomfort and fear on the beach.

Physical Burdens

Her physical burdens came from pregnancy's awkwardness and children's weight. Each pregnant woman quickly learned how heavy she was on land. Regardless of her jungle adventures when a girl, as a woman she learned to stay in the water. This lesson, flowing inexorably from her sex, had the force and effect of instinct though each generation learned it fresh.

A bipedal pregnancy was torture. Spinal-femoral alignment began evolving early in our aquatic transition, starting bipedalism while we still slept on beaches. As full-aquatics, that alignment achieved its final state; our long legs and pelvic changes allowed no four-legged walk. Pregnant females were the first to sleep in water when we became aquatics, and the same discomfort held them back as men returned to land.

No body structure had evolved to support bipedal pregnancy. Water buoyed swollen bellies for millions of years with no sling-like ligaments, no strong backs. Fossils suggest both men and women were pot-bellied in water. Though eventually life on land gave us slimmer bodies and stronger backs, the first terrestrial females didn't have them.

Pregnancy plagues modern women with circulation problems, morning sickness, and compressed bladders, yet we have been back on land for three or four million years. If being sick and slow makes living more difficult, then they are an evolutionary force, and we have had several million years of improvement. Aquatic pregnant females must have suffered greatly, trying to walk on the beach.

Even unpregnant women were slower on land than men. Women evolved as better swimmers, since they had to tow children, and had proportionately longer legs than men. Women have more weight in their hips, store more fat in their thighs, and have to move all that when walking. They would have been the slowest runners when they came out of water, as they are today.

Pregnancy magnifies these problems. A quadruped's swollen belly swings freely from her backbone like a sack from a beam; cows on our farm could run even in late pregnancy. But in bipedal humans the fetus is cradled in the pelvic bowl (which evolved not to cradle but to anchor muscles), and must rock left and right with each stride. Men could walk bipedally with effort, but the woman's gait became a laborious waddle.

Her first infant's birth made staying in water *imperative*, because women were hairless from the neck down. We had retained body fur only until fully aquatic, when children shifted to a light hold on scalp hair. On our way up the riverbank we had no chance to re-evolve body hair, and no mother on land could support a baby's weight with her scalp hair alone.

And babies couldn't dangle from what hair she had. A terrestrial primate's hands and feet can usually support its weight from birth, though ape mothers help their young for the first few weeks. But aquatic life made our feet so flipper-like that they couldn't grip at all; our hands evolved for quick grabs and lighter strains, not for supporting full weight.

In the sea our babies had also become heavier. A chimp or gorilla baby is a skinny little thing, but a human baby starts to add fat shortly after birth. Fat insulates and gives flotation, but reducing the load in water means increasing the load on land. Even if it had prehensile limbs and fur to grip, a human baby couldn't support its chubby body unaided.

Finally, babies had evolved large and heavy heads, which they now cannot lift until several months after birth. Brain expansion began during our aquatic life for head flotation. Out of water this makes heads heavy, but it helped prevent drowning by elevating the nose. As we moved up rivers, infant drownings increased because swimmers float lower in fresh water. Brains slowly evolved larger to compensate, making it more difficult for a riverine baby to lift its head on land, making it need its mother more.

Aquatic life revolutionized our babies. Infants entered the sea riding on their mothers' backs; the infant who came out had to be carried. Any intelligent woman stayed in water, where babies and bellies were weightless. When women finally did move to land they carried heavy loads, and a modern woman with a child shows the solution. All human cultures share a universal tool, used by no other species, vital for life on land: Something to carry the baby. The baby sling is our single most important tool, outweighing any weapon. When women finally did move to land, and wove crude baby slings from riverbank grasses, they cut us loose from water and made us mobile.

Pleasures Of Companions

Beyond awkwardness, being on land was distasteful for women. Remaining in the water gave no reward by itself, any more than simple walking gave men a thrill. Instead, being on land gave men the ability to move and dominate, supplying emotional rewards they had long missed. In the same sense, being in water created conditions females found emotionally rewarding and prevented things they disliked.

Women stayed in the river partly because their friends did. Female primates find allies in long-term relationships, and tend to stay near other female family members. (They'd be glad to have well-behaved male family members too, but young males prefer to join older dominant males.) Primate mothers and daughters often keep company for life.

Any group contained some pregnant females, some with new infants. These felt comfortable only in water; walking on the beach exhausted them. Unencumbered women had no strong desire to be on land and liked to stay near friends, so burdened women made the decision for all. Distaste for the land and a tendency to mingle kept most women in the water most of the time.

This lesson repeated in each generation, with each little girl seeing many women in the water and few on land. Without terrestrial role models, she too habitually stayed in water. She then became the next generation's guide as an aunt, friend, or mother, showing that women stay in the river. There women sympathized about the time men wasted on the beach as modern women gripe about the time men waste in front of televisions.

Women stayed in water also to compete, for feminine competition depends on allegiances and lineages. An instinct to form allegiances is a female competitive asset, just as a propensity to thump brings advantage to a male. (Females also use force, and males make strategic alliances. We are speaking here of small differences.) Female alliances may usurp food supplies; a female's status may (in yet-unknown ways) affect her ability to procreate. (Primatologists were slow to see female competition because subtle rivalries can elude detection.)

Rarely does female competition demand thumping someone, but male competition depends on the threat if not the act. Life in water had little effect on female competition, but destroyed classic male competition completely. Usurping food supplies (oyster beds, say) had as much value in water as ever on land. So women's tendency to form alliances remained an important trait, and women stayed in the river partly to compete in a way men couldn't even see.

Attraction To Children

Playing with infants also confined women to the river, in another competition invisible to men. Women get all gooey over babies, and primates generally show great interest in infants. More-interested females make better mothers; less-interested more often lose their young. Culling the uninterested over millions of years has evolved excellent mothering skills in most mammals. Primates, with the added dangers of arboreal life, reached a peak of parental excellence.

Deep interest that leads to play becomes *practical* experience. The kitten who stalks a spool of thread will make a better hunter, gaining from practice pounces. Little girls with dolls, or young women who baby-sit, gain as the kitten does: They wouldn't do it if they didn't care, and they'll do it better as a result.

Women may satisfy their mother-hunger with any object, pet, or kidnapped baby. Childless females often obsessively focus on others' children. Young primate females try to steal babies and mother their siblings; all ape females apparently compete to babysit. Younger infants cause stronger mothering responses, and a newborn evokes the strongest urge of all. Women and chimp females kidnap infants, not ten-year-olds, and humans value younger adoptees the most-highly.

Since new mothers stayed in water, where babies had no weight, a child might not stand on the shore until the modern equivalent of age three or four. So mothering happened in the river, and a girl's location was a form of competition. Any female who liked children enough to stay in the water and play with them was more likely to become a good mother. A female who spent her time exploring on land showed less maternal instinct, and might have been more likely to fail as a mother.

A young woman exploring on the beach did no *harm*, nor did one who stayed in the water *guarantee* her maternal success. There was no sex-specific process encouraging an exploratory nature in girls, and there was at least a minor selection *against* such behavior. So babies gave one more reason for women to stay in the river.

Fear On Dry Land

Finally, real fear kept women off the beach, because women saw danger where men saw none. Predators carried more risk for a female, smaller and less able to defend herself if surprised. With pelvic geometry evolved for better swimming, women run more slowly than men. A woman fifty yards from water might be as vulnerable as a man thirty yards farther out, and surely knew to stay closer to safety.

Primate females are less boisterous than males, and show less bravado. Their new handicap happened to coincide with their milder play style, nipping adventurousness in the bud. Caution is certainly the correct response to vulnerability, as young females who ignored their risks fell victim to predators.

On the beach men knew they could run and thump; women learned they could get thumped. As the smallest sex, each female learned that the beach gave rougher males a chance to hurt and frighten her. Water protected her from all that.

Finally, males who pummeled her were not the courteous suitors her ancestral mothers knew. Males now exercised dominance they had missed for millions of years; they were more aggressive, more intimidating to young women. Any weaker member was a convenient butt for assaults, with females the surest victims.

So young females clustered at the riverbank while males forayed inland. For late-aquatic women to stray from shore was akin to a modern woman walking down a dark alley. Land presented dangers both real and imagined; the female had reason to fear both predators and her own kind. By comparison life in the water was secure and rewarding. Who could resist?

Aquatic Confinement

Women were confined to water, chased by fears men didn't share, lured by rewards men couldn't see. But men were evolving to survive on land and women inherited the changes from their fathers. No animal can adapt perfectly to both land and water, and though women remained in rivers they could not remain unaffected. We were one species; either men or women would shape an evolutionary course that could go only one way.

When males began exploring the land, our feet and legs evolved to help them run. When short toes and lack of webbing became helpful, when knee hyperextension became limited, these changes spread through the population. Men's evolution took away the flipper-feet and flexible kicks women still needed for swimming. All returning aquatics face this evolutionary ridge, where the sum of prospects could roll us down either side but none could balance forever.

Foot and leg changes partly immobilized women who stayed behind. With swimming ability eroding, women couldn't catch fast-moving prey. Men (who had once helped with children) were now up on the land, and women did all the baby-work. Evolved to tow one infant, and now a poorer swimmer, a woman had to drag two or more while her mate stumped up and down the beach. With a cluster of dangling infants, our burdened and immobile female carried her species' future; her failure meant extinction.

Female need should now (it would seem) create supportive males, should create the foraging male hunters and more-sedentary female gatherers that most anthropologists describe. A country lush enough to drag us over the transitional ridge should provide enough extra for men to bring some to their mates. Sharing is easy to imagine, simple to evolve. It might have happened in any group, and surely happened in many, but it didn't happen for us.

Altruism is unusual in primates, though possible in many species. (In one case, pelicans fed a blind companion for apparently many years.) Male predators bring food, male ducks guard ducklings, male primates protect their own young. A helpful male supports his genes by supporting his mate; sharing genes commonly gives rise to cooperative behavior.

A returning aquatic could gain a reproductive advantage from only a little assistance, while miserly men would leave fewer children. Altruism could benefit not only the individual but also the group; tribes of generous men would grow faster than groups of less-helpful souls, whose children were malnourished. Mighty Hunter theories flow from this very premise: We supposedly evolved to carry home haunch of wildebeest on a biped's shoulders, to hunt because it fed our children.

Both male and female chimps beg for meat when adult males capture prey, and the possessor sometimes shares it. Newly terrestrial males, scrounging a predator's kill, might return with loot. Though not initially intending to feed a female, they might respond to her begging. Perhaps it could start with children, running back to the river clutching eggs for Mom.

A specific female request also might start the practice. We had used language for a couple million years by this point, and women surely could list the groceries they'd most appreciate. They could agree in advance or dicker after the man came back with meat, negotiating with language and bargaining with sex. So the theory goes, though usually framed in a terrestrial setting.

An ex-aquatic male could have brought food not to just any female, but to his current sexual partner. In water we began forging emotional bonds to overcome fear of frontal sex; she would be on his mind when he returned. Once men saw their value to women, they could use food to compete for mates; food gifts proved a man's prowess and promised future meals.

Food sharing had value only because females became less able to feed themselves while aquatics climbed the transitional ridge, while bodies restructured for running. Land-based foods became more important as they got better at collecting them, and they could finally get all they needed from the forest. Helplessly evolving along, with declining success in fishing, females finally would wait for lunch to arrive while they tended the children.

At Home, Barefoot And Pregnant

But it didn't happen that way for us. *Pay attention here.* We are turning quickly on a pivot of chance, trekking in pain over the evolutionary ridge. In some places men did help, and all died at the water's edge. But our fathers did not help, and wandered off, and women loved and followed them. We are children of men who walked away.

The limited duration of lush environments marks the outer limits of a time span. The place must be kind enough to support a returning bipedal ape (crippled by aquatic life) until the animal can become marginally competent. Legs and feet must adjust; women must invent baby-slings. Before the pocket of green fades, the ex-aquatic must learn how to live on less-generous turf. The more aquatic the ape, the lusher must be its first land home.

Within that time span, women's burdens mark the inner boundaries of our move. The more aquatic her people, the greater her doubts. If she were self-supporting, or well-supplied by her mate, the group could go on half in water and half out. But if women slowly starved in the water, dwindling in each generation, then only their rapid move to land could save them from extinction. In individual terms, only if aquatic females suffered greatly would a land-going female gain advantage (relative to her sisters) despite the new hazards. Evidence for a rapid move by an advanced aquatic therefore suggests high pressure on females.

We could move no faster than the quickest female, but we had to move before the pocket of green could fade. With men and boys on the beach, the group teetered on the ridge, nourished by different worlds. If a female's immobility in water placed her at risk, then moving to land would help despite her distaste. Scanty evidence suggests that females suffered in the water and made a rapid move, and such a move determines the group's success after the move.

Populations able to move quickly to land might do so even in habitats lasting only a few thousand years. High selective pressure can push species to evolve very quickly. Once terrestrial (and mobile), they can then move to follow opportunities. Stable invitations (lush jungle next to a so-so marsh) can shepherd dozens of such pliant breeding pools out of the water like buses loading and departing in quick succession.

Slower-evolving groups occupy the bus stop for a longer time, reducing the total flow. They might begin a land-move in a transient pocket of green, only to go extinct or revert to aquatic life as green turns back to brown. Any lush spot lasts a short time before it exists a long time; there are more new Edens than old ones. Slower-evolving groups will more rarely find themselves in a long-lived Eden; fewer such groups become terrestrial.

Any half-aquatic group might evolve extensive male support, allowing that group to move more slowly to land. So easily evolved, it must have been a common evolutionary path. While stable groups do not necessarily imply male support, highly supportive males do imply stable ridge-straddling groups.

If many groups populated the continents, then most came via the method that builds terrestrial groups most effectively, just as most cars must come off assembly-lines regardless of whether that method makes the finest cars. Of all returned-aquatics, most made the move rapidly through the available bus stops. With females the most reluctant travellers, returning happened most efficiently when women were pushed hard onto the beach. None of this means that filling the continents was an evolutionary goal or even a desirable end.

We can now see a lottery in action, where a population improves its chances by not helping females. Without any plan, the landward move accelerated the less-helpful and slowed the cooperative. Most continental colonizers were pools of minimally-helpful males, because females in those tribes had to evolve quickly, and therefore those groups moved more quickly through the bus stops, out-colonizing slower-moving and cooperative ones. Absent specific selection for a more- or less-helpful style since then, men should have remained about the same.

Men in modern hunter-gatherer groups contribute only a third the food that women do — about what one chimp can beg off another. Many men are capable of walking away from their mate and children, apparently without a backward look. Both of these suggest that men have no overwhelming instinct to aid their mates. It seems wildly improbable that such an instinct faded after our return, for life got no easier. So modern male behavior suggests (to me, at least) that females got little aid in water, and made a rapid move to land.

Women were confined and burdened in water, and afraid to go on the beach, yet *something* moved them. Within any one lifetime the situation seemed stable; for any one gene pool the process might reverse. But the move to land happened most reliably when women moved quickly, so the lottery favored any incentive that pulled women out.

Chapter 6 - Birth Of Love

Love is delicious terror, beloved imprisonment. Passion compels women to do what they'd otherwise avoid. When staying in water made sense, love disconnected sense and brought women out despite reasonable fear. Love records men's abandoning women and women's running after; it shows our land move's success despite women's distaste. Not evolution's goal, passion is the survivor's scar.

Love is life's peak experience, exhilarating and terrifying, shaking us to our foundations. Despite apparent disorder, love works perfectly today as it first evolved to work. No happy accident of sophisticated brains, passion is a primal force pure as terror or anger. To understand love we need to imagine life without it.

What do women see in men? A man might be a social vehicle that a woman learns to drive; he might offer protection, or be her best exit from her parents' house. These cannot explain passion; what else does he offer? The average man is bigger and stronger than the average woman, and might pose a real threat if angered. He's bristly in the morning, and sex with him usually won't give her orgasms. Men often cannot voice their feelings or understand hers. What does she get from him?

Love measures successive subtractions; women who saw no value in men avoided them. Women who had a weakness for men, who found subjective rewards to replace orgasms, made babies and passed on their weakness. Women's view of men as valuable became stronger as generations of uninterested women were subtracted at the water's edge.

When leaving the water gave advantage to the group, natural selection favored any mental trick making the land-move seem worthwhile. Women who found value in men had a reproductive advantage, creating daughters to inherit that mental trick. Women who joined men on land loosed their tribe from the river, to migrate and multiply. Unable to dissuade men from land-based entertainments of dominance and status, women survived by following them. Any woman who resigns herself to watching football games for the sake of male companionship does the same.

There are costs when women buy company. In the feminine mind, two distinct whips propelled them from water and left them scarred. Women evolved passion to make men seem worth the effort; women evolved sexual repression to leave themselves no other choice. As a man these whips seem distant to me, but I think they are real.

<u>Evolving Love</u>

Love burgeoned in an evolutionary instant as we moved to land, balancing new female fears it had to overcome. Love shows men had little practical value and no objective measure; if they had either, women wouldn't have evolved love as a substitute. (The love I'm exploring here is the fear-thrill and obsession that first appeared, mild and tepid, in aquatic frontal sex. We include many emotions when we speak of love, but fear-thrill and obsession will do for this chapter.)

I argue that love evolved in stages, not as a single event, because each stage has separate evidence. Minor hesitations in aquatic times made a mild thrill adaptive, and mild desensitizers evolved. Later, only the potent impact of the land move could transform mild thrill into addictive passion while keeping the basic form intact. Without our bringing fear-pleasure from the water, we could not have tolerated frontal sex on land.

Love was weak and tepid in water, a child of fear and grandchild of frontal sex. That pose gave a better grip and surer impregnation, making beards and permanent erections useful as desensitizers. Beards held infants on men's fronts, permanent erections gave pleasure as well as a mooring. Meanwhile women evolved frontal eroticism, and their childhood masturbation desensitized them to close contact.

Only fear's existence made dispelling fear useful, and gentle persuaders would help as long as fear hindered a close approach. Yet fear remained, because desensitizing evolved only enough to make frontal sex endurable. Fear is also dominance's partner, and an instinctive female sexual cue. Women's fear-thrill began evolving in water along with frontal sex. Liking the thrill let her keep her innate sexual trigger, and tolerate the stress of a new coital pose. But water suppressed dominance, while singing and language show that men turned to gentler competitions. With fear minimized in water, our fear-thrill was low, defining the first stage of love's evolution.

On the beach love exploded into awesome power in a few generations. After the land move we had no desensitizers. Sexual contact of a man and girl-child changed from a game to abuse. This wasn't just a cultural switch (voting to make it illegal), but a shift in men's motives, making such laws necessary. Meanwhile the man refused to tolerate a child hanging from his beard, nor did children any longer want to. Beards became ornaments again, free to wither.

Of the female lures to sex, only frontal eroticism and the fear-thrill remained. She had lost automatic orgasms long before. Even if nothing else changed, losing desensitizers meant other rewards had to carry more of the load, and greater passion must necessarily evolve. Meanwhile, new fears appeared. Dry land made her particularly vulnerable to predators, and men became more dangerous as dominance reasserted itself on land. A weakness for fear-thrill exactly met this hazard. We needed no new emotion, because women could clear the heightened barrier by strengthening an already-existing instinct.

Finally, women who moved to land sacrificed the company of friends and children for love's sake. Women today do the same, ignoring friends when a man appears. Women's groups built our aquatic society; women and children were islands, with men flowing past in hopeful streams. But during our land conversion, the women who made the move did it better if they cared less for friends and children, and left the matriarchy behind.

This defines love's power: Love induces women to couple without lifelong desensitizers. Passion lures women from friends and children for the sake of a man. Love blinds a woman to hazard when her delusion makes him seem more important than the risk. This is a brutal force; that we seek it desperately does not make it an unalloyed blessing.

When we left the water, love grew from mild thrill to overwhelming obsession. Love evolved when we needed it for our survival, like any creature's teeth or claws, by subtracting the failures.

Where The Men Are

When we started our terrestrial return, sex happened only in water. As men and children began to frequent the riverbanks, water remained a place of refuge. We were more adapted to water than to land, felt safer there, and knew the river's dangers better than the jungle's. Men went on short expeditions, scavenging forays brought in food, and reborn hierarchies gave men emotional rewards. The more time men spent on land, the better they liked it. From life in the river with a few surreptitious moments on land, males switched to life on land with a few panicky rushes to the water's safety.

All men returned to the river for sex (because that's where women were), but men became increasingly uncomfortable in water. When a man entered the river he lost the ability to race over and thump someone. As in the original aquatic move, his aquatic status came from the fact that others knew they would eventually get thumped, once they all were back on land. The more men enjoyed dominance and hierarchy on land, the more they gritted their teeth when forced to give it up. Women probably liked seeing men in the water, out of uniform.

Regardless of his distaste for water, a man needed a woman's cooperation and voluntary move to land. Never able to outswim her, he could not capture her and drag her out. Any man who once attempted such a terrifying act would be shunned by all. Women who first came up onto the beach for sex did so willingly, but accidentally pushed their sisters to a new level of competition.

The First Time

In each breeding population straddling the evolutionary ridge, some pair coupled on land for the first time. They had no shallow-water transition as when we entered the water. Back then, rear-entry sex was the norm, easily done standing knee-deep just as on dry land. The movements remained the same in hip-deep water with genitals submerged. But frontal sex needs support, and modern couples usually lie down to make love. Some do it free-standing or leaning against a wall, but these seem improbable in early experiments.

Our transitional couple might have used rear-entry sex, in the shallows with both parties standing, or with the woman on hands and knees. But they were already being adventurous to do it near or on the beach; it's unlikely they'd also try a pose unused by humans for millions of years.

We can guess their identities. She had recently played on the beach as a child, but had no children of her own and had not learned pregnancy's awkwardness. He was her sexual partner and her reliable entertainment.

She was hungry for his company, compelled by the innate fear-thrill and obsession that led aquatic women to copulate without orgasms. He was a dominant male who didn't like being in the water; her following him acknowledged his dominance. She would sit through a football game, if he wanted her to. Her mind held the key difference (her susceptibility to being impressed by him) but she *thought* the difference was in him. She thought him the most lovable of men, but in fact she was the most loving of women.

By modern standards he was indeed impressive. Shortly after we moved to land, a series of events depressed human intelligence, and this couple marks an early peak of human intellect. He was more gently-persuasive than any man who has lived since. Heirs to a million years of verbal probing, they shared thoughts with an expressiveness and intimacy most married couples never achieve.

They went up on the beach as two teenagers might make love in a cemetery, giggling the whole way, saying "I can't believe we're doing this!" She was laughing and confused, a little intoxicated by her audacity, happy with him and trusting him to keep things safe. She loved him.

They coupled with her on top, in the pose that most closely copies aquatic frontal sex, where she clasps him with her legs and is slightly above him. In this position she felt safest. Women's hesitation barred terrestrial sex in earlier generations; he now needed her willing participation. He'd do it standing on his head, if she'd do it on the beach. So she coupled with him on dry land, both giggling at their own daring, and if they liked it they'd do it again.

Selection For Love

Each population needed someone to do it first, but thousands of women made love on land before the last left the water. Once the first woman tried sex on land, natural forces spread it rapidly through the female population. The first woman (in each population) was deeply involved with her mate, more susceptible to his appeal, and at the right stage of life to try the move. When other men saw this, they began persuading *their* partners to come up on land; other women saw men's increased interest, and tried it too.

Until women moved to the beach, men went back to the river for sex, like married couples sleeping in twin beds. A woman who mated on the beach intercepted that male traffic, that slightly grudging commute, at least for the man who interested her. So for some men part of the time, sex was where they wanted it.

Women who conceived on the beach bestowed on their young a more-intense emotional involvement, diluted and unmeasurable. Whatever it took for a woman to do it, her children had a chunk of it, her grandchildren had a fragment. Dominant men disliked the water, where politeness outweighed muscle. A woman who mated on land allowed her partner to avoid suspending dominance, and it meant most to the alpha males.

Such a woman would have been the subject of stiff competition. Each woman's willingness to swive on the beach immediately made her a more desirable mate. Though she might not do it there every time (and might not actually conceive there), men quickly noted her sexual enthusiasm. Coupling on the beach, like wearing a low-cut dress, advertised possibilities.

Dominant males instantly courted these women while intimidating competitors. The woman who came up on the beach became suddenly the center of attention, like a plain girl pursued by the whole football team. Once the first woman made the move, all women saw how quickly the desirable men competed for her.

Sons of these women inherited dominance; daughters conceived on the beach inherited their mothers' weakness for being thrilled. After a dozen generations, every person had an ancestor or two conceived on the beach; after two dozen generations, many forebears. Sex-on-the-beach was not an inheritable trait, but men's preference for being on land gave a reproductive advantage to daring and passionate women. Though some remained in the river, genetic dilution meant that *all* shared a heightened response to love.

Initially, women criticized those who met men on land, fast girls who chased too eagerly. Everyone knew that women belonged in the water. Over generations, however, beaches became respectable places to get a mate, and finally the *only* place, as women intercepted men before they reached the water. Genetics and competition combined: All women inherited powerful emotion, and any woman wanting a mate had to go to the singles bar.

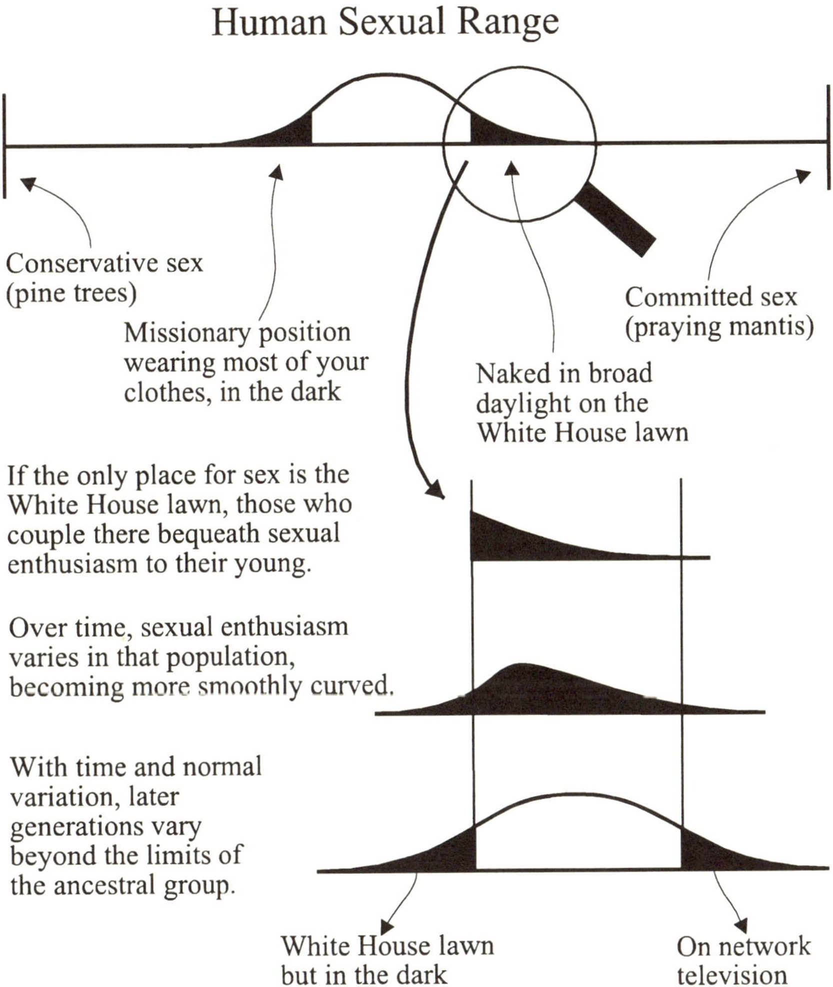

Figure 14: Inheriting Sexual Bravery

Time on land became a regular part of each woman's life, no longer for just childish play. Women with infants were there awaiting their partner's return, or seeking their next mate. At first they ventured onto the beach only while the infant was easily carried, or walking on its own. Later on, women invented baby-slings, and waddled about while pregnant. They'd rather have stayed in the river, but early daring females unknowingly raised the stakes and forced them all out.

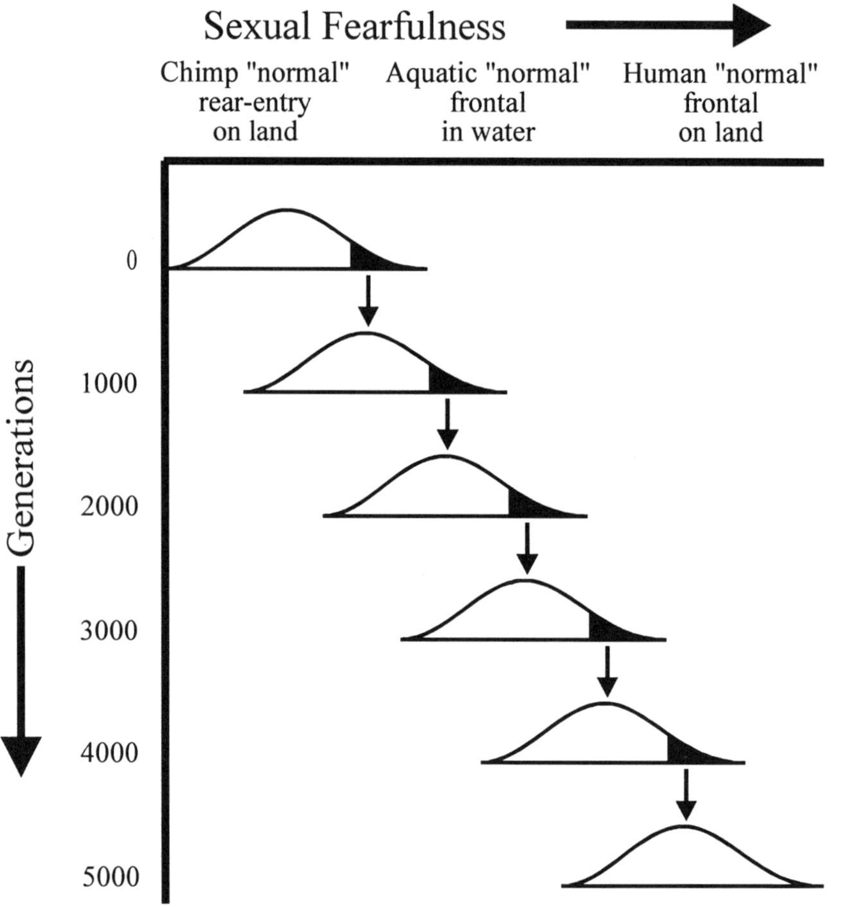

Figure 15: Evolving Sexual Bravery

Land offered women new nutritional rewards, increasingly valuable because men's terrestrial evolution hindered women's swimming ability.

Men didn't feed women in the river, and they didn't help much on the beach, either. Women's main advantage was in sharing the transient produce from their pocket of green. With infants trussed in a sling or parked nearby, women could eat land-based foods. When women brought their children over the evolutionary ridge to land, they came because they felt passion, but they survived because they gained food and mates.

With all women on land all the time, love's evolution continued unabated. Men consolidated dominance hierarchies; physical threats were a part of daily life. Women's fear of men, no longer subdued by water, quickly grew. Overcoming fear and confronting a dangerous male favored the woman who most enjoyed the delicious terror. Women evolved to think men wonderful, not because men were, but because that's the feeling their mothers had required in order to overcome fears of terrestrial frontal sex.

Defining Love

Passion's definition will form through the rest of the book, but we have enough for a first sketch. Women don't have a monopoly on love; mothers bequeath passion to their sons, too, and I will later discuss its effect on them. People embrace for more than love; love is more than a fear-thrill. But women seem the obvious place for passion to evolve, and our return from water seems the obvious time.

Love is to affectionate respect as diamond is to carbon; we left the beach carrying an emotion so fierce it continues to shape us today. It leads us to pursue our beloved despite logic; it feels much like fear because it evolved from fear of frontal sex. Women began it as a mild thrill to counter mostly imaginary dangers in water. They then evolved passion to an intense obsession, when their fears rose on returning to land.

Females didn't *imagine* danger on the beach, but faced real hazards and evolved a fear-thrill to match, tightly focussed and all consuming. Confronting males brought risks for a hundred million years, leaving us with instincts to help avoid it. Along with instinct come life experiences, teaching each individual how confronted males react, teaching each girl that boys play rough. For each breeding population to survive, females needed a hunger (instinctive or not) powerful enough to overcome well-founded fears.

Disconnecting intelligent hesitation was a key adaptive response for women. By evolving a hunger for certain kinds of danger, while clouding their judgement of their immediate best interests, obsessive fear-thrill allowed frontal copulation. Staying in the river had such clear advantages that only women unresponsive to logic would go up on the beach. So women have a better chance of making babies if love makes them a bit crazy.

Some may argue pairing is simply sensible: It's logical to cling to a loyal and supportive mate; it's reasonable to stay with one who respects you; it's less confusing to remain with the other parent of your children; it's practical to dwell with your house's co-owner. Humans marry because they are so sensible, some might suppose.

For lack of logic, nothing matches falling in love. (I don't demand that love be logical. Mine never is. But I want to see why being *illogical* apparently carried no evolutionary penalty.) This most powerful of human emotions often opposes what look like our best interests. We routinely fall in love with terrible partners, despite warnings from parents and friends.

If a woman found rewards in every man's touch, she'd need no obsessive hunger. She'd have a mate as we have cars, when and if convenient. Her hunger evolved because sex benefits her genes without her individual payoff. From the average man the average woman gets limited orgasms and limited understanding. Love, concocted from hungers she inherited from her mother, makes a man tolerable, necessary.

Love is not about pairing, it's about overcoming doubt. A woman must embrace a possibly-dangerous man in a confrontational pose with no reliable orgasm. Doing this on land means she must carry her pregnancy and her children, with no guarantee he will help. She can see her own likely fate in failed relationships and bitter women all around. Only narrowly focussed insanity can explain her willingness to copulate.

Despite examples, warnings, and personal experience, women keep trying. We may be reasonable and responsible in all other aspects of our lives, but when it comes to pairing, our inability to learn from past disasters ensures that we'll go on making babies. Love works perfectly today as it first evolved to work: It makes us periodically insane for our species' benefit.

Sexual Repression

Women evolved to suppress their own sexuality, the second force driving them up the beach. Self-repression came from bad sex. When sex became less rewarding for women, they evolved a reluctance to masturbate because that made copulation better by comparison. Self repression is not the survival of the fittest but the culling of the quitters.

Mixed blessings, love and sexual repression still guide women today. Like love, sexual repression bloomed from an aquatic seed when we moved to land. Like love, repression reflects an evolved instinct, not a choice. Women bear no more responsibility for self-repression than basketball players bear for being tall, but repression, like exceptional height, improved their chance of success. Awkward for the individual, good for the team.

Masturbation teaches a young animal how its body works; women who don't masturbate as girls are often inorgasmic as adults. Those unable to find a mate (or to conquer their own fears) may masturbate exclusively, and are skimmed from the gene pool. Their sexual relief has no *direct* evolutionary effect, for none inherits their genes, but their *act* evolves. If masturbation is too rewarding, even minor mating problems may dissuade weaker individuals from seeking partners. So masturbation evolves as a guaranteed orgasm, but never quite matching sex with a partner.

Primates in particular demonstrate masturbation's importance. The vagina's numb lower third discourages exploration, reflecting deaths from excessive masturbation and infection, hence the numbness' evolution. Reluctance to masturbate also would have protected females, but did not evolve except in ourselves. The lower vagina's insensitivity (uniform across the whole primate order) shows that masturbation has consistently worked better than self-repression for female primates.

For a female aquatic ape, masturbation was an important means of desensitization. The male's permanent erection gave girls a masturbatory tool, part of their sexual growth. For girls, the penis' *genital* identity was trivial; they used it as a girl uses a table leg or the arm of a couch. But its availability meant they masturbated on men, not furniture, and grew accustomed to the position. Masturbation in water led to better reproduction, not poorer.

Sex grew rough on women when we moved to land. No woman could separate her fears from the sexual act; she had to conquer all her doubts at once to couple on the beach. Any woman in her right mind stayed in the water and masturbated for relief. Right-minded women skimmed themselves from the breeding population, and babies came to the remainder, women who preferred dangerous sex with a man to safe sex alone.

Sex got worse when men later evolved a reduced intellect, and self-repression became even more vital. Solo sex is awfully inviting when men are both dull and dangerous. Hand-in-hand with male stupidity, self-repression set up human sexuality's final collapse. What we commonly call human superiority reflects instead a courting disaster. A pretty story this is not.

So women evolved an instinctive reluctance to masturbate. Their only relief was with a male partner — not because men were good lovers but because men were *bad* lovers! While a female chimp freely masturbates in public, a female human may devour romance novels, stroking her emotions but leaving her crotch undisturbed.

Self-suppression is entirely instinctive and involved no change to genital anatomy. A numb clitoris serves no purpose for a woman who already lost the deep-touch response — she needs all the coital incentive she can get. Women probably have evolved *more* clitoral response than aquatics, and more skin eroticism; women who climax better reproduce better. But she needed to stop using it on her own.

This all had value only because moving to land had value. Female fears built the final barrier for any primate on its way up from the water. When coupling got worse, masturbators stayed behind. Women who eventually made the move would (by default) be successful procreators, who prefer unsatisfying sex with a male to sex with themselves.

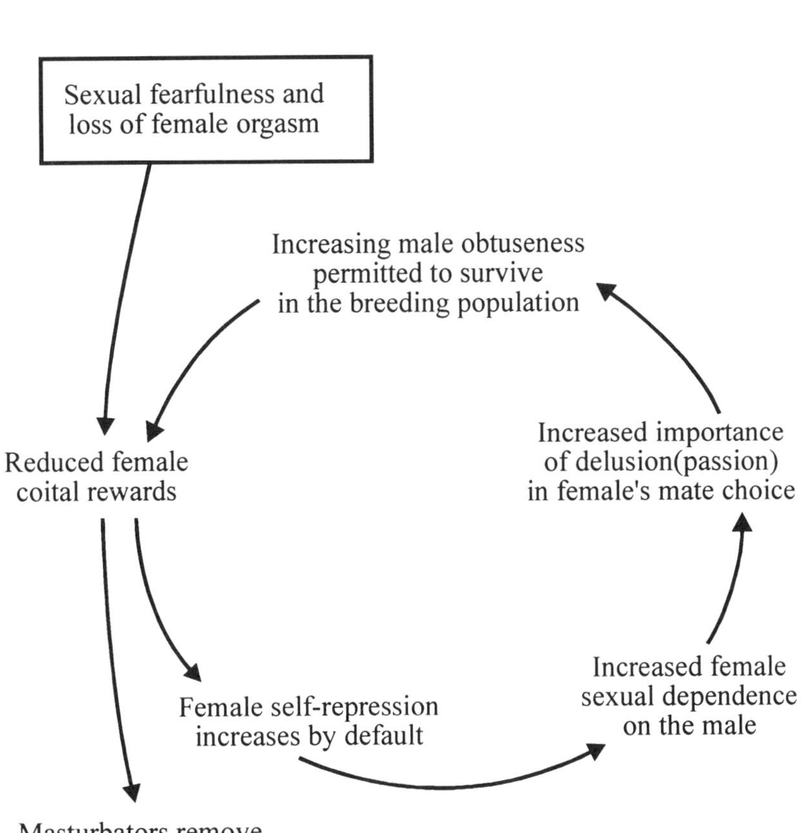

Figure 16: Evolving Self-Repression

As women retreated from the erotic common-ground, they left men standing there in confusion. Women endured a sexual revolution in the last few million years, while men merely became a bit slow. Men aren't sure where women went; women can't see why men didn't come along. Only wary in the water, after returning to land the sexes were suspicious. With this new estrangement, both men and women began to see their sexual partner as a retreating goal.

Least-Sexy Primate

Women's self-repression created a species with deep inhibitions, though we cherish a myth of wonderful human sexuality. We rape each other, use sex as a threat, shun public displays of coitus and protect children from seeing it. Some spend their lives hiding sexual feelings, for fear of public response. Many marriages are made bitterly unhappy by sexual differences the spouses cannot even discuss. Yet we claim that human sex is superior.

Men staunchly support the myth, since their sexual contacts usually include orgasms, and as the saying goes, the worst they get is great. Women might better say "human sex is good cuddling", more often an especially intimate hug than an orgasmic experience. Indeed, western science has only recently recognized female orgasms.

Those who define human evolution by a hunter's needs (hairlessness to sweat, bipedalism for weapon-use) define human sexuality by men's desires. According to the Mighty Hunter school, a woman's body evolved to lure men because snagging a hunter ensures survival. Her shape alerts him to her sexual maturity. Hairless skin evolved to be touched, lips and nipples darken when aroused so he can tell that she wants him. Kissably full lips mimic genital labia to entice him. Women even lost estrus to permit sex any day of the year. As Morgan notes (1972), men enthusiastically defend this viewpoint.

But the Centerfold Theory of Evolution fails to match reality. Most people find sex either elusive or unsatisfying. The chimp's happy promiscuity became the human's fearful faithfulness. Far from being the sexiest primate, our sexuality malfunctions so badly that only our emotion makes it tolerable.

Self-repression focussed on women because it made them reluctant to masturbate. Women's magazines even carry articles *encouraging* masturbation, describing techniques. The same article in a men's publication would cause laughter, for men know what works, and don't hesitate to do it. Telling someone to masturbate is like telling them to eat; being anerotic is like being anorexic. Nancy Friday (1977) describes female reluctance to masturbate as an unconscious bargain little girls make with their mothers, and until she was an adult she didn't realize she'd made it. Barring telepathy, she's talking about an instinct.

Self-repression covers acts short of orgasms. Men go to sex shows alone, but women go in groups — even to shows designed for women. Women give each other permission to be there, absolve each other of guilt for solitary sex. Friday suggests that women also give each other permission to have erotic dreams. Until women know that other women have sexual fantasies, they either will not admit them, or perhaps do not have them. Once they know that other women fantasize, women fantasize freely.

Men share self-repression, inherited from their mothers. But men also evolved aggressiveness for dominance-competition. More important, men need *sexual* aggressiveness exactly corresponding to female hesitation. Delicate ground. We did not evolve to rape, but a couple cannot mate until the sum of their enthusiasms exceeds the sum of their fears. If women evolve reluctance, men must evolve pushiness. We will look at boundaries where men pushed.

If a good sex life were evolution's focus, we'd all have one. Instead, we are the least sexy of the great apes and probably the most inhibited of all primates. The chimp as our model makes our sexual behavior all the more striking. We show repression while otherwise paralleling a species with gleeful sexual abandon, misplaced as a shy child in a family of clowns. We are preoccupied by sex, but we are not sexy.

None of this implies women's guilt; women did not choose to become fearful nor to create repressive cultures. Instinctive fear of sex is the shortest evolutionary path to stopping masturbation, making women turn to men for sexual relief. Not elegant, but effective.

Men And Sexual Control

When we came up from the rivers, sex became harder to find. Female fears grew as we returned to land, and each woman consented less readily. A woman became a valuable possession; male threats became a possessor's weapon.

In a sexual shortage, many males resort to "pasha" behaviors to monopolize females. Ungulates often hold mobile harems, while pinnipeds control territory and chase away interloping males. A single dolphin cannot control another, so dolphin males often team up to subdue one or more females. Primates, too, keep harems. Dominant baboons limit sexual access of underlings, while some hamadryas males exclude other males completely. None care about female rights.

Even distant relatives use precisely the same acts: The hamadryas male bites his females and attacks strange males just as wild stallions do. Pasha behavior is likely part of all mammalian instinctive toolkits, available at need. Perhaps *any* male, if he has trouble getting sex, will restrict females he *can* get, even in species that normally enjoy easy sexual access and never display pasha behavior.

Pasha behaviors include limiting female movement, chasing away strange males, and forbidding sexual acts. Harem-keeping results from limiting female movement, though the harem may have only one female. One animal (a shrimp, I think) goes so far as to capture females and cut off their legs, to keep them securely in his burrow. Territorial males create de facto mateships by excluding other males from their territory while allowing females free range. Chasing away strange males may present difficulties if a sexual shortage appears in an already-social animal. A man can't chase away all the others who ride his bus.

The final pasha act, forbidding or interrupting others' sex acts, gives a lot of bang for the buck. Each interruption may prevent an impregnation, while (presumably) gaining access to an ovulating female. No male can copulate while even a smaller male charges, so interrupting sex works without combat, though the interloper may have to then back up his threat. This male reaction parcels out sexual access according to dominance. Men today restrict public sex partly due to their personal lack of access, partly to exercise dominance for its own sake.

When we left the water, men began doing as other species do to endure a sexual drought. Unfortunately the shortage came not from other males or too few females, but from women's hesitation. Men pulled out and dusted off instincts that didn't match the job; women need persuading, not controlling. Modern men keep wives and daughters at home, chase away strange males, and prohibit sexual displays by women they consider their own. Men don't conspire to suppress women, but individually respond to a sexual shortage, instinctively reacting in ways that seem to have caused the shortage. Women exert more repressive force, but men are easier to blame.

Women And Feedback

From mother to daughter, sexual denial makes a loop, with the most-repressed teaching the most-repressible. Faced by her father's limits and her mother's sexlessness, a young woman's sexuality faces potent opponents. It's amazing that women have babies at all.

When self-repression became useful, masturbation culled liberated women. For thousands of generations, the most-repressed women (whose sex life came only through men) bore the most children. Some cultures have reached the point that their ideal mother has no sexual feelings. Mimicry is our most powerful teacher, and a mother's asexuality can easily form her daughter's template for life.

Further, mothers actively intervene in daughters' sexuality. A mother tolerates an infant son's masturbation because she cannot comprehend masculinity. With some hesitation she may remove his hand from his penis, but she doesn't want to interfere in his becoming a man — the sex she instinctively over-rates. Not being a man, she was never a boy and lacks the confidence to control him.

She has no such doubt with a daughter. If sex is danger, and mothers shield children from danger, what could help more than removing a daughter's hand from her crotch? The mother's instinctive fears lead to a confident, though often unthinking, response. Physical interventions become verbal as the daughter grows, telling her not to rub on the furniture, pointing out what happens to bad girls. Women raise daughters to hide and control sex.

Women's *instinctive* self-repression prepares them to *learn* self-repression, just as the innate language abilities of the brain make learning languages automatic. Even if both parents taught precisely the same lesson, young women would learn self-control better than would young men. But infants do not get the same lesson, because mothers teach it most emphatically to the daughter who reminds her of herself.

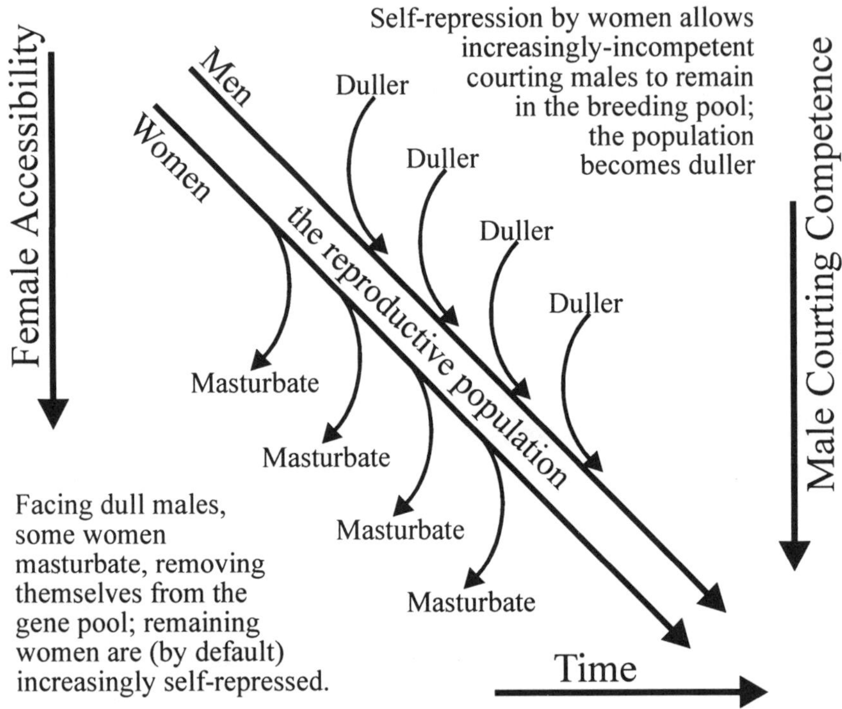

Figure 17: Self-Repression Grows Powerful

With mothers as the primary nurturers, we might ask how much self repression the daughter needs, if avoiding masturbation is the goal. The answer is (I suspect) not very much. Her small innate repression is the raw material her mother can work on, creating powerful repression with little work. The combined effects then guide her training of her own daughter. When I speak of women's self-repression, think of innate repression as inevitably enhanced by the mother's effect. The mother-daughter link is a flywheel of combined nature and nurture, easily spun up and then self-sustaining.

Forget Racial Memory

Self-repression and passion grew in women's minds as we left the water; a woman who went up on the beach for sex did so because these two forced her. They evolved in feminine minds only because of the relative value of life on land and the mental state it took to get us there, not from a goal of evolution or intent of the participants. Though self-repression feels like fear of sex, it is not a racial memory of women having been frightened or forced.

Freud was first to suggest that rape formed our modern mental state. From histories of women who told stories of rape, who got little joy from sex, Freud developed a theory of female natural masochism. In 1924 Freud described masochism as normal sexual maturity for women. Expanding on Freud, Helene Deutsch described rape as a necessary act that transferred eroticism from clitoris to vagina. She believed that women *need* to be overpowered, that female genitals evolved to first accommodate and finally *require* the act for their normal maturation (Brownmiller, 1975).

In this view, love itself began as an act of rape. Sex was originally a violent act, with a female's struggle against an unwanted embrace and violent penetration, ending in pregnancy. Those who fought less reproduced better, and over generations the assault and defense became half-hearted, almost ritualized. Women evolved to accept the process as sexual enjoyment, leaving them with rape fantasies and a masochistic disposition.

Brownmiller argues that rape fantasies reflect sadism instead. Men have always raped, and men as the cultural controllers get to choose each sex's roles. Since men have defined women as the inferior sex, they have also defined what a healthy feminine attitude should be. Casting women as sexual victims is convenient for men, so when women fantasize about rape, they are merely imagining within the mental space men have allotted them. (Brownmiller, 1975).

But what sort of minds would we have, if habitual rape formed them? Torture destroys us; abused children suffer for life. A rape victim's heart is left raw, as if the healing place were sandpapered hour by hour. Years after the attack, she will still wince at it every day, and weep for what was taken.

Any trauma can render victims incapable of managing their lives. If trauma reduces one's chance of survival, and sex were traumatic, then we would evolve minds that feel rape's trauma less. If erotic transfer were a requirement, we'd want help with it as we might need help changing a tire, bothered but not traumatized; our minds would accept what our vaginas required.

If rape formed our sexual template, I can't comprehend that in its diluted form (making love), the act would bring our greatest pleasure, but in its pure form (a rape) would bring our greatest pain. If rape were our reproductive model, then at its very worst it should be equivalent to a relatively intrusive physical exam. A raped woman would adjust her clothing and walk on as if she had simply stopped at a bathroom.

Women have no difficulty distinguishing rape from loving sex, yet Deutsch would have us think that one graded gently into the other. (That courts may have trouble separating them does not mean that women have any doubt about the difference.)

A woman's mental state matters only if she can choose what she will do. Female self-repression has value only because female cooperation is necessary. Only if females are sexually self-repressed does sex with a man feel better than sex alone. So self-repression (if you accept my thesis) shows that females have been consistently willing (though hesitant), and therefore rape is not our formative event. I will later suggest that our evolution (after the return to land) specifically shows we reject forceful sex, and have shunned rapists for several million years.

Rape fantasies do not signal a desire for rape but evince repression. In fantasy a woman can play both victim and perpetrator, subduing fears to fill her needs, playacting resistance while her control of the imaginary rapist makes him safe. Women did not evolve for rape, and have no inherited memory of sadists. Women like to contemplate dangerous men, but at a safe distance.

Women need gentle help in overcoming innate hesitations. It's not their fault, but they evolved a growing inability to do it alone. Their self-repression creates the thinnest line we walk, between saying "no" and meaning it, and saying "no" but meaning "help me".

<u>Appearance Of Humanity</u>

These people grow familiar. It's hard to imagine an aquatic society of singers and poets, but love and self-repression make our grandparents recognizable. In the next chapters I hope you will see them.

Mary Leakey found at Laetoli a double trail of footprints left in mud, now hardened to rock. The couple walked close together, possibly arm in arm. One was likely a female, several inches shorter than the other and with smaller feet. She paused to look back, then they continued together. Nearby another set of smaller prints may show their child's path. Almost four million years ago a family passed by — perhaps the woman turned to look for her daughter.

The walkers were similar to Donald Johanson's Lucy, and perhaps our direct ancestors. They were human-like, intelligent, newly returned from the rivers and with less body hair than a modern human has. They spoke fluently, wore clothing, and walked arm in arm because they loved each other. Almost four million years ago they built families from love.

Laetoli holds a tiny marker from a commonplace event. We found only these; how many more walked in mud later smoothed flat, or on firm ground and left no mark? The trail, like any small sample, like a single tooth, shows the larger picture. From all the humans who walked arm in arm over three million years ago, we found only one couple who wandered off the path, on a patch where footprints might survive, in a place where Mary Leakey happened to dig.

This couple feels familiar, walking arm in arm, balancing desires and fears. When we returned from water sex became discordant. Men strove for dominance, driven by instincts that had always found dominance useful. But women bypassed dominance, and measured men on the emotional scale they had evolved in the water. Men and women played different games on the same playing field. The couple at Laetoli loved each other but did not understand each other.

Distrust and suspicion hardened the points where men and women touched. Each began to see the other as possessing the key to sex, dangled just out of reach. Each started to misunderstand, to become angry. Men and women had evolved in ways that worked best at the time, but evolution does not plan for the future. With our return to land these facets of dominance and passion had to co-exist, though they clashed and drew blood. Our Laetoli couple stayed close because each had worked hard to find the other, and had much to lose.

Skirmishers in the battle of the sexes, they were only a few thousand generations out of the water perhaps, as close to aquatics in time as we are to Neanderthals. No longer hairy promiscuous picnickers, they seem familiar because we can empathize. Like adding a key piece in a jigsaw and suddenly the picture makes sense, when we add sexual troubles these people begin to *feel* right.

An emotion transports us as in a time machine to become again the creature who evolved it. When we feel terror, we relive the precise reaction of tiny mammals hiding from the first dinosaurs. When we love, we share the feeling that brought women up the beach. When we walk arm in arm, we recreate the warmth of this Laetoli family out for a stroll.

Love and self-repression found a place in the primate toolkit. They joined our repertoire as assets but are baggage too, and can weigh us down. Anger, for example, is aggression's hammer, the necessary tool making dominance work. Yet often we must control misplaced anger, and subdue the impulse to clobber someone.

What does a primate do to a shiny new gadget? What does the gadget do to the primate? Passion and self-repression are not the inevitable result of perfect evolution, merely adaptations helping us to survive and reproduce. I outlined how they might have evolved; we should pause and see how they work.

Passion And Bonding

First we should consider whether our new emotional tools could have evolved to create human pairing. Passion links self-repression with fear-thrill and obsession, all focussing on another individual. We want only that person to help us do what we'd rather not do alone. We feel this as a single response, but it draws from varied sources. Did it evolve because we needed marriage?

Morris (1967) says that strong sexual imprinting is a requirement, helping us stay together while fulfilling prolonged parental roles. Our tendency to bond evolved because bonded parents cooperate in raising slow-growing children. Fisher says that our mating duration evolved to match the resulting child's dependence. Love can wither safely only after the infant can get about on its own. But if parental needs guided love's evolution, something went badly wrong to leave so many single mothers.

I believe our pairing-duration is purely accidental. Our love's power (and our reluctance to finally give up) combine to create relationships with predictable spans, as last year's nest will stay on the branch for a predictable period of time. We love for a given period not because children's needs timed our passion to that clock, but because we evolved powerful emotions that take a few years to wear off.

Love evolved not to form bonds but to surmount hesitation, to overwhelm doubts and fears. Love can destroy the very pairing that many suggest it evolved to support, if one who is not our mate catches our heart. If we needed to bond for some period of parental care, passion seems a fickle emotion for the task. We need a mutual attraction which, once triggered, becomes quiet contentment until death do us part. Instead we have discord and uproar, usually non-mutual and notoriously temporary. Only with care and luck can we steady our passions to support our pairings.

We did not evolve to marry, but found it accidentally. Over countless generations we worked out compromises between our hungers and our fears, and found a crude pairing in the slippery space dividing what we want from what we can get. If there were only one man and woman there would be no word for how they act together. But with thousands of us, we learn to recognize this special huggy-kissy-staying-close thing, because we see it often, and see some without it, so we invented a name for it. Since we presume our superiority, we presume that the huggy-kissy-staying-close thing is a superior way to mate.

Not a mutually-triggered reward, the new gadget in our toolbox is entirely a creation of each individual's mind. It evolved to overpower one person's hesitation, not to weld two together. One I love might not love me back. Passion is *individual* behavior, not a manifestation of an evolved pair-bond.

Love As Delusion

Love makes us stupid. Delusion has helped women procreate for several million years, but our new tool also makes us foolish and damages women. Love's trajectory cut down women's status as an innocent bystander.

It goes like this: Male dominance is always a female sexual cue. When water muted dominance, women evolved obsession to conjure alpha-ness from thin air. By simply focussing on one man, a woman could attribute undeserved status to him, creating a sexual hair-trigger. While anyone can attribute too much status to another, only when a woman did so to a man could natural selection favor the error.

And there's more: Only women highly susceptible to obsession could find enough worth in men to leave the water for them. Only passionate love (outrageously over-valuing him) could persuade her to go. So our terrestrial move consistently fostered in women an inability to judge men's real value.

The plot thickens: After we moved to dry land, men became dull. Soon I can tell you why. Already committed to over-valuing men, women's misperception grew as men's practical value decreased. Only women who over-rated them could find dull men desirable as mates.

And the punch line: Women could compare men only to women. Women's erroneous attribution began when male hierarchy collapsed in the sea, leaving women unable to compare one man to another. A woman had a reproductive advantage if she automatically rated men higher than women. So women were bred for a predisposition to value men more than the evidence warranted, and inevitably at women's expense.

Self-repression aggravates the problem. Her instinctive reluctance to masturbate (on average) means she requires a man for any sex life at all, while a man requires a woman for only *one* of his sexual options. Masturbation is basic body care for nearly all animals. By requiring a man for this, women in effect view themselves as partial persons, like feeling an itch and having no arms. She feels incomplete alone, while a man feels not so much incomplete as lonely. Women recognize other women as partial beings too, and value them less; feminine alliances (the core of other primate societies) have withered in us because women misunderstand their own worth.

The aquatic dominance collapse first made misattributing value useful. When we moved back to land we brought love and self-repression along, and sex became a one-to-one affair. A female chimp could handle twenty males, but women evolved emotional rewards keyed to frontal sex and the fear-thrill. Each woman could tolerate only one partner, though his coital stamina could hardly keep up with hers.

Pairing arose by default. Few women had access to alpha men, yet women could not help but prefer dominance. The average woman was courted by only an average man; she was more likely to be impressed by him and make babies if she failed to recognize mediocrity. Leaving genes in less-than-perfect descendants beat leaving none at all, so most of us have average mothers who couldn't judge men.

In fact, there are no good predictors of mate adequacy and we continually misjudge them. We often marry badly because whom we select matters less than that we mate at all. If mate adequacy were important, we would have evolved the ability to see its reliable sign. An individual male may have little value, but over-rating him is women's vital trait.

We are a species fated to misjudge both danger and value. A chimp would never go over Niagara Falls, for example, or inscribe the Gettysburg Address on a grain of rice. Neither would a chimp be interested in space flight or sailing across unknown seas. We often misunderstand our own best interests, and allow emotion to pull us where reasonable fears should hold us back. Sometimes we accomplish remarkable things and sometimes we get hurt, and it owes much to the self-delusion women need for tolerating men.

Love As Addiction

We are passion addicts. Love's high value makes us exquisitely sensitive to it, just as our eyeballs are exquisitely sensitive to pain because even a tiny injury carries a great penalty. Unfortunately, our heightened response to passion means that we can never get enough of what we want so badly.

Passion bridges the gap between fear and reproduction, and continues to grow stronger because female doubts have no fixed limit. Love began as female-to-female aquatic competition, giving its early owners an edge in mildly scary frontal sex when mild love sufficed. The gap widened when women feared sex on the beach, and with men's later dullness the gap widened yet again. We needed no brand-new instinct when merely strengthening an existing one continued the species. We cannot have too much passion.

In the pursuit of passion, even smart women seek to be "swept away", forgetting contraceptives or taking other foolish steps. They may rue it the next day, and live in dread at the thought of pregnancy or disease. And when their fear fades they likely will do it all again. Passion-addiction, not personal choice, carries women on a desperate search for a reward that often disappoints. When a man takes a woman's number it is her addiction, not his words, that ties her to the phone for the next three days.

Love evolved to accomplish exactly this: Not to bond but to overwhelm; not to build commitment but to ignore failures. The woman swept away does not consider risks because women who paused to think did not make babies. An act that seems irrational may help the group, though the individual dies in the attempt; the species steps over women's hurt feelings as casually as an army over corpses.

Addiction pushes us not by how good we feel with it but by how bad we feel *without it*. Her later disappointment doesn't affect the army's progress, as long as her passion-addiction permitted coitus. A man might give little support, and his attraction might wear thin. Yet she evolved to keep trying, for each eventually disappointing man might make a baby or two on the way. She might consistently choose terrible mates, but her emotional addiction will still have done its job.

Reciprocity is a happy accident; nothing demands that a lover get love in return, for passion does not marry us but moves us. We spend much of our lives without a fix, addicts with the shakes, compelled to action, ignoring restraining orders and refusals. The stalker, for all the terror he causes, is gripped by the same addiction the lover feels.

Addiction evolved to not give us peace of mind. The woman who, having met an interesting man, lies awake all night, angry that she cannot quiet her mind, illustrates a perfectly functioning instinct. Her emotional addiction means she will much more readily accept him than would her sister who sleeps soundly. Restless nights are an accidental side-effect of having an instinct for passion and obsession.

Training, as well as instinct, led women to addiction. (You may place more or less weight on instinct or training, as you prefer.) Cultures may define married life as the only acceptable state for a woman, and little girls would then naturally focus on that state (and its emotional implications).

When in socially approved channels, we treat addiction as at least acceptable if not a virtue. Juries often ignore wives testifying to their husbands' innocence, assuming women will lie for their men. Many will. We vilify the chemically addicted who betray trust for a fix; we accept the wife who lies for her emotional addiction. When we wonder why a woman remains with a bum, someone might respond "but she loves him." Her addiction loses some moral luster if we change our answer to "but she's a drunk."

Men feel addiction too, inheriting passion from maternal ancestors. But men have little fear of frontal sex; they avoid frontal contact with another male, but easily face a female. And men evolved heightened sexual aggression to counter female hesitation. Men had no inner barriers for addiction to overcome, so men fall in love much more quickly than women. Their feelings overpower and terrify, for they have no internal brakes. Women seek to be swept away, but men more often are.

We take addiction for granted in others. Passion captures our lovers' hearts, and we know that owning hearts gives power. I don't need to bind her body, for if I win her love, she willingly binds herself. Men did not need to drag women from the water, for women evolved passion and repression, both compelling them to go. My lover and I can part with a kiss and a clear mind. If she loves me she will return, having been with no other man.

Passion captured women when it freed them from fear. Love didn't evolve to create monogamy, but love cannot tolerate promiscuity. A man wants to know that a woman loves him, for then he knows that her hidden warden will hold other men at bay. We want love potions not to enhance our own love but to imprison another. Her love, if I can evoke it, ties her to me. I don't need to be a pasha if she is a pasha to her Self.

Leaving The Beach

The hairy promiscuous picnicker died in the water; a hairless monogamous biped walked out onto the plains. Moving to land was a multi-generation walking-away by men and running-after by women. Though not all pretty, the picture does hold treasures.

I've done nothing more than interpolate from known end-points. Very few apes lived in water, but we have evidence that our forebears did. Primate society will respond to aquatic life, just as to any other change in environment. We can discern our ancestors' response because their instincts remain inside us; we can imagine their major choices based on debris in our minds.

No aquatic species inevitably must leave the water. But if we were once aquatic then we did come back to land, for here we stand. I mostly discuss what happened to only our lineage; we haven't space to explore all the possibilities. If we had stayed in water (as most aquatics did), powerful love need not have evolved. Mild thrills would accompany frontal sex, with tepid affection on the female's part. Males would still compete with singing and speech, and would not have grown dull.

Love evolved in women only because the environment urged men to leave the water, not because passion automatically aids a species. Women who felt no special love for a man stayed behind. After our return, women less susceptible to love mated less frequently, while women powerfully drawn by the fear-thrill and obsession made more babies, and so have continually reinforced a misperception of men. Women love men because their mothers needed love to make babies, not for men's inherent lovableness.

Women are otherwise sensible, and find it infuriating that irrational needs plague their lives. Women keep trying to find and hold men who don't satisfy them. Women have evolved instinctive delusions about men, and sexual self-repression to keep themselves from masturbating, to keep them searching for a mate. We might say women need men for something they can't quite put their finger on.

Our walking-away and running-after were net effects of aquatic limitations and a transient pocket of green, experienced by a normal primate who had fled to the sea to escape a drought. When the chance came, men moved to the environment suiting them best. Sensible women stayed behind, but those few who did go along have survived. So lush riverside brought us out of the water, and created impressionable and passionate women.

Courage To Go

We all grew braver when women evolved passion on the beach. Women initially hesitated not from cowardice, but because they were no braver than any other ape. When women finally came onto the land, they brought courage that we all inherit.

Suppressing fear drove passion's evolution. Desensitizers, obsession and passion have value because they subdue the inclination to retreat. When women evolved all this to deal with fear, they could not avoid evolving courage, too.

Primates evolved for team sports, standing shoulder to shoulder and squalling at neighbors across imaginary boundaries. As we evolved from marginal scavengers to hunters, teams of males stood shoulder to shoulder clubbing the victim. Nothing in this demands special bravery. Prudent hunters select prey they know they can kill, and weapons that give them an edge. They never pick on an animal more powerful than the total of their company, or larger than their vehicle.

But women, to reproduce, had to conquer fear. Without modern medicine, any woman risks death in childbirth. While we may blame culture in part, does it not also demand courage to put one's life on the line? Is not a risk faced alone more frightening than a risk taken shoulder to shoulder with willing companions? Often lacking feminine alliances, each woman faced alone the most dangerous animal in her world — a man. It takes more guts to face an aggressor than to be one.

When frontal sex in water became a useful pose, courage became a female asset. Men didn't need it then or later, but women have needed bravery ever since. In our transit to land, females evolved a repertoire of fear-controlling responses, with courage an inevitable part. Women evolved to accept danger as a part of life; they evolved courage to tolerate what men never saw at all.

When men moved to land, women hung back, and we stretched as a rubber band on two points. Men were not braver, but only had less to fear on land. When some women left the water, the more-fearful ones stayed behind and the rubber band broke in half. The timid did not change their minds; they were subtracted like slower gazelles. And as the shortened band snapped free, it caught us up and hurled us onto the plains. We all became braver together, when only the brave women came out.

The terrestrial transition was our species' rite of passage; the very act of leaving the water made us stronger in spirit by subtracting the timid. It didn't take bravery to let scavenging males go exploring, but inherited bravery let them explore farther and risk more. The human species would not be nearly so daring, had not a few women first been brave enough to follow.

Go Fission

As aquatics, we ranged from the Cape of Good Hope to China. These are temperate waters and continuous coastlines, and generally match fossil evidence of primitive bipeds. We may have gone up Africa's west coast to southern Europe, though no fossil evidence shows such a move. In the sea we used water routes, not land bridges.

We spanned a broad temporal range too. The first aquatic entered the water in the early Pliocene, perhaps twelve million years ago. The last may have gone in almost eight million years later, near the start of the Pleistocene. If it stayed in the sea long enough, each population saw similar changes: Bodies streamlined, digits webbed, hierarchy collapsed. Some newer populations mixed with more-ancient aquatics; others remained isolated. A few tendrils of each mob found new places on the land, and resumed a terrestrial life.

Returning is risky, and an awkward biped needs a green place to live while it learns to survive. Small variations in a shore's lushness locally encouraged or discouraged the move to land. As generations of fishermen try the same places, so returning aquatics repeatedly migrated through the same narrow departure points. Favorable locales shepherded many populations out of the water, spewing them onto the plains in quick succession.

The entire process may have run continuously. A steady trickle of aquatics moving upriver could join previous arrivals at a favorable site. Men would see others on the beach and quickly move to land. Women hesitated, but in a few generations their female descendants would join the men. So a lush riverside may have siphoned aquatics out of the water and pumped a steady stream of bipeds onto the plains. Once you see the picture, thematic variations easily appear.

Though a mob in the water, moving back to land through narrow portals split us into small populations, like water droplets oozing from a canvas bucket. Rivers and mountain ranges then held us apart; forests and plains guided our separate evolutions as we left our riversides.

More than topography, social pressures fractured us into small groups. As our hierarchies had softened when we entered the water, they hardened when we returned. Miocene feeding routes sustained us in small and mobile squads or platoons, but in water we were amorphous, with no group movement and no leaders. Back on land, we formed teams when we suddenly had to cooperate. Crude tribes formed within the mass, from pairings, extended families, and collected friends. Our primate interest in friendships and conflicts, long impeded in water, at last had free rein. We had food to fight over, allies to help us, and the neighbors to be angry with.

Once women made the transition, mobs fractured of their own weight into optimum sizes. Hundreds might live in a half-aquatic troop, but only a dozen or so could agree where to go next. Groups of friends and relatives coalesced around their dominant members as children clump together on the playground.

With irritating neighbors and scarce food, each real or imagined slight could end in the harrumphing departure of another splinter group, just as other primate groups divide when too large. Each returning mob birthed dozens of small bands spreading out from their natal riverbank, angrily waving grubbing-sticks when strangers came too close. The annoyance of repeated contacts slowly drove them all apart and outward, an expanding universe of crippled bipeds. And when it grew quiet again at the riverbank, the next aquatic ape could begin to climb out.

Terrestrial Evolution

As we scattered and seeded the land, we began shedding aquatic features, and evolving new ones to meet terrestrial needs. Our self-sealing aquatic nostrils had no purpose without water to exclude: They evolved permanent openness instead, leaving us a small ability to widen them a bit further at will. We no longer needed beards, since children couldn't dangle at a man's front. Some populations kept them and some lost them.

Beards and noses were trivial, but how aquatics slept became deadly on land. We had learned to sleep in water, first through habit and later by instinct. We began by balancing neck-deep in quiet coves, and progressed to treading water in our sleep. Evolved as a useful skill for sleeping in deeper bays, the instinct also let us make long passages from island to island. Late-stage aquatics didn't care much whether they slept while standing on the bottom or treading water. Children (being shorter) nearly always swam (or just floated) while sleeping, holding on to adults.

When men came up on land, they slept standing near the river, while women slept standing in the water. Fresh from four million years of sleepswimming, they would no more lie down to sleep than would a horse. Perhaps they paced slowly though the trees, half-aware of their place. But a sleeping biped shuffling in the bush makes an easy meal, and sleepwalkers quickly died. Within a few dozen generations, everyone slept lying down. Some still sleepwalk today, more often children than adults but the instinct is nearly eradicated. Few die sleepwalking, so there is little culling of the instinct's remnant.

We also evolved to dive in water, gaining a metabolic switch and instinctive breath-holding. Powerful from birth, these protected a baby (dozing or awake) while its mother swam. With continual practice, an adult's dives could last many minutes. Children still show this most strongly, and a child may survive over half an hour when submerged in cold water.

When we first returned to land this instinct killed countless children. While asleep in the water and rocked by waves, each half-aware child knew when to breathe. Water gently touched its cheek, and the baby could inhale whenever it felt no liquid stroke. As infants first began to sleep on land with their mothers, the gentle stroking ended. Slumbering babies felt no liquid touch, and did not know (still fast asleep) that the water was gone. Obedient to instincts aquatic ancestors had needed, the baby held its breath, and died of mistaken immersion.

Today crib death seems mysterious, but early terrestrial humans understood perfectly. Mothers watched sleeping infants more carefully than wakeful ones, alert to long pauses, ready to intervene. But children must have died of it by the thousands. Those who survived (to have children of their own) had a faulty breath-holding instinct, or mothers ready to save them. Deadly for thousands of generations, phantom-diving has become rare, continuing now like sleepwalking because few die of it. We prevent it today just as the first terrestrial mothers did: When the infant stops breathing you merely touch it. Still fast asleep, the baby feels no water on its face, and breathes again.

Changing Bodies

We were hindered by aquatic anatomy, too. Salty tears handicapped us when we moved away from the sea. Evolved to purge salt, our tear glands continuously dribbled concentrated saltwater when we lived in the ocean. We may have once drunk pure seawater, expelling salt in tears, or salt may have come from marine foods. But when we moved up rivers, salt largely vanished from our diet.

Salt became critical when we moved to land. Forced to survive long periods without salt while sweating in the heat, people evolved to retain it. We now weep only when stressed, with tears far less salty than those of sea birds. Our ability to purge salt has faded, and drinking seawater will kill us.

In water fat gave warmth and flotation, and aquatics were probably pudgy. Still useful on land for storing energy against hard times, it is also a burden we have to tote. Thin people moved faster, and stayed cooler in the heat. Fat people died in droves as we left the rivers, yet under the right conditions we still easily gain weight.

Walking, more than any other terrestrial change, shaped us after our return. No selective pressure urged us to become quadrupedal again, leaving us bipedal by default, so we evolved bipedalism as best we could, though it is a terrible way to move. Walking requires left-and-right pelvic tilts, balanced by the spine and shoulders. Our backs grew stronger to take walking's strain and to support the full weight of shoulders and head. Delicate bones, for which we named our ancestors "graciles", became thicker. Even today, after several million years of improvement, back injuries often cripple us. Graciles were crippled more often just after moving to land, and we reflect the success of the stronger ones.

Walking concentrates pelvic strain in small zones. With each step we lift and swing our pelvis and leg, using muscle power alone. Over a lifetime we develop narrow pillar-like thickenings of pelvic bone. These run from your hip socket to the top ridge of your iliac crest (where your belt rides). Absent in australopithecines, the thickenings help transmit strain between torso and leg. Some suggest australopithecines' lack of pillars proves our superior evolution, but it means only that they swam instead.

In water our pelvis expanded to the front and sides to anchor torso and leg muscles. With our bodies sheathed in fat, a broad pelvis was no bother and gave better leverage for swimming. But on land pelvic width limited our speed. A small pelvis took less energy to rock left-and-right, while larger shoulders provided a more-efficient counter-balance. So the large pelvis and small upper body of *Australopithecus* gave way to a smaller pelvis and larger upper body of modern men.

Women too needed smaller pelves for walking, but women needed a broad pelvis to bear children. In newly returned terrestrials the pelvis functioned for the first time as a bowl — a role often proposed as one of its primary values. With viscera pressing down, uterine prolapse became a new danger. Women evolved a smaller birth canal in response, and then started to die in childbirth. Women began for the first time to thread the tricky path between a pelvis narrow enough to live with and wide enough for reproduction.

Aquatic Death Sentence

Facing new hazards in our story, we should see hazards that passed us by. Life in water eventually failed nearly all who tried it, and aquatic adaptations condemned most returnees. We primates make lousy aquatics because we feed with our hands. Primates can grab food in pools easily, but never evolve into good swimmers. Harsh terrestrial conditions may force a primate into a fully aquatic life, but wherever it finds the chance it will return to land. And if it doesn't return soon, some aquatic predator will notice this vulnerable new food, will evolve to hunt it, and the primate will vanish.

As returned aquatics, we would not inspire confidence. Hairlessness makes temperature control difficult and increases the chance of skin injuries. Bipedalism made us one of the slowest terrestrial animals. Marine foods caused our canine teeth to shrink, disarming us when back on land. And muscular geometry giving quick grabs in water also makes us weak.

We find advantages in our bodies only because we see them from the inside, so to speak. If we had the perspective of another animal, humans would seem no more impressive than hairless Chihuahuas. We wouldn't admire a call system so vague and corrupted that one sound could have several meanings. Very few animals carry things, and if we felt no such need we'd hardly applaud one who could, any more than humans much want to resemble camels, though a camel (if it could) might exult over its hump.

And love would baffle us, foreign as a homing pigeon's skill. With no other animal on earth seeming to pair by passion, with ourselves total strangers to it, we would not point to it as an asset. Most aquatics, however, did not evolve passion, and *passion*, more than language, intelligence or physical attribute, saved us from disaster.

Our greatest handicap came not from aquatic changes but from our ape heritage. Long before our aquatic time, we and our near relatives shared a fatal flaw; in the remarkably successful primate family we apes are poor cousins. Aquatic life causes devastating changes, and nearly all populations that return to land quickly go extinct. If an aquatic primate is flawed, then an aquatic *ape* has the poorest prospects, because it starts from the wrong side of the tracks.

Apes are a dying branch of the primate order; our near relatives hover near extinction. Gorillas, chimps and, orangs survive only where men have not yet cared to go, or where we take special steps to protect them. I suggest our cousins feel heartsick, burdened by an instinct of loneliness, and we share the feeling.

We four Great Apes separated up to fifteen million years ago, in several dividings, and traveled our own paths. And each species is haunted by an imaginary ghost, as if alone in deep woods we hear a stick crack. It has frightened us for so long that we molded our hearts around the fear, consider it normal, and lost sight of it.

I found our ghost by accident as this book grew. I never went seeking it, for I had no clue it was even there. It had no form but became visible by its effect, as an astronomer may find a tiny moon when it perturbs its planet's orbit. I will try to show it to you as I found it. The implications stun me: Having found our circling ghost, we might see our own path clearly for the first time, and the valley that guides our evolution.

I'll assert this now, and hope to prove it later: A returned aquatic ape would feel even more depressed than today's gorilla or chimp. This has a physical root with mental repercussions, all implied by aquatic life. Primate brains enlarged in water to help babies float, and large brains make frightened primates, through a tragic evolutionary link. And when the primate felt lonely and frightened to begin with, as our cousins all do, the added burden can destroy the species.

But we were different. On our way out of the water we evolved a mental trick to overcome fear and malaise. Our mental trick now enables us to survive while the other apes will die; an illusion continues to rescue us today, as if we drink real water from a mirage oasis. We survived a cosmic accident, an epic tale, but I need to paint more background before you view my thesis.

Any thirsty primate may enter the water, and will return to land whenever better conditions beckon it. But we have no evidence that any other primate survived the double move. Our cosmic accident is this: The flaw that is a wound for our cousin apes is a bandage for us. We alone among primates survived the aquatic journey because being from the wrong side of the tracks, under the right circumstances, can give an advantage.

The sea was only a refuge during the Pliocene drought. As a drowning person clutches at anything, the drought-stricken primate can move quickly to water, not seeing the long-term futility. It cannot stay in the water forever, and it usually cannot come back.

Infinite Number Of Apes

Water rinsed and concentrated us like gold in a miner's pan. Over millions of years, predictable forces repeatedly collected some of the most-evolved aquatics, like the heaviest grains of ore, and concentrated them where they could survive the return. Primates will leave the water when they can, and the longer an ape remains aquatic the better we can predict where it might return. By chance, a fully-aquatic ape tends to return in the ideal place for colonizing continents.

This chapter closes with a simple sketch of our survival. The next chapter discusses our early emotional evolution on land, evolution which is essential for any successful return. The next chapter's summary adds depth to this sketch which is, for now, cursory and shallow.

A returning aquatic faces no obstacle more intractable than female fear, exacerbated by frontal sex. So why don't aquatics just revert to rear-entry sex during the move? Rear-entry coitus seems clearly preferable. Seeing other species copulating would give the hint, and as intelligent primates and sexual acrobats we could easily have tried it.

Sexual pose feels free to the individual, but reflects a species' evolution. A breeding population will consistently use predictable sexual choices, as an age group prefers a predictable dance style. Normal primates prefer rear-entry sex as a stress-free path to easy climaxes; passionate and inorgasmic apes prefer frontal sex because their hunger for the embrace replaced the orgasm.

Any species which hadn't already evolved love would prefer rear-entry sex when back on land. This implies a short aquatic stay and orgasmic females. Since apes might constantly flow into and out of the seas during the Pliocene, we are speaking of an individual species' aquatic timespan. At any time, some populations would be more adapted to the ocean than others; returnees would vary in their aquatic fitness, compared to others returning to land at the same time.

Mid-stage aquatics (i.e., bipedal but orgasmic) can quickly move back onto land. Females will readily join males, if male contact is orgasmically rewarding and not threatening. Since it takes little to coax such females out, it takes only fair conditions on land to make it worth their evolutionary while. This means that early- and mid-stage aquatics could move to land in only a generation or two, wherever land-based opportunities presented themselves. More-adapted aquatics, by contrast, might need many generations of slow acclimation and passionate culling before they could make the move.

Many mid-stage aquatics would find adequate resources even on small islands. Perhaps nine out of ten aquatics who scampered up an island beach found a workable life, even without emotional bonding. They hadn't undergone vaginal reorientation, and orgasms still lured females to copulate. With few island predators and little to fear, females could return to land without special incentive. But they were still bipedal, hairless, and burdened by children.

Thousands of populations could grasp this opportunity, and tens of thousands of islands could fill with ex-aquatics. Yet each island has little room for growth, geographic separation, and then further evolution. So each little population on each lonely island would eventually die from some local illness, crop failure, or climate change. And with no permanent mate to help them, females died first.

They also occupied rivers, quickly moving onto land in marginal pockets too lean to lure our own lineage. Less hesitant, they found suitable places soon after entering the river valley, close to the sea. Barely adequate habitats filtered barely adequate bipeds from the water before they could move far upriver and compete with our ancestors for choice sites. Lack of female hesitation rinsed them from the water too soon — lighter grit washed from the miner's pan.

We can finish the sketch with broad strokes. Rear-entry sex would be preferred by mid-stage aquatics, evolved enough to be bipeds but not enough emotionally for love-based pairing. Unhindered by the late-aquatic's fear of frontal sex, females would readily make the move to even a marginal habitat. Their short-sighted rush to land may have repeatedly confined them to tiny land-masses and limited resources. Rear-entry sex might *correlate* to extinction without *causing* it.

Love evolved with frontal eroticism to enhance frontal sex, and became vital when female orgasms vanished through vaginal reorientation. So a species with love (if you accept my scenario) implies a species of relatively inorgasmic females. A female (on average) can endure frontal sex on land only with love. This implies that women who copulated face-to-face on the beach had powerful emotional responses and therefore few orgasms. So apes who used frontal sex after returning to land were late-stage aquatics with reoriented vaginas.

Late-stage aquatic females, pending affection's explosion, have less reason to join males on land, making their people slower to move. They did not move up rivers *in order* to move to land, but because rivers were places where an aquatic ape could live. And their reluctance to leave the water meant that they filled the rivers from mouth to headwaters; the more advanced the aquatic, the more riverbank real estate it touched.

Rivers are to ocean shores as muscle to skin. Late-stage aquatics migrated up hundreds of rivers and thousands of streams, moving on aquatic arteries deep into the continental meat. For every mile of seashore they found a thousand miles of riverbank. For every mid-stage aquatic who scampered up an ocean beach and into the trees, a hundred populations of late-stage aquatics slowly injected into the heartland. Perhaps more apes lived in the oceans, but every ape living in a river pressed against the land. Their reluctance to leave the water filled the rivers to saturation.

Of thousands of rivers and their hundred-thousand branches, perhaps ninety-nine in a hundred offered no lush pocket of green for a returning aquatic. Those rivers' inhabitants had no path back from the water. They didn't know it and didn't miss it, for becoming terrestrial was never a species' goal but merely the evolutionary result of local conditions.

Of those populations who found a pocket of green, perhaps only one in a thousand could make the move. Though males played on land, females could not muster the courage to follow, or the garden faded before the aquatic could become a hunter. In most, males brought food to females still in the water, and they all survived that way until climate changes erased their little Eden. Of those who came back to the land, perhaps most died before they moved out onto the plains.

Yet this predictable pattern left thousands of mobs up on the land, budding mobile tribes from birthing gardens. The most-evolved aquatics made the land-move least successfully as individual populations. They predominated because women's reluctance always compelled them first to saturate the rivers, ensuring they would contact all possible places, and make the move in the best locations, the most-forgiving schools.

Again in broad strokes, I propose that we seeded the continents from rivers because rivers have more mileage, more ape populations touching more beach. And late-stage passionate apes filled the rivers by default, because early- and mid-stage apes moved to land before they could take over the waterways. This means that most returned aquatics were late-stage with inorgasmic females. Frontal sex and love might *correlate* to our survival wthout causing it. (The next chapter suggests that love actually *ensured* our survival, by forcing monogamy on men.)

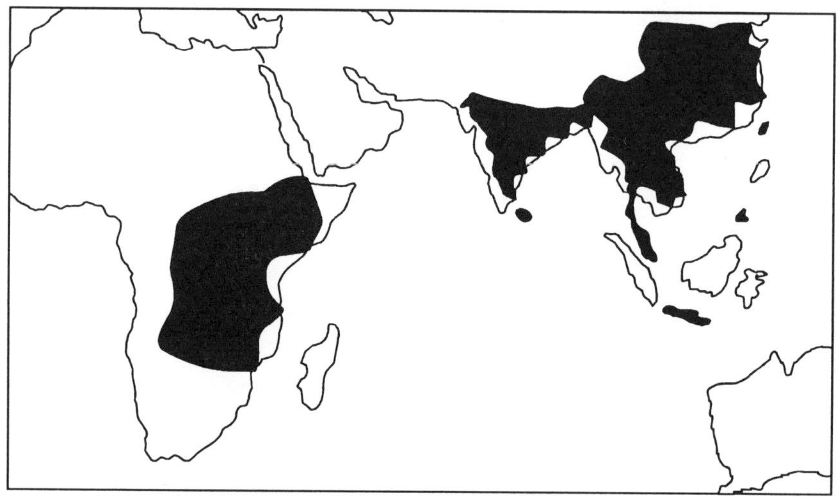

After evolving the body adaptations and emotional ties needed to live cooperatively on land, ex-aquatics spread out from the rivers.

They merged with other ex-aquatics spreading from other rivers.

Figure 18: Populating the Continents

These are only trends, briefly sketched. Between the extremes lay many populations whose choices would vary. Even humans go either way; many couples use rear-entry sex and thoroughly enjoy it. Frontal sex forces love's growth; love does not force frontal sex. You do not love someone less just because she occasionally faces away when you couple.

Our ancestors were late-stage aquatics who retained frontal sex. This caused us to evolve powerful emotional bonds but also made courting slow and difficult. In the next chapter I outline how this combination caused a courting collapse. Coupled with the apes' family flaw, our sexual collapse eventually became our survival's key. Nothing makes the human species inevitable, but predictable forces will likely create something resembling us if an ape returns from a long aquatic sojourn.

Chapter 7 - Death Of Sex

Our sexual instincts became impossibly unwieldy as we moved back to land. When water first masked estrus signals, males evolved to see sexual possibilities without estrus. Back on land estrus became obvious once more, and again triggered male courting competition. The first terrestrial men followed each estrous female as they had before we became aquatic. But new female passions opposed classic male behavior, making estrus hurtful instead of helpful. Courting slowed with each increase in women's doubt and anger, leading eventually to estrus' collapse.

Estrus died when staying with one female gave men more offspring than competing against other men every Saturday night. Estrus on land did not disappear in women but in men, when they lost the ability to sense it. Women unknowingly bred that numbness into men by the way women chose mates. Numbing men to estrus required that their intelligence decline. This decline avalanched, leading eventually to child abuse and the brain's rapid growth. In brief, men evolved to be dull rather than smart, and evidence for stupidity includes our large brains.

Estrus' death left us with new and brutal behaviors. The time of murder and rape fast approaches. Yet this collapse also set the stage for our key possession; we are pre-eminent because of the mentality we evolved when sex went bad.

Brink Of Collapse

Temporary lush local conditions lured us to land, but new pressures faced our ancestors there. To understand our sexual collapse we first need to see how early terrestrial life worked. Our prospects were grim.

Few tribes kept possession of the riverbank Eden where they left the water. Each newly terrestrial band was continually pushed on by other new tribes moving outward from the riverbank. No clear boundaries said where to stop and say "Far enough!" Population pressure, changing climate, and skill at finding new foods steadily pushed mobile groups to the frontiers, as once they had squeezed up the rivers.

Women had the worst of it. They evolved love on leaving the water, but love only enabled frontal sex and had nothing to do with cooperative pairing. Love acted as a license, letting men stay on land, play at dominance, be careless. Love appeared in women as a useful instinct when men's interests happened to match environmental possibilities. Love forced women to follow, by persuading women it was worth it.

Neither sex can evolve in isolation, but each can reflect different selective pressures. Each evolutionary change slides into the next; each alteration adjusts the rules of the game. In evolving passion on the beaches, women had been pushed to their reproductive limit, and men began to evolve in response. Our sexual evolution passed from man to woman and back, as a square dancer swings hand-to-hand around the ring. We look at men's changes in this chapter, and women's again in the next.

Ex-aquatic females could not survive without male help, and female hardships came mostly from children. As soon as hominids moved to habitats less than lush, females carried the increasing burden of infants' needs. Not constantly starving, women continually fell victim to hard times. Many little bands vanished, with no infants surviving to adulthood.

Infants were weak, cumbersome and helpless. A mother had to carry her child for at least the first year, since it couldn't hold on by itself. We can only guess how long it took to invent the baby sling; for many generations, each new mother had one or the other arm committed full-time to carrying a baby. A child could not gather significant food for itself before the age of three or four. A woman might only carry a single infant, but be feeding two or three, plus herself.

It might have eased her load, but female bodies rarely grow as large as males'. When male competition became more important, each man competed better by being bigger, but women gain no such advantage. By contrast, the more her food feeds her own body, the less feeds infants. Though a larger woman might have an easier life, her genes compete best by reproducing quickly in a small body. This has worked so consistently in mammals that a small but efficient body characterizes nearly all female mammals.

Hardened by necessity, tough women could raise one or two infants while heavily pregnant; women who couldn't do as much left few descendants. As women who survived to the end of the Oregon Trail were stronger than sisters still in the east, so savannah-dwelling females could do more than their aquatic ancestors. But they got tougher only by subtraction, as the weak continually died away.

Each woman might have suitors bringing some food, but her hesitations limited most women to one partner at a time. Language was by now an ancient skill, and she could ask for a share of a his scavengings, perhaps promising sex for food. Sex clearly interests female chimps even when not in estrus, and in the water males evolved arousal in estrus' absence. Such bargains are possible, but sometimes unreliable.

A man had little interest in feeding her children, though also perhaps his own. Like some divorced fathers, early terrestrial men felt no great responsibility, for they had never evolved any. Instead, tribes had made the landward move best if males gave *no* support. Helpful males in solicitous tribes fed females at the water's edge, never completed the terrestrial transition, and died there when the climate changed. It was male callousness that booted women up onto the beach.

With our ancestors now on land, evolving solicitous males seems a useful and simple adaptation. Males who reliably brought food would have helped raise more young and so passed on the trait. The Mighty Hunter theory rests on exactly this premise, but we evolved along a different path.

Male Competition

Males, of course, were chasing other females as they had evolved to do. Despite a woman's pleas for help, men could not resist other women in estrus. We can't tell what estrus felt like; we can only guess from analogy. In mating season male pinnipeds may fast for weeks, not to miss any chance to copulate. Estrus' scent is an infallible aphrodisiac, inducing even exhausted bulls to mount. A lion male copulated 170 times in two and a half days, by one report, and male primates may approach that rate for short periods if the female insists.

Drug addiction gives a likely equivalent, measurable by laboratory experiments. Rats will consistently choose cocaine over food until they starve. Many humans will discard careers, families and possessions for the same chemical. We might roughly equate estrus' pull to the addictive power of cocaine, though no living human has ever felt estrus itself.

When women returned from the water, they offered this drug-like lure for the first time in several million years. Water had masked estrous odors, but the male olfactory instinct had remained as intense as ever. Just as several million years have not stopped us throwing out our arms when we fall backward, so sexual scent retained the power to steal men's attention. In water men evolved to be aroused without estrus, to live without the drug — quite different from ignoring it when it reappeared. The female mobbed by men on the beach was not merely available but in heat.

For men, love missed the point, inherited from their mothers like nipples but not necessary for sex. A male could love, but (in our story so far) love meant monogamy only to the female. He might honestly love the woman (who *required* that emotion to accept him), and yet he could with a fairly clear conscience go catting around. Men still do, enraging and baffling women who love them.

Men could no more ignore estrus than they could ignore the need to breathe. (Women read this, I suspect, remembering boyfriends who claimed that it was medically necessary they sleep together.) Female primates cannot detect their own estrous odors, though they recognize estrous swellings. The male who probes a vagina and smells his fingers is not conducting an objective test, but sniffing the stew and feeling his mouth water. Women never could smell the dish, and men lost the ability several million years ago. I can describe it only by showing what it took to make it go away.

Men couldn't leave estrous females alone; if they could have ignored estrus there would have been no need for its evolutionary eradication. Women's pleas for fidelity couldn't stop men, nor could a man's pleasure in one woman's company. Though we now have virtually no idea what it felt like, estrus once was the most powerful social force in our world.

Every woman in estrus collected a gaggle of hopeful, grinning men. They hadn't a prayer, for her hesitations (on average) limited her to one partner. Men jockeyed for position, tried to dominate the scene, tried to intimidate her single mate into leaving. Not yet rapists, they could do little more; they were hopefully drawn while helplessly frustrated.

Her mate meanwhile took full advantage, copulating as many times as he could. She would have preferred more action than one crummy erection supplied, but her heart belonged to this one alone. When her estrus ended he left with the rest, sniffing after the next female. She couldn't keep him; he couldn't stay.

The Subject Is Female

We are seeing the birth of women's rage. Women alone had evolved love to overcome fears men didn't know, and acquired hungers men didn't understand. Once a woman accepted him, a man felt overwhelmed by her needs he had not evolved to fill, though he might echo her feelings. Women were trapped in boiling pain and anger, desperate for the company of men who wandered, evolved for passionate attachments men shared only in part.

Women's response (discussed in this chapter's second half) led to less-intelligent men. This in turn made women's wisdom vital to each tribe (acknowledged or not) roughly balancing men's muscle with women's sense. But women didn't design this to counter helpless anger, and the rough alignment succeeds only a little. If your boss is callous, knowing that you're smarter is small consolation.

Women's subjugation is human evolution's great tragedy. Pain comes not from mere difference in role but from resentment and helplessness. Most species evolve divided roles, often with physical distance between the sexes. The stag does not help raise the fawn, but neither is he in the area. His competition doesn't concern the doe; her nursing the fawn means nothing to him. Chimp females raise their infants alone, largely unaided by paternally-vague males.

Roles may differ even when the sexes cooperate. A female bird is no parasite for taking food from the male while she warms eggs. We have no reason to think either resents this. In mammals even the closest cooperation leaves at least one essential difference between them, but female mammals don't seem to resent nursing.

Our tragedy comes from *subjugation*, not difference. We have intelligence to see differences, language to announce how we feel, and predispositions in both men and women to value women less. Tragedy comes from the smarter sex being trodden on by the dumber, while the victims know it and unknowingly contributed to it. We probably could not have stopped the disaster even if we had seen it coming. Both sexes initiated it by doing the best they could, without ill-feeling toward the other. Even now, if I can make you see what I see, solutions are not apparent.

Modern relationships descend directly from tentative pairings of early terrestrials, each sex groping blindly to fill raw needs. We have not changed much since then. Not mindless matings of shaggy man-beasts, these are fragile pairings by intelligent people. Ancient pedigree does not justify women's subjugation, but it does clarify the depth.

Status Of Women

Early terrestrial females had almost no individual power. No one planned it that way; men did not conspire to build a male-supremacist culture. Instead, our preceding evolution set up barriers that no person could either see or surmount. In the culture that resulted, a woman when alone was less a person than an opportunity.

Her low status came partly from common primate female traits. Smaller than men, women are vulnerable to the physical force (threat if not act) underlying much of our daily behavior. Females compete less overtly, using alliances and subtle displacements. Though female struggles have powerful long-term effects, these don't make a single female especially imposing.

Second, the female desire for dominant sexual partners led to an inherited tendency to over-rate males. A woman who instinctively over-valued men would more likely find an alpha-seeming male, and gained a reproductive edge. Women reduce their own value by erroneously giving too much to men. This helps procreation but it's hell on female status.

Most important, female sexual self-repression is social suicide. Sexual activity normally corresponds to dominance in both males and females; submissive animals have little sex. This holds true in primates, and dominant females have better reproductive rates than subordinate ones. We don't fully understand the phenomenon, and it may reflect either poorer nutrition or greater stress in lower ranks.

When status grants sexual activity, sexual activity announces status. The copulating bull elephant not only impregnates the cow but also challenges any male to stop him. The alpha baboon enjoys sex and also advertises his supremacy. Female primates need not compete for sex; it comes without competition during estrus, yet the female proves her control by occasional refusals. Though not physically dominant, she can claim status equal to any male she scorns, and he will get the point.

Asexuality kills status. Women's self-repression, necessary for replacing orgasmic masturbation with inorgasmic copulation, costs them their animal validity. Women compete with sexless allure, eroticism they can plausibly deny. Allure is second-hand power, the difference between saying "I can attract a strong mate" and "I am strong." Through instinct and training, women come to believe that they are not fully-functional (that is, sexually functional) without a man.

Each woman was defeated by a lifetime of sexual non-control, evolved to need and over-value men. Such a compliant cheering section delighted men, busy beating their own drums. Blatant male competition and female self-defeat easily created a male-supremacist culture with no plan or conspiracy. Once started, subjugation then passed by cultural means, learned best by the sex most vulnerable to the lesson. Early terrestrial women had almost no chance for equality, and could only avoid being the most subordinate of their sex.

Females had changed. Women's Miocene forebears served no one and suited themselves; they respected male muscle but did not value males. An estrous female focussed the community, coupled with all males and created a network of pleasant memories. Her promiscuity caused paternal doubt, leading to male cooperation or at least less infanticide. Her occasional refusal, in a world without rape, said she was to be reckoned with.

But women could not tolerate promiscuity after obsession replaced orgasms. Obsession's power multiplied as we left the water, for only the powerfully obsessive could make the move. And no feeling strips self-respect like obsession, leaving its victim aching in hopeful agony. Passion ensnared women on our return to land, a hidden chain made her trail along behind her chosen man. For most of her cycle he ignored her; in her estrus he mated with her. Unable to ignore her fears, she was unable (on average) to accept any other male.

Women allied with men because men had visible power, while other women seemed equally vulnerable, useless as allies. By announcing her alliance with a male (whether he wanted her to or not), a female claimed the right to ask his help and share his place. Primates often identify each other through associations that become part of each member's personal assets. If you mess with me you have to deal with my friends. We announce alliances in our uniforms, gang colors, crucifixes, recognizing them almost without thought. Women didn't invent it, they just put all their eggs in the only basket they had.

Human monogamy started when women allied with their obsession's target, gaining emotional rewards and power that otherwise eluded them. Women compete by pairing, scoring each other by whom they marry or whom they recently discarded. But this also leaves each in danger of being discarded herself, for her potential loss outweighs her mate's. A woman's position would be less precarious if she defined herself in terms of herself alone, if she starred in her own life.

I use "wife" here for the first time to mean willing monogamy by women. We can predict monogamy for an inorgasmic and passionate female with no other path to power; we can also foresee her anger at having no better choices. Her monogamy coincided with men's limiting female sexuality, which they attempt whenever access to females is limited. Our story has not yet arrived at "husbands" — willingly monogamous males.

Emotional Wrecks

Each newly-terrestrial female experienced life as a series of emotional disasters. Women obtained sex, love, and identity only through men because women felt repressed, obsessive, and without status. To think that men had them was only a mental trick women evolved on their way up from the rivers. A woman sought all three in an alliance with a man she loved and respected, but men remained promiscuous and unreliable.

Any species will inevitably evolve mental devices where needed to get sperm and egg together. If men had been closer to women's original desires, rather than merely necessary for reproduction, women could reproduce with fewer delusions. Men seemed to supply sex, love, and identity only because women evolved to think so; by disavowing their own right, women made men seem worth the trouble.

Men (and the terrestrial life they followed) posed such clear danger that women could reproduce only if they found enough value in men to outweigh the hazards they faced.   Being required for reproduction, women's misattributing value to men inevitably grew to the point where most women would consider most men as mating candidates.

Women had to overvalue mediocre men; any woman who couldn't get the alpha male could still reproduce if she couldn't tell the difference. Except for the woman who got the best man, each woman's genes gained by her misjugment.   By expanding (in her mind) the pool of eligible mates, this error made more babies than the alpha pair could make alone. So most of the population had mothers who could find alpha-ness in any man.   I remember my father as a wonderful person, but I have no doubt my mother overrated him in their courtship.

Women's misattribution is necessary in the individual but tragi-comic in wider application.   The woman needs to dislike masturbating; she misapplies this in thinking sex is distasteful.   She needs a focussed fear-thrill to make frontal sex tolerable; she misapplies this in thinking that men are necessary possessions, that women are disposable.

The ruckus going on in women's minds alternately bemused and irritated men.   Men put women on par with children who followed behind on scavenging forays.   Women's importunings annoyed men who wanted to go with other guys to the game.   A woman's inability to tolerate more than one partner gave him a clear path when she was in estrus, but the rest of the time he likely felt bothered by her.

A woman could only take so much of this.   After several months or years of dogged persistence she realized he was not as wonderful as she once thought.   Passion that overcame her own fear had no hold at all on the object of her affections — a hard lesson for anyone.   So eventually each woman mourned and buried her love.

But life goes on, and her needs compelled her eventually to seek again.   We evolved to learn little from these mistakes, because those who learn to avoid pain also stop making babies, and so we go out to find another and hopelessly pursue them in turn.   Each woman's life was (and often is) a series of searches, filling permanent needs with temporary men.

Women's Anger

Repeated injuries to the heart built women's rage.   Without status or power, addicted to what hurt them, bruised and bloodied, women had gone as far as they could. Though men had not planned it, to women it seemed that men had caused the problems.

Men loved, too, inheriting passion from their mothers, but love is not automatically fidelity. Love in men stirred strong but fickle currents. With no hesitation to allay, no need for focussing to evoke a dominance-triggered response, love captured men without enabling sex. Love in men was genuine but loose, tumultuous but useless, thunder without rain.

Love never evolved to make pair bonds; a woman had no assurance that a man loved her in return. Men had little to fear if they wandered; the woman was bound by her emotions, not he by his. Small and trapped, she could not punish him for transgressions he did not understand. Each time her estrus returned she welcomed him back, as she would have done any other day if only he showed interest.

For men monogamy had no meaning, foreign as color to the blind. When she pleaded with him to stay, weeping, telling him she couldn't live without him, she spoke a language he couldn't hear. I picture him gently pulling away, the day her estrus faded, in search of the next fertile female. His buddies went ahead, having gotten nowhere with his mate, all looking for another who might be more compliant. He patted the sobbing woman's hand saying "Honey, I like you too, but I really gotta go ... ." He left not because he was cruel or shallow, but because estrus was strong. If it weren't too strong to ignore we would still have it, and we would be ignoring it.

How many episodes like that could a woman take? She couldn't detect estrus, and she surely didn't understand the grip it had on him. With each visit she knew he would stay forever; she scorned all others for him. Though he may well have loved her, he came to her more for her estrus than her personality, along with his buddies. Men couldn't see why women picked only one man for sex, and made such a big deal about it, when she could tell they all wanted to join in.

Women dressed men in illusions, stitching princes from the material at hand. But men never fitted their role, and each contact betrayed women's expectations. In each woman's life this coalesced into deep anger. Yet each time their emotional addiction struck, they were once more swept away, secretly seething, and at a loss to understand why they were trapped anew.

The Courting Wall

Terrestrial passion was hell on women. Their susceptibilities and men's behavior together made courting increasingly difficult. At the same time, men's courting skills declined, and our entire mating process approached the Courting Wall. I have tried to depict their emotional life in some detail, but now step back and see it again in broad strokes.

Late-stage aquatics (more highly-adapted) hesitated to leave the water, and women made the move last. The lottery of rivers, habitat, and population meant that most returning apes were inorgasmic. Women made the move for love, and passion's power multiplied during the transit. Male promiscuity began causing profound stress for females, and their pain then became an added barrier to mating.

Passion is not a byproduct of human intellect but a hunger that outweighs hazards. Passion and self-repression evolve together and for the same reason, pushing away one's own sexuality while pulling toward sexual relief with a partner. Anything that causes love to increase will also increase self-repression. Passion was so vital to our success that women now rate love higher than the simple sexuality it evolved to assist; repression carried equivalent weight.

Normal men infuriated passionate woman. The very emotions making men desirable made men's absences intolerable. Repeated injuries discouraged women and led some to avoid men altogether, removing themselves from the breeding population. As relations between the sexes became strained on the beach, successful mothers were those with heightened passion and self-repression.

This defines the feedback loop that endangered our courting process. Self-repression grew because sex often disappointed and pairings often failed; only repressed women could prefer sex with infuriating partners. But self-repression feels like sexual doubt, and makes women harder for men to court. Passion grew stronger, to both overcome women's own repression and to make sex worthwhile with an annoying man. But growing passion made male promiscuity more intolerable and increased the sum of pain for women. Pain makes celibacy more appealing and so requires more-intense self-repression.

While women's emotional world collapsed of its own weight, men's courting skills declined. In the water they had no other outlet, and evolved conversation as a competitive tool. This competition drove human intellect to a peak we would not see again for many generations (sketchy evidence follows) and men's verbal skills overmatched women's. Back on land, men had better things to do than talk; they showed off on silent forays instead of in poems. An actual decline in innate verbal ability appeared soon, but first men began to simply lack practice.

Male fluency grew rarer just when female doubts increased. Men already found women perplexing in their adamant monogamy, for men don't instinctively link monogamy and love. So a man might love one woman and yet hopefully follow another in estrus. Men had trouble being eloquent, and any conversation was risky when he didn't understand her emotional state. For men, courting began resembling dance lessons for a small boy — hard to guess the right moves and easy to stumble.

Humans hit the Courting Wall. Women's needs built upon themselves to create a higher barrier for courting men, while men's abilities to surmount those barriers declined. Men could not climb over; women could open no door. But long before the last angry woman rejected the last careless man, we began evolving in a new direction.

Estrus Obscured

Women's practical need for men now fueled evolution's engine; any trait advanced if it made men better providers, because ex-aquatic females could not consistently survive alone. Women's passion became evolution's paring knife; women's anger culled wandering men, regardless what otherwise-useful traits men had to shed along the way.

Women bred men who couldn't detect estrus because oblivious men made more-tolerable partners. Competing for favors of doubtful women, the least-annoying men won. An excellent way to avoid irritating a woman is to pay attention to her. A man who couldn't detect other women's cycles paid more attention to the woman who last accepted him (because she loved him).

Mate-selection always affects evolution. It urged male aggressiveness because choosing a dominant male is an excellent female strategy. With females' innate response to dominance, courting made men evolve singing and language in water when they had nothing else to show off. Less combative than butting heads, these were still competition.

Verbally adept men left more progeny, competing against other males a woman talked with, though perhaps the men never met each other.

Inability to detect female cycles is similarly competitive behavior (or non-behavior). The man who fails to notice the estrus of a woman down the block won't chase after her. Less often running off, he less often angers the woman whose emotional addiction requires his presence. He gains an edge over other men she may like more, but whose behavior makes her angrier.

By preferentially mating with such men, women evolved males numb to estrus, giving us an astonishing and defining human feature. From elephant to rabbit, mouse to buffalo, males infallibly notice ovulating females. Without this sensitivity they couldn't hope to reproduce. If humans had equal sensitivity, a man could identify the state of each woman he passed, distinguishing immature from menopausal, or first day of ovulation from second.

Modern men haven't a clue when their female friends ovulate. Couples trying to make a baby have to maintain a calendar and record her temperature, to find auspicious dates. (Newly measurable and subtle patterns don't change the evolutionary result. We lost estrus when men couldn't detect it, regardless of whether modern instruments can.) We cannot see what other species find obvious.

Some think estrus-loss came from female competition and proves our superiority. Humans can make love any day they choose, so this must be a superior sort of sex. Since men think women always look sexy, women must have evolved to continually entice men. So (the theory goes) women evolved all this to make a man think fondly of her, next time he trudged home bearing haunch of wildebeest. Enticing hunters with constant sexuality had enough survival value to cause estrus' disappearance in women.

But estrus' death came from shutting down *male* responses, not from female competition. A man waits near his latest mate, unable to detect her ovulations or anyone else's, constantly in hope of some action. The willingly monogamous female bred an uncertainly monogamous male. Her urge to pair (caused by her inability to tolerate multiple males) crudely matches his inability to do much else.

Aquatic evolution accidentally prepared us for estrus' loss. Water had masked odors, and men evolved sexual arousal without them. Of course, males can always masturbate without sniffing a female crotch, but in water we became accustomed to copulating without estrus, which wouldn't interest a chimpanzee. Once we could couple without estrus, being entirely numb to it was only a short (but momentous) step.

When women left the water, men could again detect women's cycles. Men's instinctive pursuit of estrous females infuriated females themselves, who had evolved emotion-based sex in the meantime. The more male behavior angered females, thc more females preferred males who couldn't detect estrus, and so oblivious men evolved.

Disappearance In Groups

Estrus vanished through a ponderous trend, not an isolated and freak mutation. On its face, losing estrus seems unlikely; no other primate has clearly done so. Yet two long-term patterns made it nearly unavoidable in us: Emotion's value in late-stage aquatics, and late-stagers' predominance as land-colonizers. Losing estrus is not superior, merely probable.

Female passion sets a clock to time estrus' collapse. The more important a woman's passion becomes, the longer (on average) it takes for a man to couple with an indifferent woman. Her ancestral mother needed no courting at all; modern women may require months or years to accept a man. Or minutes, for some. If only estrus can trigger men's courting, then courting has no chance if it takes longer than the time between the start of estrus and the death of the new egg. If estrus *does* trigger courting, then a wandering mate hurts his prospects by stepping on his mate's feelings.

Love in turn was the result of frontal sex, not its cause. Mild love evolved when purely mechanical problems made frontal sex advantageous though slightly scary. Love grew more powerful when long-term aquatic frontal copulation removed automatic female orgasms, and stronger still when we moved to land. Love's present power reflects what it took to lure our mothers into a frightening and inorgasmic pose.

Frontal sex on land implies late-stage ex-aquatics, who coupled face to face in water long enough for vaginas to aim toward the body front. So estrus' loss coincides with a returning late-stage aquatic, inorgasmic and emotionally-paired, addicted to the frontal embrace. Early returnees reverted to the less-personal rear-entry sex they found more tolerable, where a male's promiscuity hurt no female.

Of all aquatic apes, late-stage groups made the most-hesitant and most-numerous land colonizers. While early aquatics could move to land over an ocean beach, late-stage aquatics required slower transitions up rivers. Early-stage aquatics who moved into rivers made the land transit near the sea, while late-stage aquatics occupied vast drainage basins filled with innumerable favorable sites. So the lottery of rivers and time meant that most returning aquatics were late-stage and passionate, primed for estrus' disappearance.

A few primates achieve monogamy while keeping estrus. If monogamy helps parents reproduce, it can certainly evolve as an instinct in both sexes. It might have evolved that way in us, given more time. We surely benefit from monogamy, because only that benefit gave women the power to breed oblivious men. That is, women's preference for dependable men could force men's evolution only because women's preference coincided with practical need. This echoes our return to land, where men's choosing to be there could foster women's passion only where life on land did bring advantage.

But we are not instinctively monogamous, and a few humans can be wildly promiscuous. Women, on average, prefer monogamy as simply the best solution to a dilemma of need and fears, though their daydreams may reveal lustier impulses. Whether or not we lived monogamously in water, women quickly chose it as fears multiplied on our return to land. Once they arrived at that solution, they began breeding a monogamous man from the men at hand by simply preferring as mates those men who stayed near.

Widely separated populations carried the same emotions, consistently forcing estrus to disappear. The longer a species evolved in water, the more passion they carried back to land. Women's love made male wandering intolerable; love's power caused pain when a man walked away. The larger passion's role in mating, the more desirable a man oblivious to estrus became, and the more likely men would evolve an inability to detect it.

Men who couldn't detect other women's estrus couldn't detect their mates' either. By selectively mating with those who annoyed them the least, women removed men's sexual confidence and left them constantly hopeful, constantly confused. Men had already evolved in water a heightened sexual awareness. With estrus scents drowned there, men became aroused even without odors. So the loss of smell sensitivity did nothing more than return males to the courting mode women had found so gratifying when aquatic.

Each male's best reproductive strategy then became the same as it had been in water. Each man looking for a mate scanned the population of females, looking for one who returned his glance. Partly like a fisherman hoping for any fish, he also felt passion's tug inherited from his mother, drawing him to the few who hooked his heart. One he liked he would court continually, focussing on her as his mother had focussed on his father. He recognized her cycles no more than aquatic men had, but she had a reliable and trusted partner, confused though he might be. An excellent chance of sex makes a terrific lure for a male, though the timing may be uncertain.

Human estrus vanished as the simplest way to evolve a tolerable male. It accidentally caused constant confusion in men — not a result women sought, but women couldn't see long-term effects. Women merely preferred to mate with men who hurt them the least, and women's preference consistently took control when passion made their anger last longer than their cycle.

Imaginary Partners

When men stopped following estrus, they gained by falling in love. More than mere survivors of a culling, men now competed against each other in a newly restricted arena. The more important passion became for women, the less value promiscuity had for men. A passionate woman who wanted him would accept him immediately; one who didn't already want him took longer to court than her estrus (and therefore his estrus-driven interest) lasted.

Suddenly love had value for men, because it helped maintain their interest in the newly delicate relationship. For as long as women had loved, mothers had been bequeathing the instinct to both daughters and sons, and as aquatics men had already gained by inheriting it. Just as her obsession helped a woman accept a mate, the aquatic male's focussing had helped him court her long enough for her to notice him. But men never required love for sex, and can mentally disconnect the two as women cannot. Even so, some early terrestrials surely shared deep mutual affection.

Once women's anger was a predominant evolutionary force on men, men gained from their own obsessive focussing because it carried their interest past women's estrus cycles. The same concentration that helped an aquatic man court his mate past undetectable ovulations now sustained his interest over years on land. Not callous, early terrestrial men were drawn away by an irresistible instinct. They could not (on average) ignore estrus, for if they could we would not now be numb to it. But once numbness had freed a man from estrus distractions, passion helped make him into the attentive mate women hoped to find.

A man maximized his reproductive chances by not disrupting the one relationship he had. Any trait helped him if it helped that relationship, whether or not the trait ever before had value. So when estrus-awareness became a disadvantage for men, passion became as much an advantage. No longer inheriting love willy-nilly from mothers, men began evolving it in their own right because faithful but numb men raised more children. Husbands appeared.

Women didn't know when men grew oblivious, for women didn't share men's estrus sensitivity, and the differences in any generation were too small to measure. Men didn't know it either, for they had no way to compare themselves. Men who wandered left fewer children; men who stayed bequeathed the trait to their sons. Men on average felt less drawn toward other women (because their fathers had been the men who stayed near, numb to estrus). And they felt more drawn toward their mates, as passion grew in them, reflecting the new benefit of monogamy.

A woman didn't always get what she thought, when she chose an oblivious man. We can sustain delusions about those we love on very little real evidence, as flowers in the desert. When a woman did find a man who'd stay near, she could tell herself he loved her though he might not. Men never needed to be perfect, they needed only to give women something to work with.

Components Of Estrus

What then did we lose, when we lost estrus? Smell is the simplest part, and an adequate definition for now. Shedding cyclical sexuality first seems a bargain, traded for merely our sense of smell. But we also paid by losing intelligence, which then adjusted our eye for beauty, enlarged our brains, and finally ensured our survival. That discussion can wait.

Estrus is an interaction of male and female, not just smell; a communication between the sexes, not a female attribute. Most use the term too narrowly, to mean a female's ovulation-related preparations for coitus, announcing her state to the male. In the classic view, estrus is a purely female property. Women evolved constant signals (breasts, touchable skin) to keep the Mighty Hunter interested, and men respond helplessly; Sneaky Eve hid her cycles. But we can't define "estrus" without the male response; estrus cannot evolve in the female alone.

Male recognition evolves to exactly match the female signal. She emits odors to announce her state; the odor evolved because it brought males; male odor awareness evolved because it led to fertile females. Odors, colors, or neon lights would work equally well, so long as females signal when they ovulate and males detect it. Whatever the exact form in any species, female features and male recognition evolve together because the vital message is heard.

Sexual signals began with smell and taste — the most-primitive senses. Eyeless males in primordial seas found females by following chemical trails, and most animals still signal estrus with odors. Though smell and taste imply estrus' primitive beginnings, *any perceptible change linked to ovulation* may signal estrus for a species, regardless of whether any other species shares it.

Estrus varies widely in primates. Chimps display extravagant genital swellings, as external labia become a firm extension of the vaginal barrel, engorged and lubricated. Swollen pink skin stands out from darker fur, and Goodall reports males racing several hundred yards to get to a female in this state. But she must have estrus odors when they arrive, or they lose interest. Arboreal life weakened smell sensitivity in early prosimians. Scent doesn't linger in trees, and is less useful than sight for detecting food or enemies. Our widely scattered, chimp-like ancestors had to announce estrus over some distance, so female swelling, first evolved for purely-mechanical reasons, increased over time perhaps because it attracted attention from afar.

Female chimps recognize their swollen state and associate it with male attention. One female at Gombe was confused at the conclusion of her first estrus when her swelling subsided. Apparently she thought she'd damaged it, and during her next cycle she continually protected her vulva with one or the other hand. Though conscious of their swellings, females cannot smell estrus odors and do not sense what excites males.

No living man can describe estrus' pull. In chimps it incites sex-to-exhaustion; it triggered our bodies as automatically as a meal triggers saliva. Like a blind person attempting to describe color, we can barely guess how it felt. Immediate, all-consuming and imperative, estrus has no modern equal. Ogling a pretty girl is to estrus as food's picture is to food.

Every facet of estrus must be cyclical; any non-cyclical feature cannot announce ovulation, though it may announce sex, status, or condition. A woman's beauty is *not* estrous because it does not change during her cycle, but we can use the analogy. Any female in estrus, though the most haggard of the troop, would seem breathtakingly beautiful to any male. (Goodall reports the oldest female chimp was unusually popular as a sexual partner.) The youngest and most-nubile female, by contrast, would seem plain if not ovulating.

Though they loved their wives, our fathers followed other women in their estrus. Not shaggy man-beasts, they were intelligent people with a powerful language. Women pleaded for fidelity; men tried to explain their side. As estrus faded, husbands increasingly stayed home, but in early days men could not ignore it. We know they couldn't ignore estrus, because we evolved obliviousness to it instead.

Men became numb to estrus by first losing sensitivity to estrous odors. A man whose smell sensitivity was fractionally less acute than another's was slightly less likely to notice the odor of a female down the block. He less often chased after her, and faced fewer tearful recriminations from his mate. Each straying might require days or weeks to soothe his betrayed and angry partner, a time when they would conceive no child.

Men had already evolved to couple in the water without estrous odors. With estrus now causing emotional turmoil for women, those men with a less-acute sense of smell seemed to women more attentive and reliable. Chasing another woman was largely futile anyway — the male might be inescapably drawn to *her*, but she probably ignored *him* completely. She selected her mate by personal preference, and while a few men might find many females, most men would find only a few.

Like any other trait, smell sensitivity varies among individuals. Natural selection needed only that tiny variation to work on, in encouraging scent-dulled men. Each slightly scent-dulled man gained a tiny advantage, beyond quantifying, repeated over uncountable generations, keeping each father's sons just slightly closer to their own hearth.

Insensitivity to estrous odors also implies insensitivity to *other* odors. When suppressing response to a single odor became useful, the easiest evolutionary route was to decrease awareness of *all* odors, so men now have a poorer sense of smell than do women, a footprint of sex-specific selection for olfactory numbness. If we had needed scents for hunting we would have kept the sensitivity, but hunting shaped us less than courting.

Estrus died because women's natural preference happened to erode men's awareness of it. It vanished not to make our lives better but as an accidental by-product of women's attempt to minimize their pain. It reflects not superior sexuality, but men's inability to ignore once-useful instincts.

Female Competition

Men evolved to ignore estrus, rather than women to conceal it, because women couldn't afford the loss. Other women's estrus caused pain because emotion-based pairing made male promiscuity intolerable, but her own estrus remained vital to a woman's competition. When women bred men dull to estrus, women accidentally removed their own competitive asset.

Female competition is a fairly new idea, while blatant combat has always typified male competition for sex. Though primate females are less aggressive than males, their competition may have larger effects. A male primate often mates for only a few years in any one troop, possibly an instinctive trait preventing inbreeding. But a female often spends her lifetime near blood relatives. While a male competes for sex, a female competes to advance her family.

Females compete more subtly than males, matching the subtle gains they win. In any one day the advantage in displacing another female may be small, but a lifetime of little victories may give better health and reproduction. Generations of such advantages, coupled with familial alliances, extinguish low-ranking lineages while stronger dynasties thrive and multiply. Female competition is a deep ocean swell shaping the primate world, while transitory male combat only ripples the surface.

One female competition depends on estrus. By signalling ovulation to the male, her estrus triggers his sexual interest and competition. Being competed for guarantees each female the best available mate. Any female who emits no signal fails to attract males sensitive to it. Reducing her odors reduces the male crowd and their competition, and her chance to pick the best.

In many primates (including our near relatives) infanticide's danger led to promiscuity as another competition between females. A promiscuous female induces paternal uncertainty in all the males she swives with; the more-promiscuous, the less-likely her child will die at their hands. Estrus helps promiscuity; the more powerful her attraction, the more males will know her and be unsure of their own paternity. Any female with weaker estrus signals couples with fewer males, and bets her baby's life on the kindness of strangers.

Skill at choosing mates is still another competitive arena. When a female selects healthy and dominant males, she bears young who start life with an advantage, while indifferent females raise mediocre infants. Primate females may be less choosy early in estrus, with their preference apparently linked to hormone levels. A female may mate more casually before ovulation, like greeting all at a party to avoid offense. With rape virtually unknown in primates, she can then choose only alpha males at her cycle's peak, though she coupled with many to protect her infant-to-be.

The right to refuse incidentally grants her status during her estrus, when a hormone-enhanced preference for more-dominant males narrows her choice. Refusing lower-ranked males combines with *all* males' interest to raise her status for a time. In effect, a female appropriates a temporary slot in the male hierarchy just above the highest-ranked male she refuses.

For many years primatologists thought estrus-dependent status was the only status available to females, and widely reported it. Estrous females compete directly with males for food, while non-estrous females don't — behavior that catches a primatologist's eye. Estrous females may also dominate other females, possibly because they can temporarily enlist male aid if needed. Estrus-dependent status is probably no more than the short-lived celebrity of a socially-adept primate. But choices available to an estrous female (flowing from the competition her estrus evoked) were vital to her lineage's success. A smart female in estrus can guarantee her child an excellent father and many helpful uncles.

Losses Of Estrus

Not only did women need estrus; they had no way to shed it. Even if estrus' absence made women's lives easier, no selective pressure could guide women to that end-point. Natural selection operates on tiny variations between individuals. Any weakening of a female's estrus odors would hurt her, not help her, and any evolution in that direction would extinguish the lineage involved.

In the classic analysis, women evolved hidden ovulation to keep men interested and at home. Not knowing when she ovulated, a man coupled frequently with his mate to ensure pregnancy. But this works only for one man and woman alone in the wilderness, not for a social primate. An anestrous female can keep a male interested only if he already ignores estrus.

In a group, any one woman with reduced estrus odors would lose men's interest. Other females with normal scent signals would attract males in their cycle, while more-subtle females would attract fewer. Modern women know well that blatancy gains male attention more reliably. So a less-odorous female reduces her chance to pick from suitors, reduces her prospects of procreation, and weeds herself from the gene pool.

Not only did a female lose chances at mating, her lineage lost genetic advantage. Without cyclical allure and status, a female had little chance to select the best males. No group surrounded her, no one was breathlessly interested in her acquiescence; her refusal impressed none. She had once chosen mates freely, cycle by cycle, as male rank surged and ebbed over years. Now her emotional limits required that a male court her more slowly and stay with her longer. An alpha male, if she got one, might not stay long if other women were after him. A bad one could be hers forever.

Perhaps females could lose estrus if the breeding population evolved it as a whole and without any individual variation. Over generations female signals could decline, with no one female perceptibly different from the next. Evolution teaches, however, that variations are the norm, and any female with stronger odors could bring down the whole structure, destabilizing every hearth she passed. Only if every woman had exactly the same degree of attraction could estrus vanish by *women's* change.

Women concealing ovulation runs against evolutionary currents. Women could lose estrus only if there were no variation, and only if men felt no competing tug from other women. Suppressing estrus has value only to the extent that estrus lures men, but women cannot suppress it as long as it *does* attract men. A lineage cannot shed estrus as long as any other lineage has it; a woman gains not when she loses estrus but when other women do.

Estrus is nearly universal and uniformly important, not some trivial trait to misplace, yet we clearly have shed it. Estrus became inactive in humans when each less-sensitive man gained a reproductive advantage. This mechanism depends on responses we still display, and works *with* individual variation instead of *against* it. Women lost a vital competitive tool when estrus ceased to function, and men gained one. Good evidence that men became oblivious, rather than that women concealed estrus.

Intelligence To Trade

Our cyclical sexuality was an iceberg, with smell the visible tip and intelligence the hidden part. Men paid smell's small and obvious price to shed estrus awareness, but also paid with large, obscure and crucial intellectual decline. Humanity's survival owes much to mental tricks we evolved when men became dull. That men (not women) shed estrus matters because only a male event fits the rest of our story.

Smell had already become trivial; I discussed its loss first only to show the process. Life in trees weakened the primate sense of smell, we evolved to find odorless mates in the seas, and we shed scent-awareness easily enough when estrus caused new problems. Perception and intuition surpassed smell in the courting game, and men's aquatic courting competition lifted human intelligence to a peak. Obliviousness suddenly gained value after we returned to land, and we started back down the slope; dull men made more-tolerable mates.

Aquatic reproduction focussed on male competition for hesitant females. Vaginal restructuring, vanished orgasms and the fear of frontal sex made emotions vital to female procreation. Men evolved new courting styles because aquatic women mated with the least-fearsome and most-persuasive men, and men had little to do but compete to persuade. Men vying for female attention gave us singing and language, competing by displays instead of force. As complex as language eventually became, it began as simply a new way to show off. Physiological changes to support language (Broca's area, Wernicke's area) show the competition's importance. The male's toughest obstacle came from female doubts, and brains grew language as peacocks grow tails, both to impress mates.

Male competition also depended on knowing how women behaved. If winning a hesitant female was the goal, then accurately discerning her responses let a man gauge his progress. As males moved between family groups of aquatic females, men who quickly saw their opportunities did better than men who wasted time when a woman wasn't interested, or men who gave up too quickly when her mouth said "no".

The better a man understood women, the better he reproduced. Men who talked women's language touched more hearts and found more mates; their children outnumbered the children of duller men. In water men had only speech and perceptiveness to compete against other men, and both grew to counter female doubts. When sex got mental, smart men got more sex. So in water men evolved to discern women's subtlest signs, and no man had to wonder what a woman thought.

Feminine signals were partly estrus-linked, partly not. Courting times lengthened in water as female hesitations increased, and affection gained importance as her sexual trigger. Yet a modern woman's sexual interest still depends partly on her cycle, and her welcome for the same male varies through the month.

Early aquatic males visited several females regularly and courted them all, like a trapper working a line of snares. Women grew less orgasmic and more passionate as their pelves deepened, courting slowed, and a man spent more time with each of fewer women. Men's longer stays also owed something to the affection men inherited from increasingly passionate mothers, as all aquatics felt love's impact. At all times, men competed in their ability to discern women's responses, to know which woman was the most accessible on any one day.

Men watched women's faces, since water limited body language. Each smallest facial flicker, each eye movement, gave a clue to her feelings. Primates are intently face-focussed, with a broad range of expressions. We can tell from considerable distances whether another animal is looking us in the eye. Without knowing it, we can detect expressions lasting as little as a fifth of a second, and routinely measure another's response to us by the size of their pupils.

But men could also converse at a level now lost to us. A single word can speak volumes, when long-time friends catch each other's tone. When we pick the exact phrase to pluck another's response, and hear their answer in kind as if hearts directly touched, conversation can thrill us. We are lucky to find one we can talk with so well, but this *typified* aquatic conversation when conversation was all we had.

Several million years of aquatic courting fostered ever-higher human intelligence, and we peaked in the sea. Men were poets at a level we can barely guess, perceptive and intense, subtle and intuitive. Men in water evolved mental traits we now think of as feminine, though men now often have trouble talking with women at all. Women had no special need of mental skills, but inheritance from fathers carried women along just as inheritance from mothers gave men love.

Back on land men tried to maximize their sexual contacts, as they had always tried to do in water. When estrus scent lured them they went, for that infallibly led to sex. When more-subtle signs told of female availability, they followed them too, using the masculine hyper-perceptiveness they'd evolved in the water. Men could recognize an available woman clearly as if she were on fire.

Declining Male Intelligence

Male intelligence declined after our return to land, because intuitive and perceptive men understood women too well, and could too easily see prospects beyond their own hearth. Male promiscuity wounded passionate females, and women by their cumulative anger culled male traits that angered them. A woman didn't care whether her mate had a good sense of smell, she cared only that he stayed home. She didn't mind if he ignored subtle signs if he would just *be* there. Her new passion demanded his presence, opposed his water-born ability to understand women. She'd rather have a faithful clod than a fickle poet. In the end she got neither.

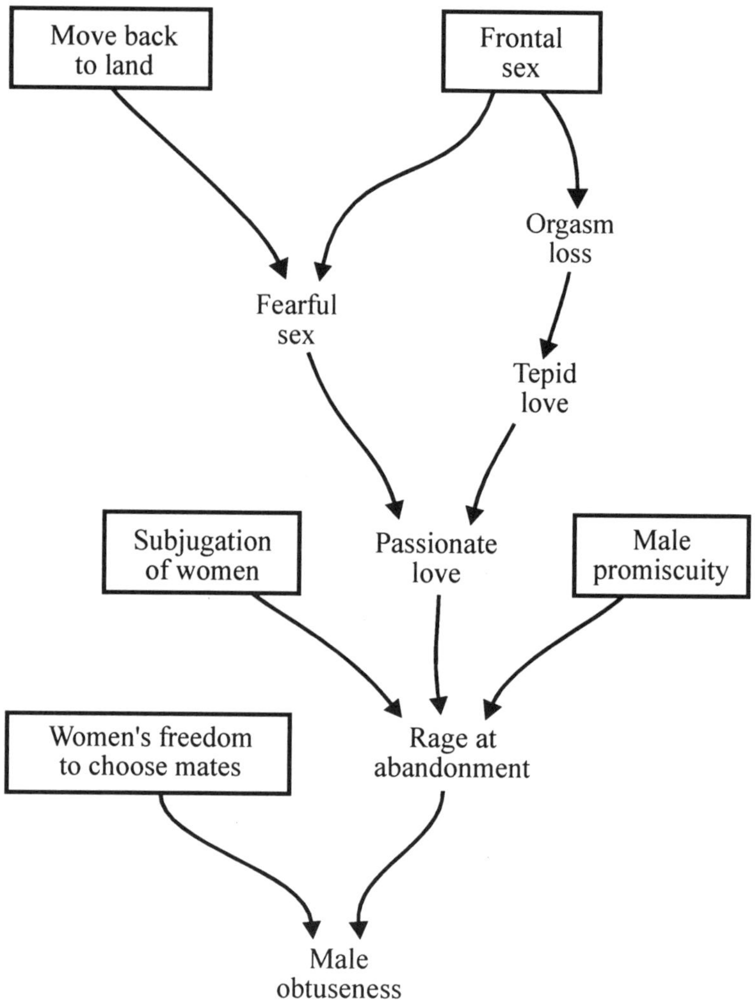

Figure 19: Declining Male Intelligence

Any male able to detect female signals used them to pick his short-term sexual strategy, to stay with his latest partner or to follow another who beckoned more clearly. His wandering came from his perceptiveness, and left long-term distrust in his mate, hurt and enraged by a wanderlust she didn't understand. The woman he might chase had her own needs; if indifferent to him, she made his sexual foray unlikely to bear fruit.

A tolerable husband was one who found no added lure in another woman when she ovulated, no less attraction in his mate when she did not. A man could not simply be indifferent to women, or he wouldn't mate at all. We were a social animal living in tribes; women were all around. Men's obtuseness focussed on *cyclical behavioral* changes in women; the less a man noticed how women acted when they ovulated, the less likely he would enrage his mate.

Evolution is never more precise than necessary. Cumulative rage of betrayed women eroded wandering men's procreative payoffs; more young were born to men who paid less attention to a nubile woman swaying by. The simplest way to evolve an oblivious husband was to suppress *all* male perceptiveness, just as ignoring estrus odors came with suppressing all smell sensitivity. A man could not avoid seeing another woman, but he gained in the long run if he could not tell that today was the day she would have accepted him.

Not all men wandered. Men inherited passion from their mothers, and love evolved in men by helping a man court a woman long enough to win her . Most couples had genuine and reciprocal love, as couples do today, but temptations occur, and any who strayed evoked pain and rage at home, and reduced his chances of having more children. Being a little duller hurts no man, though not all needed the hobbling.

As women's anger culled men who were both perceptive and disloyal, perceptiveness dwindled in the male population and we all inherit the lessening. By preferring stable men, women bred duller ones for their daughters. The resulting man stayed home because he detected nothing luring him away, and his wife could interpret his nearness as fidelity. Men accept monogamy partly because women unknowingly reduced men's ability to stray. It is often an uneasy acceptance.

Perceptiveness and intelligence come from neural complexity, from the ability to sort loosely related information. Intuition is perceptiveness without attention, digesting details until the gut reaction alerts us. Men recognized available women by intuition; they *knew it* but didn't need to know *how* they knew it. When women preferentially mated with less-perceptive men, women stripped from men an intellectual process; the intuitive leap had to stop for a man to be a tolerable husband.

In breeding faithful men, women altered the genetic recipe for men's brains. Just as demands for greater intellect shape a more-complex brain at the cellular level, demands for reduced awareness strip complexity away. Stupidity has no value by itself; a woman gained no competitive advantage from breeding a mate she could outsmart. In the process she acquired a mate who could not read her moods any more than he could read the ovulation signals of other females. But at least he stayed home.

Women now trust intuition while men tend not to; some think it is women's exclusive terrain. Some deride intuition as a soft and sloppy way of thinking, while clear and linear logic translates more easily to print. Having lost intuitive acuity, men often feel overmatched by women, and shy away from dangerous ground.

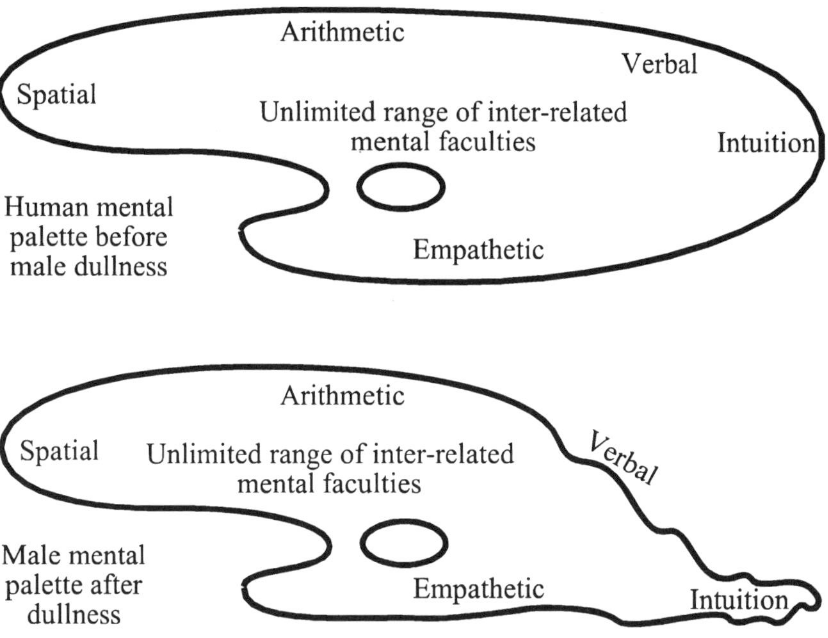

Figure 20: Palette of Human Intelligence

Linear thought remains when once-powerful brains lose subtlety and intuition. Linear thought evolved not as a better way of thinking, but as the skeleton of more-complex mental processes after subtracting intuition. We did not evolve a powerful brain to become a Mighty Hunter. Instead we eroded an earlier intellect leaving behind enough, as it happens, for hunting.

Declining Language Skills

Women weakened men's courting skills when they stripped men of subtlety. Men had evolved eloquence in aquatic courting competition, with complex brain circuitry to support it. When women reduced men's brain complexity by culling the disloyal, men's conversation was cut down by accident. This answers the problem that earlier arose by implication: If men evolved language, why do they now suffer a verbal disadvantage? Because long after language arose, new forces simplified the sophisticated machine that first created it.

Courting created language and intelligence, both dependent on the same complex brains. Eloquence depends on more than just brain anatomy. In an impoverished culture, a brilliant mind may have no words. Spartan speech prevails where fancy talk is disdained and action admired. Yet mental processes surely *limit* fluency; speech doesn't hold more than the speaker puts in.

Language was the inherited record of women's preference for interesting males, just as intelligence was the record of both language and perceptiveness. Under constant selective pressure men evolved an easy mastery of speech, for the man who struggled with words lost advantage. Specialized brain areas made eloquence easy. Competition was so strong we began it in infancy, as a kitten plays at pouncing, honing adult skills.

At its height language came to substitute for physical touch and grooming, and still can. Grooming is a primate's central social act, more important than sex in forming relationships, so vital that primates evolved skin eroticism to encourage touch. But through language we stroke each other with phrases instead. These phrases feel like actual grooming; the receiver feels slighted if too few phrases are offered. That mere noises can replace skin contact shows language's importance to us.

When women began to select loyal men as mates, intuition and subtlety hurt men by leading them astray. A man had to *be* oblivious to other women, not just pretend it, and that required a physical change to reduce neural complexity. Men began shedding mental high-performance parts and fuel-injection, until their brain functions reached a level low enough to keep them consistently at home. With brain complexity cut, verbal ability eroded.

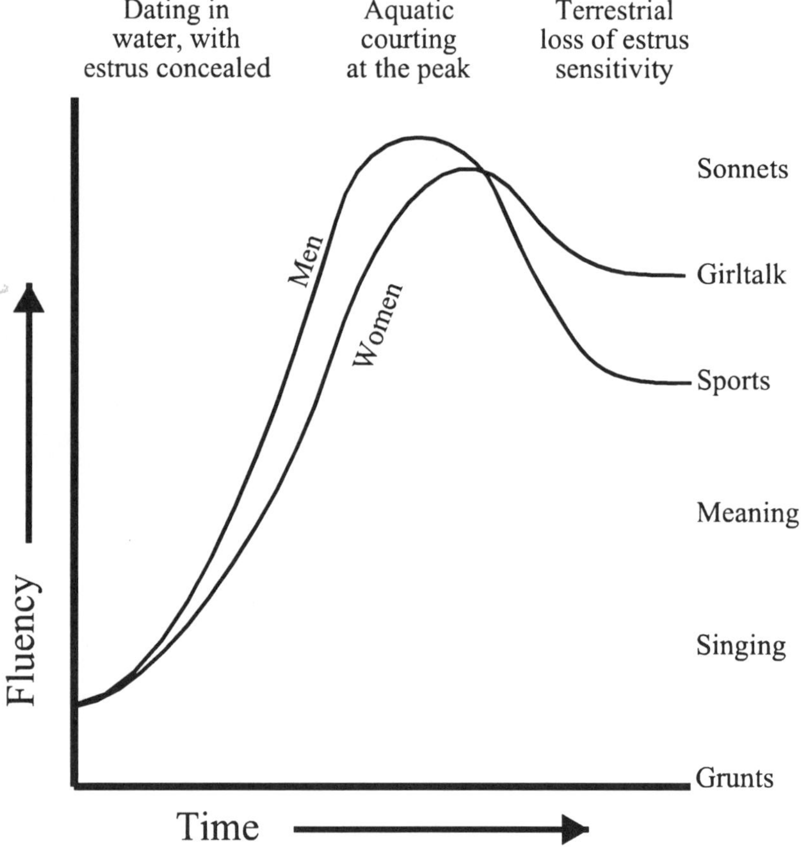

Figure 21: Rise and Fall of Language Skills

Just as men inherited from mothers the ability to love, and as evolutionary followers are less-adept at it, women inherited from fathers the ability to speak. In the water men were more fluent. A women could keep up, of course, but it was her *admiration* of male skill that helped persuade her to mate.

Now men's facility was accidentally cut back, leaving women out in front. Suddenly women saw men as verbally *inferior*, and conversations became awkward. Where once men had easily held feminine attention, now men struggle to guess "the line" which might evoke interest. Men's powerful courting skills fell away because women were willing to live with the results.

Reduced language skills did no great harm, though women hear the difference. A modern man might seem wonderfully facile to an early-aquatic female, but dull beyond belief to a late-aquatic woman. Men alone were selected for mental regression, but women's language skill also declined, as they inherited less-complex brains from their fathers. Men sliding back down the verbal scale left women a bit higher up as they had once left them lower down when skills were rising.

Men retain a compulsion to speak but have lost facility; men seek vocal contact but have lost verbal depth. The gap between the sexes widened as a byproduct of reasonable female strategy. A woman's choice of the male who was less-annoying to *live* with made it harder for her daughters to *find* one who could hold up his end of the conversation.

Measured By Women

Evidence for men's intellectual decline comes first from women, and more clues will accumulate as the rest of my thesis becomes clear. We can never find direct evidence because key people are long dead, just as we cannot take a dinosaur's temperature. As with dinosaur bones, we can make sense of clues we have only by comparing how alternate theories fit larger frameworks.

We are starting to come over the brow of a hill, with a broad valley taking shape as fog clears. Early hints will become the focus of our tale; there will be less talk of what we do, and more of what it means. Sex and courting are trivial until we use them to see who we are and how we arrived. In the end, this will have been a book not about our individual reproduction but about our species' just barely completed birth.

Comparing speech of men and women makes us stop and notice; one of many trails in a wilderness that seems to not match the map we brought. Women are the clearest measure of men's minds, just as men are the clearest measure of women's. For such subtle differences no other species will do, and the only non-male human we can compare men to is a female. I don't imply superiority in either sex, because intellect is not necessarily valuable. Here in a verbal forest, men and women left trails just a little apart, and their small separation tells us that those who made them went in a different direction than we had thought. The direction matters more than the difference.

Women's verbal edge is well documented and shows early in life. We typically explain the female advantage as a fluke, or perhaps an enhancement for women's cooperation. That is, perhaps women evolved verbal skill because they had no other power in a male-controlled world. Men had strength and tools and didn't need words; men could invent weapons and needed no fluency. Men, in that view, trail in a meaningless race.

I consider the evidence crucial, not meaningless, and the first hint of our real track in the evolutionary wilderness. Women's conversations show thought processes more complex and subtle than men's, evincing more-sophisticated mental machinery. Conversation implies cell-level differences. Men are verbally less-adept not because they don't need talk but because their minds no longer support it. I suggest that intellect doesn't really matter much, and we are not gaining it but shedding it.

Women in conversation resemble playing lion cubs. While some interpret pouncing games as practice for future hunts, cubs have no such intention. They simply vent energy by pouncing; having evolved for it, they enjoy it. That this play will eventually make better hunters of them in no way implies that they pounce *in order to become* better hunters. Women's conversation is similarly evidence of interest, not practice for future verbal combat; play rather than strategy for advancement.

Herman Hesse, in <u>The Glass Bead Game</u> (<u>Magister Ludi</u>), outlined an art form so subtle he couldn't describe it in clear terms. As his story begins, the game had grown from an early form where distinctive glass beads represented concepts of philosophy, science, and art. The first games juxtaposed three or four beads, such that the arrangement communicated a new message — as one note is meaningless but several notes become a tune. One bead might represent physiology of a fish; one might symbolize Darwinian theory. Carefully arrayed, beads called to mind the associated knowledge as a symphony of thought.

In Hesse's story, the game has evolved beyond glass beads to symbols of vast subtlety. To play as a novice took years of study; the game of a master called for great ceremony and might be discussed for generations. It might include hundreds of icons, and comprehending it might take more than a lifetime.

Complexity announces its presence when it defies reduction to print; Hesse declared his game's complexity by claiming he couldn't describe it in detail. I suggest that people are much like his glass beads, and to grasp in one's mind the personality of another is to see all this complexity and possibility at a stroke. The most intricate problem is not to predict a star's movement but to predict whom the astronomer will love.

Women play matchmaker for fun, playing the glass bead game, detecting subtle interactions not obvious from a cursory understanding of the symbols. Where a man may see two people, a woman is more likely to see a relationship; where a man hears two talking, a woman may hear a fight. To weigh these contacts is the most complex of games. While such mental agility may *be* a path to power, and a person may practice it *in order to* gain power, women do it also just because they like it, with lifelong use reinforcing an inborn advantage.

Women's conversation betrays an innate skill at complex thought. So much information is compressed in their talk that a man included in the conversation can easily find himself excluded from the meaning. Defying reduction to words but understood just the same, this intuitive grasp of complex reality marks a domain inhabited mostly by women. At one time we all lived there.

What Smart Is For

To admit weakness in one's own sex approaches betrayal, so prickly is the contact between the sexes. We value human intellect so highly that belittling it seems treasonous to all humanity, yet we have good reason to say that intellect did not drive our evolution, and male obtuseness had value. Finding fault with men or women gets us nowhere; differences are trivial except as evidence of larger forces we have failed to see.

We usually explain sexual differences using the hunter-gatherer myth. We assume that intelligence gave humans the advantage, so human brain size must be our evolutionary key. With hunting our formative behavior, men needed skills to track and kill. Male-supremacists accept what matches that myth and reject what does not. Feminists counter with female-gatherer theories, never doubting that brains were the point of it all.

Measuring human intelligence began with measuring brains. Craniometry depends on the idea that brain size correlates to intelligence. Men's larger brains were initially taken as proof of greater intelligence, thereby justifying women's subjugation. But larger animals normally have larger brains, and correcting for men's body size removed men's supposed advantage. Though craniometry fell from favor in the early 1900's, we hear its echo in every argument that links superiority and brain size.

To support our myths, we next perverted intelligence testing. Alfred Binet created the first tests to identify learning-disabled children for special education. He explicitly denied that his tests measured intelligence, but his warning was ignored. Smitten by "physics envy," psychologists had longed for hard numbers they could use to grade and measure people. With craniometry's collapse at the turn of the century, we embraced IQ testing and have used it ever since.

Like Olympic contests, intelligence tests measure unnatural acts. We did not evolve adding columns of numbers, but because we can measure this skill we use it to compare ourselves. We then presume that our acquired proficiency, which no other animal can match, proves our superiority.

Cultural bias permeated early tests. Questions about food products or sports figures could be answered only by people who shared the test-designer's native culture. Tragic results include rejected immigrants, involuntary sterilizations, and ruined lives. Since then we have continually refined mental tests to compensate for cultural differences. We have never wavered in our conviction that intelligence was a valuable thing, that its measurement was a worthwhile effort, and that it could in fact be measured. But the most we can say about our intellect is this: We are able to design tests such that other species and selected humans will fail them.

We don't know what intelligence is, but we can measure behavior that we think reflects it. Until we know what intelligence is, we cannot legitimately compare our own to that of any other animal. Studying other animals may give valuable clues to brain processes, but does not permit placing them and us on a scale. If a whale composed a singing test, or a rat designed a maze, we may be confident that humans would flunk.

If brains did not evolve for excelling in mental tests, then excelling in mental tests does not evaluate our evolution. To see the forces causing our mental evolution, we should study what a brain does when it is doing as it pleases — not what it does when trained and molded. And our brains, when they do as they please, jump to conclusions. The intuitive leap is scornfully simple because we evolved for it, while logic must be learned. The essence of natural thought is the instant awareness of intuition.

Linear thought is a trick of the brain as ventriloquism is a trick of the mouth. Logical structure often helps; only sentences strung end-to-end can communicate my argument, yet my thesis came to me in thunderous insights which stunned me for days at a time. It is more difficult to make *you* see than it was for me to see it myself. I use linear thought as a tool in this book, not to tell you that linear thought is invalid but to tell you it is *no more than* a tool.

Treasonous and radical perhaps, I compare men and women by how they play with their heads. Thin but consistent evidence suggests that men avoid intuition and the Glass Bead Game. Most men don't trust their brain's most complex function and avoid the world's most complex topic. Not all scorn intuition or avoid personal exploration, but enough do to let us call it a trend. I submit that this is crude evidence of an underlying mental handicap compared to women. I must use this comparison because women are the only human yardstick for measuring men.

Evolving Stupidity

Intelligence is a disposable asset. Though brainpower may help an individual compete against fellows, it apparently had too little value to keep men from sliding back down the scale. I will later present more evidence that rising intellect did not shape human brains, and that *decreasing* intelligence better explains our brain growth.

By our own IQ tests we are only about 25% smarter than a chimp, after perhaps five million years of separate evolution. Gaining so little implies that intellectual advances had limited survival value. My thesis explains the slim gain. No exercise of brainpower compares to finding and then keeping a mate; each of us knows the urgency and delicacy of that task.

We need to see brains in a larger perspective. Human intelligence has no more cosmic significance than peacock tail feathers. The universe did not await our awareness. It is not intelligence that sets humans apart, but the ability to focus on problems and a language to pass on the answers. Even this was not our evolution's goal, but grew from courting devices in an aquatic ape, an odd variant of a flawed primate family approaching extinction. And when brains became a problem, men shed them as readily as hair.

Intellect regression affects women too. Just as men inherit from their mothers the ability to love, and women from their fathers the ability to speak, so we all inherit from our fathers a degree of stupidity. Absent sex-specific requirements, we all take a common shape and behavior. If intellect became trivial, then only men's need to be dull guided our mental evolution, and women were again carried as passengers.

Over generations women would inherit successive doses of obtuseness and would finally be indistinguishable intellectually from men. Complete parity might take thousands of generations to attain. Though maternal genes would be coasting and under no selection, they are diluted and inherited over unmeasurable time. If women have skills men can't match, then men's dullness has not yet fully diluted into the female population.

One way to interpret this gap is to say that men are still regressing. If women are inheriting dullness from men, but are yet smarter than men, then men must still be getting duller. We are living in a still photograph of a very slow chase. Still under pressure to be ever more oblivious to subtle attractions, men are slowly moving farther down the intellectual scale. Women are not yet equally regressed because men are a moving target.

The other interpretation is to say that women gain some advantage from intelligence, or humans gain some advantage from intelligent women. Perhaps women are separate travellers evolving in another direction, not slowly following. Feminine intelligence might be useful in female-to-female competition, though semi-permanent pairing has probably reduced that competition's value. Or each woman's brain might give her a tactical advantage over men in her life.

Menopause may be the best evidence that women evolved to be smart. A reproductive halt is virtually unknown in primates, whose females usually bear young until they die. Morgan (1972) says that menopause probably evolved in us because the whole tribe gained some benefit from old, sterile women. This is common in deer, whose groups are often led by menopausal females. Hulse (1971) says that finding old people in a community implies they had value, and the most valuable thing an old person can do is pass on knowledge.

Each tribe gained from smart old people who could pass on their accumulated wisdom from a lifetime of perception and intuition, but each tribe also gained when men were oblivious to subtle events around them. So perhaps women became by default the repositories of knowledge. It wasn't that women won a survival advantage by being smarter than men, but that the tribe gained from having *someone* perceptive.

Menopause may mark intelligence-loss in men, much as an island appears when water level falls. Aquatic males marked a peak in our intellectual evolution, shaped by intense competition for female hearts. When we returned to land male obtuseness became valuable, and women were left smarter by default; wise old women replaced wise old men.

As men slid back down the intelligence scale, women's knowledge became increasingly important. In other species, when random females fail to ovulate their genes vanish, swamped by thc offspring of fertile females. When random human mothers ceased to ovulate they lived longer and became more valuable. Tribes carrying menopause genes learned more and adapted quicker and flourished, rescued by islands of perceptive memories in falling lakes, surpassing populations that competed by merely bearing more children.

Only our ability to pass knowledge through language made brains as valuable as babies, but sterile intelligence could not survive. A valuable woman had to reproduce valuable genes and so menopause became regular, stopping babies after genes for baby-stopping had passed safely along to the next generation. Women's age at menopause perhaps shows us how natural selection balanced making babies against community knowledge.

If women's knowledge is worth *preserving* at the expense of more births, it must also be worth *evolving*. But the mechanism could begin only under pressure; trading brainpower for births gives a benefit only after stupidity is already causing problems. Women inherited male intellect-regression until growing obtuseness started to hurt the tribe. Female timed-infertility then evolved along with sex-specific intellectual contrasts, providing a core of tribal wisdom. Once intellect became sex-specific, women's skills might differ considerably from men's while leaving male obtuseness intact. With female intelligence carrying an important social load, it may have evolved some distance back *up* the scale. Women may indeed be separate travellers, bearing intellectual responsibilities someone had to carry.

We cannot measure the adaptive value of male stupidity; clues merely suggest that some dullness *was* adaptive. We don't know whether women are now the smartest they have ever been; evidence only hints that they didn't grow as dull as men. So we cannot judge either the speed of our movement along this scale or the distance we have come. If we are declining, we apparently still mentally surpass chimps from whom we split at least five million years ago. If our intellect is advancing, still the gap remains useful.

After our return from the water our brains grew (but not to make us smarter), about which more later. This brain expansion further clouds the picture, adding some intelligence fortuitously to both sexes. The contrast between men and women arose on a moving ground, with shifting reference points. We can't tell whether modern humans are smarter or duller than late-stage aquatics.

We are at an unmeasurable point on an intelligence scale of unknown length. Male obtuseness pushed us down the scale from our aquatic intellectual high-water mark. Wisdom's value in women may be pushing up. And we cannot completely disconnect the two because we all inherit half our mental faculties from the opposite sex. And as long as humans need both intelligence and dullness, the distance between men and women will remain.

Infinite Number Of Apes, Revisited

We should pause here, and again see ourselves from a distance. Our return to land was not a goal but a symptom of a larger process. More than simple physical re-adaptations, our return measures female evolution. Our success owes nothing to our ability to hunt, but everything to our willingness to cooperate. We survived because female passions grew strong enough in water to breed monogamous men on land.

Our return came more from numbers than from superiority. In the previous chapter, with a broad brush, I sketched two pictures of returning aquatics. Less-adapted ones could quickly return to land, but by quickly returning they rinsed themselves from the sea before reaching the best spots. Our more aquatic ancestors resisted the move to land, filled the rivers, and accidentally found the lushest places.

We need not consider brief aquatic episodes, likely common in primates for forty million years. We primates dabble in water when we like, live on fish when the land is bare, and live as aquatics when droughts are long. Primates can quickly learn new ways as opportunities arise, and aquatic foods are an obvious asset. We can ignore all primates who entered the water and became quadrupedal again when back on land.

I defined a "mid-stage" aquatic as a primate who would no longer revert to quadrupedalism if returned to land, "late-stage" as one with a fully-deepened pelvis and inorgasmic females. Orgasmic females need not evolve emotions to replace climaxes, and when back on land will prefer the less-threatening and less-personal rear-entry pose, forestalling any future growth in passion. Inorgasmic females evolve strong affection as orgasms wane through frontal sex, can even relish frontal contact, and continue to deepen their passion when back on land. For orgasmic females, contact with any male brings an automatic physical reward; inorgasmic females must first focus on the male, and find part of their coital payoff in the pleasure of his company.

Orgasmic (less affectionate) females would revert to normal ape promiscuity once back on land. Inorgasmic females can move to land only if aquatic affection explodes into passion during the move, passion that then leads the female to shun promiscuity. Aquatic primates (at mid-stage or later) rarely return to land at all; defenselessness and bipedal burdens make water a safer place. Yet mid-stage aquatics must have returned more easily than our late-stage ancestors. Mid-stage aquatics needed only a fair terrestrial environment to lure them from the water, females and all. Late-stage aquatics needed a location so good that the move was advantageous despite strong female reluctance. Feminine reluctance caused powerful culling, leaving its stamp on the survivors in the form of passion.

The last chapter left an implied question hanging in the air. If a promiscuous returnee seems so easy to persuade, then why did passion predominate?

When mid-stage aquatics were rinsed too soon from the water, they lost two different competitions against late-stage aquatics. The first competition was by simple numbers, sketched in the previous chapter. Orgasmic women in estrus did not hesitate to couple on the beach, since they saw an immediate payoff in sex with any man. This quickly brought mid-stage females out of the water (absent passion) whenever local conditions encouraged males to make the move.

As mid-stage aquatics expanded up rivers, marginal local conditions let them move to land before they saturated the tributaries. Late-stage aquatics, by their reluctance to move, first filled the rivers to overflowing and then moved to the land in thousands of places. Though any one mid-stage group moved to land more readily than would any one late-stage group, too few mid-stage aquatics moved at too few locations to flood the continents with orgasmic and nonpassionate tribes.

The second competition came in the immediate evolutionary result of life on land, as outlined in this chapter. Estrus renewed its ancient power as soon as women arrived on the beach; mid-stage (non-passionate) women let men revert to promiscuity, and their population was doomed. Lured out into a lush pocket where women could feed their own infants, mid-stage aquatics died whenever local food sources faded and men's help became important.

Mid-stage aquatics differed from our ancestors in the degree of female passion. Mid-stage females preferred some men over others as a natural result of men's subdued aquatic competition, and as normal primate preference. But they were highly orgasmic, preferred rear-entry sex once back on land, and so were less particular than we regarding whose penis scratched their itch. Whatever the exact mix of emotion and sensation, there comes some point where the average female cannot endure promiscuity. The mid-stage aquatic female, when back on land, never reached that point.

It was not during the move that women needed men's help (helpful men would have *prevented* the land move), but bipedal females eventually needed help raising heavy-headed children on land. No lush habitat lasts forever; as soon as scratching out a living got tough, children began dying. Not bred for helpfulness, fathers were elsewhere when needed. The mid-stage returnee had a society with no bonds, as a spoonful of sugar has no strength or form. When women need help, promiscuous tribes dissolve and die.

Promiscuity is not inevitably fatal for all primates; many are famously promiscuous with no apparent damage. But when a short primate childhood lengthens, there is an increasing chance that at some point in that childhood a shortage will occur. When a child becomes burdensome, there is more likelihood that a mother at some point will fail. Fathers need not always be present, need not supply most of the food, if only they can help a little in hard times. When mothers are already burdened, a famine every decade can end their tribe. Moderately-adapted aquatics did not die immediately, but they all died eventually.

For primates, aquatic life is evolutionary fly-paper; any but the lightest contact becomes a trap. Once beyond minimal adaptation, once bodies are hairless and streamlined, the animal must return to land as a biped if it returns at all. Bipedalism increases women's burdens and reduces the likelihood that they will make the move. Yet we continually returned here and there during our aquatic time, where the land presented pockets lush enough to lure crippled bipedal apes, male and female. Good lands near lean rivers could lure even reluctant apes out of the water.

Returned aquatics dragged bipedalism's fly-paper, doomed by an ancient decision as the panda now is doomed by the ancestor who began eating bamboo because no one else wanted it. Once an aquatic primate evolves a deepened pelvis, it cannot avoid bipedalism and eventual extinction as a fair aquatic or lousy terrestrial. To stay in the sea is difficult; to return and survive is impossible.

It is doomed, that is, unless men become monogamous. When women evolve enough passion, women's anger at unfaithful mates outweighs men's reproductive gains from promiscuity. When males become monogamous, even with dullness as the price, the couple together can raise their children. Once pairing is strong enough to carry their child through a drought, by degrees they can learn to bear lifetimes of shortage. Feminine passion gave strength to our species by breeding imperfectly-monogamous males, mimicking the pair-bonding other species show.

But passion evolves only when needed — in the move to the beach. During the transition (as less-emotional women are culled) female affection multiplies by some unknowable number and blooms on land into obsession. Only a few women followed men onto the beach. Only when the whole aquatic population has evolved affection to a certain level will there be a few women at the fringe who feel the requisite pull. When that fringe group of women reaches some critical mass their reproduction carries the day, and passion permeates the population as it quickly moves to land. Freedom from the bipedalism trap comes not directly from passion, but from passion's tepid yet necessary aquatic precursor.

We might compare our return to making candy by cooking syrup. From time to time the cook drops a sample in cold water where it instantly hardens. The hardness measures how the cooking has progressed — vital to know, though the sample itself be discarded. Hardness is not a *present* character of the molten mix but in its *potential* nature, concealed in the melt, revealed when conditions change. When the first sample assays right, the entire batch is ready to take on a new state.

Love was the necessary ingredient in our aquatic recipe, of no great value until it hardened into passion on our return to land. Tepid aquatic love grew only when needed as a replacement for female orgasms; climaxes faded only when deepened pelves and a thousand generations of frontal sex weakened their source. Love swelled into passion in our landward move, enabling frontal terrestrial sex. Purely by coincidence, passion also made male promiscuity intolerable for women, letting women breed monogamous men who lent a hand. Terrestrial survival correlates to frontal sex because they both gauge the same female feelings.

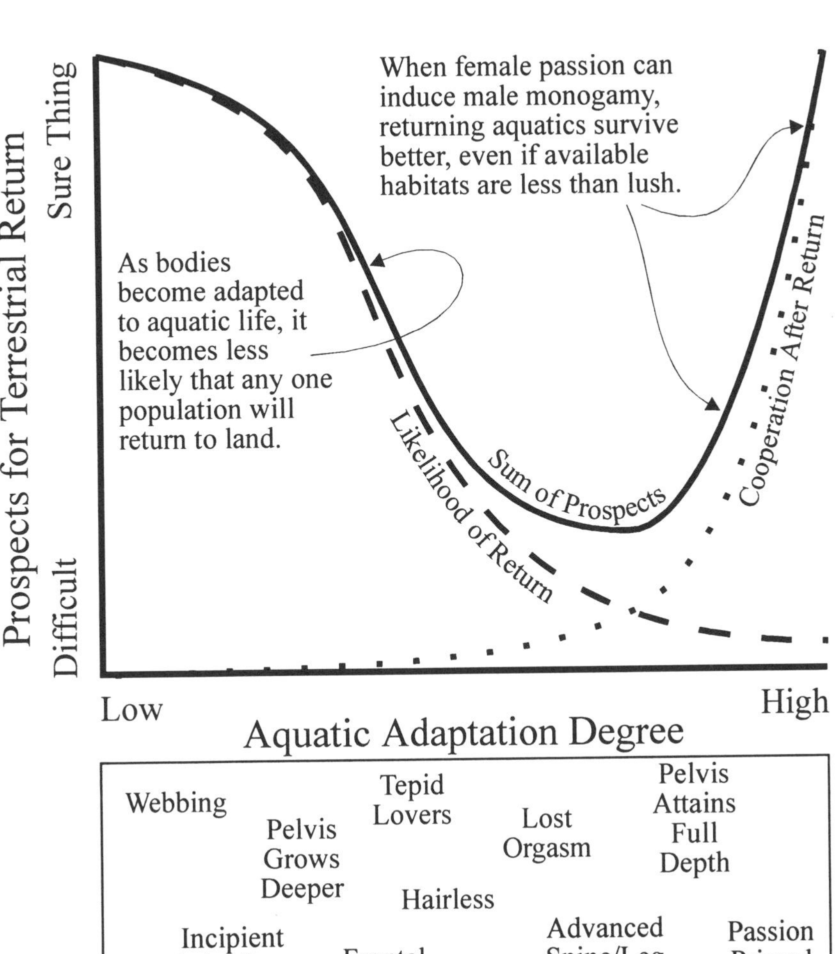

Figure 22: Prospects for Terrestrial Return

For a million years or more we simmered in the Indian Ocean, committed to the aquatic recipe, while orgasms gave way to tepid love, while samples now and then tossed themselves onto land. The longer we were aquatic, the less likely it became that any one group would make the transition, but the more groups *did* return because our numbers were filling the rivers. None survived long until the mix was ready. Until women had evolved not passion but a mild precursor that could *harden into* passion when drawn from the water.

Endlessly the pattern repeated while the aquatic population grew increasingly emotional and steadily larger, filling rivers deep in the continental meat. Wherever conditions were good enough, some men tried to return and sometimes women followed. Lacking bonds, earlier ones quickly died when times grew tougher. As aquatic affection grew stronger over millennia, it became powerful love, not orgasms, that made women follow. Women's passion made promiscuity intolerable; their anger at betrayal then began weeding out the men who understood women well enough to wander.

Passion did not make aquatic groups return, but raised returnees' prospects to near certainty. Ex-aquatics began consistently surviving when women's anger grew strong enough to make monogamy a man's best strategy, and he then evolved to suit. We lived in the water as friends, but populated the continents when women began forming families.

Chapter 8 - Battle Of The Sexes

In the battle of the sexes I've been an unwilling soldier; I've injured my share of women. Buttoned in our uniforms we take aim at each other, drawing blood from our best friend and worst enemy. If I can't stop causing pain I want at least to see what the fight is about. As this book took shape in my mind the battle began to make sense.

The sexual fight is as real as Waterloo, as quiet as fingers on electric skin, as loud as a door slammed. The raped, murdered, and divorced are casualties of this one battle and no other. Men and women wound each other like soldiers following orders, not stopping to ask if an acre or two warrants the cost. We no longer see the blood, having bathed in it so long.

Men and women meet as armies who've come great distances, firm in their resolve. Frontal sex made women inorgasmic and passionate. Their commitment to passion suppressed men's intelligence and made courting difficult. We've plodded these paths longer than written history, bent under sexual baggage.

Each army rolls in a deep channel of purpose; young lives are harvested, uniformed, handed dogma. At the front each unwilling warrior is the frightened point of an army thrown spear-like by his homeland. Vast and bloody inertia holds sickened soldiers helpless in a whirlpool of carnage. Only death or the battle's end gives a way home, an escape from the trench. No countryman at home knows the hurt as well as young soldiers in the opposing lines.

Individual stories seem small next to the saga of the massed ranks they fill. When an army turns to attack, a thousand may die under their own wheels, shrugged off as a necessary expense. But in the sexes' battle, armies turn when individual needs merge to push in new directions. We are many soldiers with no commanders, lost on the battlefield together. We are sliding down a blood-slick slope, locked in combat with our best friend who asked for this no more than did we.

Impregnation Race

Female hesitation remained the male's greatest barrier. Losing smell sensitivity and subtle perception kept a man home but did nothing to help him over the courting wall. Instead, those losses aggravated the problem; reduced verbal skills and perceptiveness made him even less able to court effectively. Evolving into good husbands made men terrible dates.

Broad patterns are the sums of individual stories, but men and women in the trenches never saw the broad pattern. A woman who by chance loved a more-obtuse man had a slightly better chance of a peaceful mateship and more children. In her eyes he was faithful and did not ignore her most of the time, and angered her less often than did other men. And his traits repeated in his sons, who when grown had to find mates of their own.

What did dull men do when courting grew difficult? They did as some men do today — pursue younger, more impressionable women. Not all men chase girls or foolish women, or pay prostitutes. Such women attract because their personality seems no barrier to his sexual interest. When female consent becomes elusive, men seek women with limited ability to choose. The more trouble a man has finding a mate, the more attractive this route becomes.

This is not automatically reprehensible, though infuriating to women. All animals seek sexual relief through the easiest means. The bull holds a harem to make sex certain, and not to give other males a chance. In many species, males chase away other males to force females' choice by default, short-circuiting the courting process. No animal intentionally makes sex (or anything else) more difficult than it has to be.

Women's self-repression shows that they also took the easier path. When sex became frightening on land, some women masturbated rather than find a mate. Masturbation was so safe and satisfying that only women who instinctively avoided it could convince themselves that men were a better source of pleasure. Sex grew worse as generations rolled by, and the repressed left more children while masturbators left fewer.

Women are self-repressed because their antecedent aunts took the easiest route to orgasms they could find. But women who took the easy route failed to reproduce, leaving that role to their more inhibited sisters, and over thousands of generations repression became instinctive. When men took the easy route they changed their sexual target. Both sexes have left us with the modern version who reproduced most successfully. Neither holds the moral high ground.

Female choice became entirely subjective when aquatic life destroyed male dominance indicators. Singing and language show men's evolutionary response when women chose partners by personal preference alone. Preferences that help mating need not be wise or even useful; singing and language began as human plumage, nothing more. Love began as women's mental trick allowing sperm and egg to meet despite fear, not as an objective and accurate measure of men.

Losing estrus similarly freed men to evaluate women subjectively. By breeding men numb to estrus, women cut men loose from old instinctive lures and let them choose by preference alone. When estrus functioned, a woman seemed sexy when she smelled and acted sexy, and all men knew it. After estrus died a woman seemed sexy merely by being female, and her *feeling* sexy had nothing to do with it.

Eroticism replaced intimacy when men became dull. Sensitive males were culled by the female need for a stable home life, leaving men who could not talk with women or detect their cycles. Real exploration of hearts often gave way to a useful *delusion* that such intimacy had occurred. To an unknown degree, the aquatic marriage of minds gave way to a terrestrial merging of mistaken impressions.

Eroticism now led the next sally of our evolutionary armies. Any man gained if he could detect and court the most-accessible females. But he could not evolve greater perceptiveness and skill at reading feminine signs, or his sons would anger and then lose their wives. So men gained by detecting accessible women by appearance alone.

Male Competition

As sex grew elusive, men learned that young women were their easiest sexual prospect. Even without subtlety or perceptiveness, men could target young women as a group because their physique made them easily recognized. Courting young women started as a reasonable competitive strategy each man learned but was not born with. Eventually men's new interest evolved into its own instinct, with thunderous repercussions that still guide our evolution.

Each man saw around him females varying from infancy to old age. Just after estrus' disappearance, men had no more age preference than a chimp does now. Each man would court women he met, probing for interest in return. Instinctively men were oblivious to both puberty and menopause, though each male learned to appreciate the female form as he grew, and to recognize stages of life.

Maturing females moved into an arena of male eyes as ducks in a shooting gallery move into view, as growing boys move into women's view. Death in the end removed each woman from men's attention. Her life would include under a hundred ovulations and perhaps a dozen pregnancies.

Each woman was a moving target. Only a child at her first ovulation, with age she grew wiser. She was easy to impress when young, with sex still the exciting stuff she'd been waiting for. Never before courted, she thrilled at the new attention. Adult males seeking her were the same ones she had held in awe a few years earlier, as she desperately awaited her nubile time.

Years later in full adulthood she bore scars of emotional wounds. Courted many times and abandoned more than once, she required more time and persuasion. An older woman is not always reluctant to mate (some are quite eager), but age usually adds caution. Less afflicted by her hormones, she takes time to judge. Less thrilled at having her own man, she searches for a good one. The older and more experienced she is, the fewer men can overcome her indifference.

In a modern woman's conversation we hear men's hurdle rising year by year. Some women find to their surprise that life is better alone. Like any addict rid of a compulsion, a woman may find an unanticipated freedom, where once searching for a fix had ruled her thoughts. Like any other addiction, recovery can start once the addict recognizes how much the habit has cost. For many women, only experience can teach that men may provide fewer rewards than expected.

Some anthropologists think we remain passionately bonded for predictable spans, to ensure our children's nurturing. I suggest instead we remain gullible and addicted for predictable spans, to ensure our children's *conception*. My sister has two sons, with a failed marriage behind her. She told me the next man will have a tough time persuading her. Having passed on her own emotional addiction and her ex-mate's shortcomings, my sister now deflects dull suitors toward easier targets.

Each woman recreates the courting wall in her own life. When a child, she is ingenuous and easy to persuade. With advancing years she becomes wary and less accessible. By the time she has children, she has learned too much for a dull man to court her, but she has also helped earlier dull men reproduce their genes. Her later wariness warns away any who seek easy access.

Each man faced a courting wall built of all women's hesitations, as women themselves flowed past him in a procession of individual lifetimes. A man cared little *how* women chose partners, but only which women were receptive to *him*. Women's rejecting promiscuity meant that each man needed to target his few best prospects. Each man sought to penetrate the courting wall at its weakest point, to pass its least vigilant sentinel.

Over his lifetime each male learned that younger women were more easily impressed. In part this reflected men's delayed maturity. By hitting puberty several years after girls, boys eventually grow larger bodies that help in dominance contests. The delay is emotional as well as physical, and a boy may be better matched to a younger girl than to one his own age. Schoolgirls are often unimpressed by their male contemporaries, and seek "men" a year or so older. Where the age difference is larger, the man's task may be easier. The girl's first important bond to a man had been with her mother's mate. This powerfully sexual relationship forms the way she will see men throughout life. A man who can evoke those same feelings can gain a competitive advantage over younger men.

So each young man learned early that younger women were the most likely sexual prospects. He most easily impressed girls a year or two younger, and older men could impress any girl they chose. Since most men tried easy targets first, young women were in great demand. Competition often prevented younger men from mating at all, as older men usurped the relatively small population of young women. Fairness is irrelevant; it was each man's most efficient way to reproduce.

Each woman's lifespan was a man's race to impregnate. As her ovulations tick by, each may result in pregnancy. Any male wins who can impregnate her, for he appropriates her body and motherly care for his own genetic purposes. But if she has any children, she is automatically tougher to persuade. Each child of a now-available woman means the man who fathered it is gone (or she wouldn't be available), and she is scarred (because she loved him). So the goal for men (perhaps unaware of it) is to impregnate a woman for the *first* time, before she has learned the price.

All men in the group run the impregnation race, each woman marking another sprint. Each man tries to gain the emotional allegiance of the youngest interested woman he can find. By winning her heart he bars access by other men, he makes her a pasha to herself. By targeting the youngest and most impressionable, he reduces his courting requirements, maximizes his short-term sexual rewards, and improves his chances of reproducing.

An Instinct For Girls

The courting wall, built of moving individuals but lasting innumerable generations, fostered a male instinct to target its weakest point. Men's instinct came from competing against other men, always seeking the same recognizable physique. An instinct had value because all men were learning the same lesson, so the prize went to the man with the lesson prewired in his brain. This parallels language, where the learning comes fastest to those with genetically embedded syntax.

When a girl became sexually active, competition swirled around her; men who waited until then to pursue her were lost in the crowd. The winning male was already courting before her first coitus, as the shooting gallery winner has his rifle pointed where the duck will appear. By courting pubescent girls, men could appropriate them before other men wanted them, before women could be bruised by love. Feminine personality having raised a barrier, men evolved to hijack girls' hearts before their barrier became too high.

Any man who found prepubescent girls attractive had an edge over other men, just as inorgasmic women who found something to like in men surpassed less-passionate sisters. His life experiences already taught him that girls of a certain age were the youngest sexual prospects, but this was not enough. He gained by evolving an innate preference for girls younger than their age of first acquiescence, so he would already be courting them as they entered their sexual life.

Men's new instinctive interest was sexual, whether or not the girl was ready for coitus. Men saw girls as sexual beings, courted them as they would any woman, and urged sex on them from the start. Already evolved to find females sexual without odors, men easily evolved a new instinct to see children as sexual before adult odors could appear, before puberty's physical changes.

Puberty is life's only physical milestone; no earlier physical change announces age, no sign says "This far and no further." An infant at six months was clearly not a sexual prospect, a girl of fifteen might be a veteran, and no clear line separates them. So the male's best avenue was to court any girl he found, on the chance he could keep her until she ovulated.

This was an ominous event, men evolving to seek the vulnerable. If mature women of all ages were equally accessible there would be no competition for the gullible, no need to be her first. Though girls had used men as masturbation tools in water, there need not have been man-girl sexual contact on land. In water *girls* initiated and controlled contact. Men evolved beards and permanent erections to make it easy, but it was *feminine* interest that strengthened through evolution. On land *men* instigated change, seeking easier orgasms. This didn't continue the aquatic pattern, but paralleled it when urged by completely different circumstances. In the next chapter I will describe fossil evidence for the distinction.

As all instincts do, it arose first by random mutation and then by selection. Most males, wooing females of all ages equally, would move on if a girl rejected them. A few males, carrying this dim and rudimentary instinct, were a shade more persistent with younger girls than older ones. Men could not know that an instinct urged them, they only knew they felt attracted to young women. Their higher interest led to longer courting and a marginally better chance of sexual access. Each man was oblivious to his long-term gain; the younger the girls he thought sexy, the better his chance of being her first.

In each generation a few more men carried this slight interest, reflecting their fathers's reproductive edge, and slowly a crude instinct spread through the population. Eventually the entire breeding pool shared the gene-driven preference, but with normal variation some men had it more, some less. The stronger a man's focus, the more time he would spend courting pre-sexual girls, and the more likely they would eventually be his sexual partners. Men with a weaker version were gradually left in the dust.

The instinct carries no moral message, it merely reflects the reproductive success of an understandable male avenue to sex. Male hamadryas baboons give us a parallel behavior, kidnapping infant females and waiting for them to grow up. It evolved in men as an instinct, instead of being passed as a cultural pattern, because it had to operate beyond men's conscious interest. Such an instinct in other species had no value at all. It was adaptive in us only because female hesitations made it a workable strategy. Beginning with a few men, the innate preference arose not because men are bad but because it helped them compete for access to women when competition grew difficult.

Wooing young women paid off reliably, with youth-obsessed men fathering youth-obsessed sons in growing numbers. The minority's preference became the majority's, and the mechanism began to avalanche, sweeping older behavior away. When they encountered the courting wall, men ducked under it.

Pursuing young women worked because it made babies, not because courting children was socially desirable. In its full flower we might see older men surrounded by harems of willing girls. Men held their charges by personal allegiance rather than force, copulated with each as soon as possible. Younger men would be largely excluded, waiting for their prime years to arrive, perhaps snaring a cast-off older widow.

Young women considered themselves victors, not victims. Each had the susceptibility to see any male as an alpha — the useful delusion of love. A male won her by impressing her, and she thought herself the mate of the tribe's best man. Each girl, in her first flush of sexual control at ten or eleven, scorned the withered twenty-year-old she replaced.

Every young woman who thrilled to her victory gave each generation of men permission to continue. Each girl-child who loved an older man passed his genes and her acceptance on to her child. I can't help but feel that his inability to court mature women should have been culled from the gene pool. But his youthful lover guaranteed that their sons would in turn seek little girls. Neither of them could see the burden they were placing on future generations.

Competition sharpened men's new instinct. A man had to outrace not only a girl's puberty but also her other suitors. If other men competed for ten-year-olds, a man could gain by targeting eight-year-olds. If others were wooing girls of six years, a man could edge ahead by winning their hearts at five. As soon as a girl could be lured from her mother, men began wooing her.

Over thousands of generations men's new instinct became focussed and precise. The man with a stronger preference had a faster erotic response to a girl-child; he elbowed other men aside to reach her. Clearer instinct bore fruit in the next generation, as men with vague preference were culled. Competition between men was unceasing, and each generation sharpened men's focus.

A man would be a good mate to his one or more child brides, until he became bored with them. Normal primate promiscuity tends to move males on every five years or so, and his once-youthful mate would in that time slide out of the girl-child state he was evolving to prefer. Without knowing why, he would become dissatisfied, and would shift his attention to younger girls. A cast-off woman became the wife of a younger man or weaker competitor, she on her way down, he on his way up. Love still lured her, but her anger grew. She might love several men before cynicism made love unreachable, and she settled for a final pairing as some young man's cook, worn and bitter at thirty.

This was a huge muddle of a clear need. Women had wanted men only to stay home and not hurt their feelings. But we have no simple monogamy gene, and the fear that stopped women's promiscuity did not stop men's. Breeding a faithful man was an accidental product of women's pain; breeding a dull one was a double accident. Competition by dull men focussed male instinct on the most-accessible female physique, and perverted women's ideal of long-term mateship to short-term eroticism. Instead of being unreliable mates for a lifetime, men evolved to be reliable mates for a limited time.

Now I must pause and confess that I paint this picture too bleakly, for male obsession never reached this point. Men dream of harems and women are often abandoned, yet many know lifetime mateships and true love. I overstated the brutal effect of male sexual obsession in order to make the forces clear. Having sketched with bold lines, we can start to add shading.

Women could not stop the injury men did them, but women *did* evolve to deflect it. In the battle of the sexes each side continually evolves for their individual good at the other's collective expense, and I will soon outline women's response. For now menopause stands as evidence that men's preference was not unchecked. If women were abandoned by age twenty or so, there would be no value in halting fertility at forty. If menopause evolved to keep older women alive, then something was keeping older women sexually active, too.

Beauty And Perception

A young woman is the human beauty standard, small and smooth-skinned, distinctively curved. We measure all women against a girl nearing womanhood. This is not an accident, not a contest men invented, not a weapon to hurt women.

Many argue that youthful beauty is how a female advertises for a male. A young woman has more childbearing years ahead; the male who wins her has a chance of more progeny. Young women have more-obvious curves, their hips may *seem* wider or their breasts fuller. To the extent these make childrearing easier, a man seeking them may gain a more-fruitful mate.

This approach has two flaws. Men often find new young wives rather than exploiting all the childbearing years of one, so chasing young women seems unrelated to the time men keep them. And there are broad variations in what men seek, with some features clearly unrelated to fecundity. Long legs do not imply fertility.

Some cultures favor large breasts, some small, some ignore them entirely. Some favor thinner women, some fatter. I know of no culture that thinks old and withered women are pretty. Though the viewable body parts vary by culture, men seem to appreciate whatever they see. And in every society a young woman's *visual* appeal, more than any feminine *character*, drives men's lusts.

This raises two questions. First, why see beauty at all? Men's scorning so many still-fertile women feels like a wasteful evolutionary turn. Second, what are the exact features that make us see beauty? Why does one culture notice cheek or breast while another sees eyes or ankles? How can such variations exist while the focus remains the same young woman? *What* we see must wait a bit; *why* we see it is my subject here.

Some suggest that beauty exists outside the human experience, and that we merely recognize it as we recognize gravity, but I disagree. We enjoy seeing beautiful things because our species evolved an eye for beauty; having evolved it we find beauty in many places, as a composer hears music everywhere. An eye for beauty, like love and language, evolved because it helped us reproduce. I briefly mentioned it as part of men's ignoring estrus in the water. Beauty, like love and language, is always inherited by both sexes, and this blurs its origins.

Act Of Perception

"Beauty" began in the sea as men's pleasure in seeing a feminine face, one of several pleasures that evolved to trigger men's courting. Beauty is an inherited ability to appreciate, not an inherited shape in females. With back-sloped foreheads and thick-boned faces, our ancient mothers looked pretty to their mates.

When we praise a meal we announce our palate's boundaries as much as measure the food. When we call women "beautiful" we trace the dimensions of an instinct born in male competition. We attribute beauty to women as a woman in love attributes merit to her lover. Yet finding a woman beautiful demands no one physical feature in her, any more than a man must be a good person because a woman thinks well of him. Beauty is not a property but an appetite men evolved in order to reproduce. Without men's competition there would be no beautiful women.

Animals like what helps their survival and reproduction. Kittens like to pounce because as adults they need the skill. We like food's taste because liking it leads us to eat and survive. We like the pleasures of sex because it is vital that we perform the act. Neither eating nor coitus would give any pleasure if we had not evolved a liking for them while we evolved the processes.

In water, men had to start serious courting for the first time. They needed a simple cue that the person facing them was a valid courting target. Women's voices gave an audible guide and would have sufficed if nothing else evolved. But men also sought out women by their faces, with nothing else visible, and evolved to like the view. For face-focussed primates this was a natural change, but it played a small role in the water.

At its start beauty was not a certain age or shape, but simple pleasure in seeing a female face. Men who found pleasure in women's company, whatever the lure, spent more time there and reproduced better. In the sea, before men became dull on land, they enjoyed hearing women's voices, probing women's minds and viewing their faces. With our social web disrupted in water, these preferences helped glue our species together.

Beauty-perception in men parallels love in women, both mental tricks helping reproduction. Love overcomes fear; beauty overcomes indifference. Love is a hunger of the heart, evolved to make one person desirable above others. Seeing beauty, like noting movement in the corner of your eye, evolved to nudge the mind awake.

Beauty was not the most important but only the first thing, pulling men in from the passing stream so real courting might start. An eye for beauty is not even close to love, but evolved to give love a space to grow. It evolved to grab attention at a distance, to draw the man in as hunger draws prospective diners to a restaurant before they have read the menu. Drawn to a female face, a man could then try to prove that he was special, could learn who lived behind that skin.

Beauty is a hunger of the eyes, holding a man captive by a face. As dominance waned, before singing and language could arise, females started evolving obsessive focussing to make their mates seem special. Males inherited obsession from their mothers and tailored it to their own needs. It was focussing that held a male in front of a female long enough to win her, while aquatic courting deteriorated. The male with an instinct for obsession was more likely to persist long enough, whatever it was that first caught his attention.

Men attached their obsession more to faces, women attached theirs to hearts. Just as a woman's focussing gave emotional isolation in a crowd and enhanced her estimate of a man she focussed on, so a man pedestalled his intended by her face. As women gained from thinking an average man remarkable, men gained from thinking an average woman striking. Finding a woman pretty gave a man focus for courting before he could know her person.

Liking faces began as such a small thing. Singing carried farther, said more from the heart, and could be answered. Conversation was worth more than either. Men noticed women's faces only because it was easy, faces were distinctive enough to separate the sexes, and primates read faces to recognize moods, conflicts, and friendliness. Men did not choose mates for beauty, but they started to see it. When a man thought of his new girlfriend, in his mind he saw her face.

Men did not choose to evolve an eye for beauty, any more than they chose to evolve singing or language, or women to evolve love. Liking to be in women's company, to look them in the eye, was the necessary but gentle inner pasha, guiding men's behavior to bring the sexes together. Drawn and held there with their fellows, success came to men who found a way to make themselves stand out from the crowd. Singing and language evolved in male groups skewered on beauty's velvet pin.

Beauty in men parallels love in women. A woman in love need not know why, she merely wants the company of her beloved. She need not see how this same feeling in her mother was a prerequisite to coitus. The male who responds to a pretty face does so because it feels right. He need not see how past men with such reactions reproduced better. Disinterested men were surpassed by men who saw any woman as a prospective mate. Beauty in men, like love in women, is a record of subtracted nonparticipants. In water the female face was enough to evoke male courting, because that was the first signal men received.

Love is the wrong emotion to catch the stranger passing by, but love worked well enough in men once the conversation began. Love makes a single person tolerable and desirable, and women needed that. But men needed to see all women, to troll for female attention through aquatic family groups, to find one to talk with. The fisherman catches a *specific* fish, but he goes fishing because he would be happy to catch *any* fish. Love attracts us to a person; beauty attracts us to the possibilities.

Beauty began as a mental tickle any woman could evoke in a man, by simply being recognized as female. As soon as a man recognized her sex, he began flirting, eye contact, a smile. Whatever heights a couple might later reach, this first verbal touch needed some impetus to start it. Aquatic men competed with other men for women's attentions, and the man whose sexual interest was more easily evoked had an edge over less-focussed men. Over time this created a population of men who saw beauty in women of all ages.

Preference Evolving

We defined beauty more narrowly after returning to land. Love grew during the move, male promiscuity became painful, and women preferred men who stayed home. Women's anger culled men who were too perceptive, and estrus-awareness vanished. Young women became the beauty standard as the impregnation race pushed older (more cautious) women to the side. We did not change young women's *appearance*, only men's *view* of them.

As men became oblivious to estrus, they grew numb to part of what had attracted them to women. For dull men, women's thoughts were beyond easy reach, while reduced verbal skills made initial conversations difficult. So what women said became harder for men to value, took longer for a man to appreciate.

But beauty could still pull men in from the passing stream. Men's ability to recognize a woman by her face had always preceded real courting. Physical attraction still skewered men who now had less to court with. Appearance was the first, least valuable and least meaningful attractor, but gained power by default, as men grew duller.

At the same time, men's courting target changed. By breeding out estrus, women accidentally made men into dull husbands; duller single men had a courting disadvantage, and the impregnation race began. Any man who saw sexual possibility in a younger woman focussed his courting on her, ignoring older women. Men evolved this preference only because younger women were in fact easier to court.

Imagine if only old women could be mothers. Some fluke of anatomy, for instance, could make childbirth impossible before age fifty. Any male targeting younger women would be culled from the gene pool. Men would inevitably evolve a powerful sexual attraction to the fertile women's age; fifty years old would exemplify beauty. Younger women would seethe, apply aging makeup, hope to lure a mate who could fill their needs.

Feminine emotional barriers caused the impregnation race, giving an advantage to men who courted young girls. A man who saw beauty in younger women was initially a fluke. He reproduced better because young women happened to make easier targets. The competitive pressure driving the impregnation race left this male obsession as its footprint. Men could not see young women as beautiful until pursuing young women was a useful strategy.

The narrower a man's idea of beauty, the more precisely he targeted girls. No man had to understand this, in the rush of sexual pursuit. Each man competed against his fellows to procreate. Each competition was not merely in verbal skills or personal dominance, but also in how well he saw his target, how closely his appetite matched his need. Men's courting choices were the stuff of random variation, tested in each generation.

Our idea of beauty narrowed as the competition weeded out males. Beauty was always the least-personal recognition of the most-accessible sexual target, evolved as only an initial courting trigger. When all women were equally accessible, all were equally attractive to men. As women's individual hurts built the courting wall, the wall's weak point favored men attracted by the most-accessible females. And the greatest gain went to men who were most-hampered by the courting wall, who had the most trouble coping with an adult woman. Skin-deep attributes of youth gave the perfect cue to those men for whom skin-deep was deep enough. It's no accident that beauty has nothing to do with personality.

This once-minor instinct began as only an eye-catcher, but evolved into a filter culling our grandfathers. Men who recognized youth reproduced better; men with no preference often mated happily, but often dashed their hopes on stouter barriers. By always subtracting a few of the men who ignored youth, the courting wall eventually made young women beautiful and older women invisible.

Our present scale of beauty is no male conspiracy against women but a record of mindless selection. Over countless generations, reproductive competition refined men's sexual target. Men's focus narrowed and sharpened, leading men to court the girl too young to give adult consent, too young to harbor adult skepticism, old enough to be coaxed from her mother.

Our standard of perfection became instinctive; men innately seek girls not yet awakened to sexuality, they seek virgins. There is no other human age or form that draws us so potently. Racial memory is a myth, but a mindset can evolve. A she-bear does not *remember* male attacks on her ancestral mothers; her high aggression reflects deaths of less-aggressive females. Men do not *remember* ancient virgins' value; they pursue them today because they are descended from the men who successfully reproduced by pursuing them in our dim past.

Our myths illuminate inherited hungers; we repeat the stories which strike a chord. Stories of heroes impress us because we are sensitive to the virtues of leadership; stories of fear grip us because we know what fear is. Our myths persist by being close to our hearts, and among these relics are tales of girls aroused for their first time by men. Women do not need erotic awakening; females are sexual from birth, and these myths come not from women's passiveness but from men's hunger. Virgins fascinate us because the men they fascinated won the impregnation race. Men kissed Sleeping Beauty four million years ago.

The Virgin Berth

Targeting young girls became the predominant male strategy, with a new instinct to support it. Women then did unknowingly the same thing that women do intentionally today. When men's obsession led them to target a small group of women, each woman gained by seeming to join that group.

Men wanted virgins. All primates have hymen equivalents, but only humans value them. Some think this shows men's insecurity: Frightened of being compared to another man, men feel safe only if a woman can make no comparison. I suggest it shows men's failure to court and the failure's effect on countless generations. Men want naive women because men with that preference reproduce better; physical proof of naivete increases a woman's value in such men's eyes.

A man's preference for young women comes from his fathers' courting deficiencies. A man appreciates girls without ever intending to; he can no more alter his visual hunger than he can ignore his need to eat. With effort, a man can avoid drooling on young women he passes, but he knows they're there.

Women's most desirable age results from their maturation. If women became stupid with age (while yet fertile) men would evolve an obsessive preference for wrinkles. Men's relative inability to court would create an innate preference for any age where women's resistance is lowest. As it happens, we're all stupider when younger.

Men's obsession is visual. Men targeted young women not because men adored childish personalities but because men shied away from adult barriers. Bypassing personality leaves physique as the only raw material for instinct to work with. Men's new instinct focussed on young women's shape, leading men to court the bodies where the least-demanding minds lived.

No quality of the feminine brain appealed to an aroused male like brainlessness did. No female facet had as much value to him as the guarantee she would couple with him. He cared nothing for her wisdom or intuition. If she rejected him sexually, even if her mind was the treasure of their tribe, he had to court another.

She had no words to counter his instinctive focus. No personal appeal could match his avoiding her personality. The best of arguments couldn't outweigh his desire for no arguments at all. In a world where men sought young women, a woman's only recourse was to disguise herself as young. When the adult female psyche became a barrier to men, women set it aside for the sake of the mateship. Some still do.

Women began to evolve in response to this male obsession, a ponderous course-change for the female army. Millions were crushed when their forces turned. We are still embroiled in this campaign, the latest (but surely not the final) in the battle of the sexes.

This Spectered Eye

The first question about beauty was: Why see it? We see it because men needed a non-estrous sexual trigger, and physical appearance filled the need by default. Men focussed specifically on young women as their easiest route to reproduction. We see sexual possibilities (and call it "beauty") because evolution wove an eye for it into men's minds. Having evolved the instinct to see beauty, we can then see it in many things.

This leaves the second question: What forms look beautiful to the male mind? What is this mental event some women can trigger, and on what curve of flesh or smoothness of skin does it depend? The battle of the sexes focusses on this question, the skirmish that determines the campaign's outcome.

Beauty is not vague, does not need a thousand words, but can be reduced to lines on paper. It is not youth or a certain figure or elastic skin.

An artist can depict a beautiful woman with half a dozen carefully chosen strokes. In a single eyebrow, a slope of nose and curve of forehead, he can create a recognizable face and we can see beauty.

What happens in our heads when we see this? If pressed, you might say she looks pretty because her drawing resembles beautiful women you have seen in the flesh. By evoking that memory, she can remind you of beautiful characteristics not actually drawn. Seems reasonable.

I believe the truth is exactly opposite. You already had the sketch in your head before you saw it on paper. When you see women, their closeness to that sketch makes them seem beautiful. When you see the sketch itself you can appreciate it for its own sake, a rendering of an image you evolved to enjoy. To follow beauty's evolution we must strip away color and texture and view this elementary mechanism ticking in our heads, this sketch in our brains.

Response By Caricature

Inherited images are caricatures tattooed on a species' brains, exaggerated yet recognizable pictures. Sketches evolve because they benefit the species, and improve until they function reliably. Images differ by species but the mechanism works alike for all.

About 66 subspecies of deermice live in various habitats. In the 1950's and 1960's two groups, one from grasslands and one from woodlands, were bred in a laboratory for twenty generations. When released into either their original environments or a simulation, they correctly sorted themselves out. Something in their little brains after twenty generations could recognize where their ancestors once lived.

Instinctive preference clearly has value. A mouse who wanders out of his home range has an excellent chance of being eaten by a predator it doesn't know to watch for. Instinctive recognition of a home range is likely to match instinctive fears of the resident predators.

Animals store innate preferences and fears as mental cartoons, fragmentary sketches of reality. An example comes from sticklebacks, a highly territorial fish colored white and red. Tinbergen (1951) reported that introducing any small red and white object sparked combat by a male defending his little territory, though the shape could be wildly inaccurate. The male's instinctive response didn't need to discriminate real from fake fish. In the world of the stickleback, the only red and white object was always another male stickleback.

When deermice chose fake grasslands, they responded to a caricature, in this case paper strips scattered on the floor. In their natural world no such substance existed, so they had no ability to detect the difference. It felt close enough, to them.

In one experiment, rhesus macaque infants were deprived of their mothers. Some infants got artificial mothers made of wire mesh and towels, and the babies clutched them for security as they would a real mother. Even a rough approximation evoked a cuddling response; towels-and-wire felt more like Mom than no Mom at all.

Nikko Tinbergen's jackdaw provided another famous response-to-caricature, when it started courting Nikko. Unable to find Nikko's mouth for the feeding rituals of daw courting, the bird shoved bug-pulp into the naturalist's ear, the most mouthlike feature he could find. Similar reactions include mothers' willingness to accept infants of other species, or a bird's readiness to hatch the wrong eggs.

We use response-by-caricature each time we pacify a baby with a piece of plastic instead of a live nipple, put a ticking clock near a lonely puppy, or fish with an artificial lure. Any instinct evolves enough precision to discriminate only the choices in the natural world; selection tunes instincts only when the current precision is inadequate for the species.

Instinct responds to a feeling of "fit", and animals do what they can to make the fit feel better. The deermouse moved to make reality fit its cartoon, to make the world match its internal template. The stickleback fights what feels like a stranger; the puppy lies nearer to what feels like a heartbeat; the baby sucks what feels like a nipple.

We all carry a young woman's caricature in our minds; each of us inherited it through a thousand fathers. We can't ignore the picture; recognizing pretty faces is an internal artifact of the mind, as clear and real as the fear of falling. The caricature isn't perfect, but close enough to make us like to look. Instinct makes a good guide but a rotten artist.

We find young women pretty because they fit our caricature, not because we objectively judge them good breeding material, not because men delight in the rage of older women. We gain pleasure from matching our sketch to the outside world, never knowing what's going on. Women also prefer pictures of women to pictures of men because women inherited the same innate sketch from their fathers. The instinctive sketch is the same in all human populations and in both sexes.

Long before modern humans, our brains evolved a modern human's caricature. Our evolution since then has bowed to many forces, one being men's preference for women matching this image. By consistently choosing these women, men whittled away others who did *not* fit our internal image. In the process we hollowed out an empty place in the human heart, and in that place carved the treasure that can save us while chimps and gorillas die away.

Drawing Caricatures

Innate caricatures arise continually by chance, to be inherited according to any benefit they bring. They begin in random mutations of brain-forming genes, expressed as crude sketches of smell, sight, touch and sound, meaningless as radio static. Conception wires them in the brain, and natural selection weighs them as it weighs claws and teeth and muscle.

An instinct evolves because an animal must respond correctly to unforeseeable events. An instinctive *caricature* evolves because animals must respond to some thing but cannot carry the thing itself in the brain. I use "caricature" to mean an object's mental model and corresponding response, as opposed to behavior evoked by circumstance alone. Throwing out your arms as you fall back requires no inner tree-cartoon, but a songbird fears a hawk because it carries a hawk sketch in its brain.

A caricature need not be visual or inherited, though inherited ones are my interest here. In the womb a puppy builds an auditory sketch of its mother's heartbeat, and we fool it with a ticking clock. Many lemon-scented soaps contain a chemical that feels like lemons we have known; chemists provide a cheaper product that fits our olfactory sketch. Neither recognition is inherited.

Responding to scent is like trying on clothes; we seek not the thing but the fit. You don't look for a shirt exactly like the shirt you have but for one that fits your body. A male doesn't have a sample of female scent but a template of how it should feel. We save not a lemon-scent molecule but a memory of how it smelled. Clothing factories exist because bodies are predictable; estrus exists because scent-recognition is inheritable.

At her first ovulation a female exudes estrus odors, though the scent molecules never existed in her body before hormones turned on the chemical-assembly process. Nor did the male as he grew have a sample molecule to grow receptors around, as a child molds many clay copies from one stamp. Instead, a male grows a nose much like his father had, and a brain that likes the same smells.

Starting from initially-random preferences, male interest quickly evolved to focus on scents emitted by females of his species. As soon as any odor led males to females, females would gain mates by exuding that smell. Males gained by following it, by evolving a nose that picked it up well, and a brain that liked to smell it. Offspring inherit genetic recipes for both manufacturing and receiving the scent, without inheriting the scent-molecule itself.

A male's brain evolves scent-cartoons to get him excited when female scent molecules hit receptors in his nose. It's important that he not miss a female signal, but he can be fooled. Some flowers attract male insects by emitting imitations of female scents; in trying to copulate with the flower, the male spreads its genes instead of his. A male never seeks a female for herself; he seeks that which fits his hunger, and his hunger is only a sketch.

Deermice similarly evolved visual cartoons, brain templates that fitted grass. Early sketches only hinted at color or shape, of no more inherent value than liking Picasso. A mouse who carried a grass-cartoon stayed where his cartoon fit the real world; a mouse without a cartoon didn't care.

Mental cartoons evolve only when they affect choices. On woodland fringes a grasslands mouse who cared nothing for grass would wander freely, and stumble into danger in the trees. By some minuscule amount the earliest grass-cartoon gave a survival advantage to mice who stayed home. When woodland owls had finished eating cartoonless mice, the remaining mice all liked grass a little bit. Woodland mice who didn't care for the woods wandered into the open, to be subtracted by hawks that any prairie mouse would have seen coming.

The fringes of any habitat will continually strengthen innate preferences of its inhabitants, where preference aids survival. In each generation some hungry souls will wander into danger, and the less a mouse cares for home the more likely he will be the one who wanders and becomes someone else's lunch.   Interbreeding slowly diffuses the preference into the prairie's heart, where it has no value. The grasslands mouse at the prairie center inherits a liking for grass from mice at the boundaries, but he gains nothing from it because he has no choices.

Each generation inherits the mental cartoons of its parents, with occasional mutations. The image gains power whenever it more closely matches the real world, when the matching helps survival. The caricature never stops evolving, but it never becomes perfect. Deermice recognize grass after twenty generations` but we can fool them with paper strips. Natural selection adjusts the caricature only when the natural world requires it.

Image Of Beauty

Advantage went to men who targeted young women.   Each individual's preference was a pre-existing state, just as each mouse's likings take no effort. At reproduction's fringes where men chose whom to court, natural selection strengthened men's preference for girls.

The first visual caricature of femininity appeared while we were aquatic, as a crude impression in men's minds. Men with cartoons of femininity (a rudimentary concept of pretty) had an edge over men who could not copulate without estrus odors. When women began preferring duller men and courting grew tougher, natural selection began adjusting men's image to more-vulnerable females. Men whose female caricature matched younger women had an edge because younger women were more accessible.

No clearer than the first image of grass-ness, the first image of a woman was only a fragment. As some people have a favorite color, some men had a reaction of pleasure to a smoother face or a rounder forehead. Fragmentary images began without intrinsic value, but estrus' collapse made matters of taste into assets. Men with a preference left more progeny to inherit and adjust the image, as woodland mice who liked their home just a bit were marginally less likely to become hawk food. Men's interest in young women is the genetic result of repeated selection for innate caricatures. Though not every man courted girls, every man who did gained advantage and all men eventually inherited their preference.

Innate images grew more powerful over generations. The clearer a man perceived his target, the more intensely he pursued her. When all men inherited a caricature of a young woman, that man whose image was more precise courted more precisely. Young women's gullibility gave value to a clear image in men's minds. We recognize a pretty face by its resemblance to this inner image, an unthinking response all humans share. Beauty doesn't advertise a woman's fecundity to men; it reflects men's recognition of the most accessible female age.

Images Disclosed

We can put our caricature on paper, incorporating in half a dozen lines the visual essence of an immature female, no closer to a real female than strips of paper are to grass. All young mammals share a distinctive ratio of forehead to muzzle; on first seeing another species we can distinguish infants from adults by this alone. Infant apes have the same large foreheads and small faces, the result of faster brain growth in infancy. After birth the muzzle begins to grow, supporting canine teeth and chewing muscles. Human males evolved to target the juvenile-face ratio because this led reliably to immature females.

Our sexual caricature includes a large head, a small face and nose with large eyes, drawn with smooth lines. Other feminine assets are sexual only because men *learn* that they are sexual, because women solicit with their own erogenous areas. Only this cartoon of femininity is innate and powerful, spanning all cultures. Different peoples can have different sexual preferences and still value the same young women because all humans recognize the same young-female face without realizing it.

Women carry the same caricature, and women measure themselves against it without thinking. When women look in a mirror, their innate cartoon appears in their minds, superimposed on the reflection like a misaligned tracing. The image in their mind's eye never quite matches the reflection; the cartoon's eyes are larger and more-clearly drawn, its face more delicate. Women use cosmetics and diet to reduce the gap between the two images, thereby emphasizing features men are looking for. But it is the caricature and mirrors, not men, that set the standard.

So far this may seem barely plausible. It's intuitively appealing that sexual attraction should operate by crude cartoon instead of detailed picture. It corresponds to our preference and behavior, and explains why other animals give apparently instinctive responses to the wrong stimuli.

And we have another piece of evidence: Most of us have seen our own caricature drawn on paper or in films, the theme of a common hallucination.

People occasionally report contacts with aliens, describing smooth faces with huge black eyes and bulbous heads mounted on slim but human-like bodies. It's laughably improbable that aliens would exist as described. To expect upright bipedal symmetrical creatures with two eyes beside a nose over a mouth, all looking passably human, is to expect quite a lot. Even bilateral symmetry may be rare in the universe, and is only one of several general forms on earth. I suspect the reporters have experienced a relaxed consciousness, of unknown causes. While it felt real to them, it has no more reality than any dream.

Believers use aliens' similarity as proof of their existence: How could so many see the same thing if they didn't exist? I propose that they saw something intimately familiar, not alien at all; in their altered mental state they dreamed an image embedded in our own minds. Alone of all animals, we are able to render on paper one of our caricatures. We have seen these aliens before; they gained quick public acceptance after the first reported contact because they look so familiar, not because they are real.

This imaginary creature, frail and skinny with no breasts or genitals, is a rendering of our sexual template. The rounded head, large eyes and smooth skin are all attributes of females and of children, carried to extremes. Like any caricature it is rough but it captures the essence; it is as close to real women as are paper strips to grass. This is the pattern embedded in men, the shape toward which we are all carried by forces making young women sexual targets.

Vision And Substitute

Men instinctively measure women's beauty using our caricature as an overlaying tracery. Being an instinct, it not only requires no thought but resists conscious probing. Like finding an unseen moon by measuring perturbations in its parent's orbit, we can locate our instinct by how it moves us. The best evidence for our caricature's existence may be not in what we knowingly see, but in what our evolution *implies* we saw.

Seeing beauty takes more than instinct. For a man to reproduce he must get excited by actual women, no matter what cartoons he carries.

Real life builds breasts and thighs on our foundation caricature. But these are not innate, for we can differ in what we like, and change our minds, and talk about it. In spite of acquired preferences, the core of physical attraction is recognizing in others something we already see in our minds.

We enjoy seeing our caricature's physical counterpart. This pleasure-reaction is as old as cartoon-making; caricatures can evolve only if fitting a pattern to reality always feels good. The deermouse's pleasure (I have no better term) keeps it at home and strengthens its allegiance to grass. Patterns and their enjoyment are the basic behavior-shaping evolutionary tool, not subject to our control or willful disconnection. Men can take many avenues to mating, but enjoy it more when their partner is closer to our caricature. No matter how bright as individuals, all men are descended from some dull ones who could mate only by courting the most vulnerable, and needed this image as a guide.

Our template is entirely *a*sexual, appreciated for its own sake. Cartoons do not know what purpose they serve; a sweet tooth does not know it is a sweet tooth. The young woman is our standard of beauty not because men command it, but because she matches both sexes' inherited cartoon.

Men can hold our caricature's embodiment by possessing young women, in their imaginations if not in fact. The caricature evolved in men to encourage exactly this. By giving the man pleasure, the cartoon led him to pursue that human who matched it most closely. Young women being reliably naive, over enough time, made young women pretty to men.

Frontal sex for women is surrender (but dangerous to say so). Women cannot avoid signalling vulnerability, when they open themselves to a man, and male dominance has long been a female sexual trigger. But women don't want subjugation, they want only to choose their own partner for a play-act surrender.

Women cannot select sexual partners to match their innate image because the frail caricature cannot comfortably trace over a large and dominant mate. For women the effects of the image are entirely accidental; it did not evolve for their benefit and has caused them much harm. They carry the image only because they inherited it from fathers who needed it.

Yet instinctive hungers demand feeding; women want to fit their cartoon to something, though fitting it gives them no survival advantage. It seems reasonable that both sexes would want to look at pictures of the opposite sex, who should be their sexual focus and potential mates. Men live up to this theory, with men's magazines consistently carrying pictures of women. Women's relative lack of interest in pictures of men causes some to say women are less visually inclined. But women no less than men want to see pictures of women, and women's magazines provide them. These illustrations are clearly what women want to look at, though women err in trying to mimic them. Women are as visually aware as men, and both like seeing copies of a caricature they bear.

Women's obsession with children owes something to this innate image. Women are often riveted by a child, staring at the infant with fanatic concentration. Women go to extraordinary lengths to acquire children, occasionally stealing them. We might agree that a man showing the same interest warrants watching, but because a woman's interest coincides with a nurturing role, we accept her fascination as normal though perhaps overzealous.

She shows more than a visual interest; this is also a hunger to mother, and her valid interests make our innate cartoon harder to see in her. By giving a reason for staring at children, mothering-interest camouflages the caricature. Able then to overlook it in themselves, women are free to label men's obsession a willful and hurtful conscious act. But men didn't plan or want what both sexes share.

No burden seems to attach to the man who seeks young women, except perhaps payments to the angry ex-wife he leaves behind. Experience and empathy can make him considerate; love and warmth can keep him faithful for life. But even the most loyal husband will occasionally stare in dumb befuddlement at some cute young thing, unable to explain himself and equally unable to ignore her. Beyond a defensive "Because she's pretty!" he cannot explain his pleasure in seeing her, any more than a woman can explain why she is riveted by a child, or a deermouse by the grass.

Women In Disguise

Once men focussed on young women, women evolved to match men's image. Men's caricature of women, just beyond the conscious mind, drives the most spectacular, misunderstood, and fateful evolution of our species. Women's competition to match men's image saved us from extinction and gave us the keys to the world. Yet it was all so disarmingly simple: Women who looked younger had a better chance of gaining suitors and then children. The pattern still works.

This reflects female *individual competition*, not all femininity accommodating men's new preference. What benefits an individual may not help the mass. Our whole species would gain sheer numbers if all women got pregnant at each opportunity. Given the need for love, we all would gain if we made permanent and peaceful mateships. We'd be better off, perhaps, if love had never evolved to overcome fear that never appeared. The battle of the sexes would not exist at all if each sex evolved en masse to some state of ideal compatibility.

Our battle reflects *individual* strategies and successes. In water the *individual* man gained by an ability to court verbally; it made him more appealing by comparison to other men. In our landward move an *individual* woman gained by passion, intercepting on the beach men who would have visited her sisters in the water. When women bred dull men, the *individual* man gained by his obsession for young women, leading him to pursue easier sexual targets. The sharper his obsession, the more advantage he had over other men less focussed.

Women began to evolve toward men's caricature, resembling what men were looking for. It was not women as a group who gained from this, but the individual who surpassed her sisters. A woman who looked younger to men had an advantage over her contemporaries, woman against woman, for men's eyes. Her attraction gave her children to carry on her trait, gradually overwhelming the genes of other women.

One sex shaping the other had long been part of our story. Women's preference for interesting but nonthreatening mates in water led to singing and language. Men's preference for life on land led to love's evolution in women. Women's need for stability at home made men dull. Following broad themes, our sexual tug-of-war has passed the initiative back and forth, with individual advantage repeatedly accumulating to change the whole.

This final contest stands alone. In no other human step did such profound effects follow such trivial behavior. Feminine youthful competition led to child abuse, murder, and fear. Yet it is hand-in-hand with the passions where most find life's meaning. It troubles me, but I cannot condemn the conflict that defines us.

Our tragedy is that our sexual armies never came to grips, circling each other in the darkness. Murder and child rape come from women's physical changes, yet men never intended such a change, they only preferred accessible women. An individual woman gained from looking younger, but a population of younger-looking women parried men's preference and very nearly killed us.

Women's army wheeled heavily to meet a new attack, crushing bystanders as it turned. For the first time humans evolved based on appearance alone. It made no one smarter, faster, or stronger. Youthful evolution's only value is in deluding men to attract a mate — vital but somehow silly. When it began, no one could have known it would eventually save us from extinction.

Becoming The Image

Women's new evolution centered on head shape because head shape was the male caricature's focus. Like a stickleback attacking a red-and-white lump, men's instinct focussed on the trivial but consistent fact that infants have relatively larger brains. Women grew larger heads for the same reason peacocks grew larger tails — it's what the opposite sex looks for.

Men courted girls because girls are reliably stupid. As girls move out of the most-vulnerable age, men's interest wanes and their attention moves to a fresh crop. An individual woman gained by looking childlike later into life; men courted her longer and gave her more children. Women evolved to look young by retaining into adulthood the brain/muzzle ratio of a child, by matching the male caricature for a longer time. Women also evolved to *sound* young by keeping the child's higher voice into adulthood.

Brains grow more quickly in the womb, with growth slowing soon after birth in most primates. We had already evolved slightly larger brains in the sea to prevent drowning as infants, because fat floats and brains are fat.   We achieved this moderate increase by continuing fetal growth patterns farther into life, a neotenic shift that also gave us sucking lips and smaller canine teeth.   Neoteny gave a larger-headed aquatic infant, who grew into a larger-skulled adult.   If there had been any other fatty deposit on the skull, that other fat might have grown instead.

Any girl who had a slightly larger skull had a childlike look a bit later into life.   By comparison with the brain case, her face seemed smaller and less protruding.   This illusion was always effective, but the appearance had value only after men evolved a caricature to help them penetrate the courting wall.   A larger-skulled woman was closer to men's caricature and evoked more courting interest.   She left larger-brained daughters, who again had this reproductive edge.

Nothing in the Great Scheme of Things demanded that *brains* grow. The only requirement was to evoke a child-impression in the minds of child-obsessed men.   But the infant's brain gives its skull the distinctive shape, and the simplest way to evolve a large-headed adult woman was to let the brain keep growing.   It would be possible but unlikely for some completely new feature to create an *illusion* of brain, when brain was so easily added.

And neoteny was again the easy path to do it.   Continuing fetal rapid growth had served us in water, and extended-growth stages were already common in the anthropoids.   It took no genetic breakthrough, but merely the continued enhancement of a pre-existing human property.   When men fixated on girls as sexual partners, women evolved large brains as camouflage by delaying their adult stage.

Large brains are indirect evidence of human dullness.   A large brain says nothing at all about its possessor's intellect, but only about women's need to attract a man; a large brain is adaptive only because humans are a bit dull.   We've linked intelligence to brain size for several hundred years, but we should discard the entire line of inquiry.   Brain size means no more than eye color in measuring typical human intelligence.

Course Of Change

Female competition becomes clearest when viewed most narrowly. Fully adult women were too far outside the male's caricature, and didn't compete directly against infant girls for child-obsessed men. Instead, adult women bear the stamp of individual competition between their mothers, when those females were not yet ovulating. Female evolution is a record of child brides.

For the moment forget adult women and the personality-barrier, and assume men avoided them completely. In a moment we will include them again. I'll take a single age for girls and label it the Female's Optimal Courting Age, marking the male's most-auspicious target population. This is the female age where courting males face the weakest obstacles.

A woman's anger grows as she ages and learns how passion can hurt, making adult females difficult to court. Motherly vigilance is born in a woman's knowledge of her sex's vulnerability and her own memory of pain. Her vigilance weakens slowly as her daughters mature and leave her control, making them accessible to men. Men learn that adult women make daunting prospects, and competition from other males pushes each man's interest down to younger girls, until he meets their mother's objections. The Female's Optimal Courting Age is the sum of these, marking the male's most-accessible route under the courting wall.

We cannot reduce the Female's Optimal Courting Age to a firm number, though we might pick an arbitrary date for convenience. We do this when we use twenty-one as the age of adulthood (though forty-year-old children are all around us), or sixteen as the age of female consent (equally questionable). For our discussion we might assume a girl was most effectively courted at age six, in our ancestors a couple million years ago, with full adulthood at fifteen. At six she could be lured from her mother and be swayed by men, though copulation might be a few years away.

So a male's caricature was a rough sketch of a six-year-old girl. His preference came not from evil intent but from evolutionary forces beyond his grasp. His caricature was a six-year-old for no more reason than that *she was optimal*, and he inherited his image from fathers who bypassed the courting wall at this point. She seemed beautiful to him because trial and error picked that age as his most auspicious sexual target.

A girl at the Female's Optimal Courting Age was the center of male competition. Her appearance evoked male interest as estrus did for her pre-aquatic forebears. Male competition, by culling less-assertive men, guarantees the better genes in humans as it does in any species. The longer a girl remained the focus of competition, the better mate she might get, and the more likely that assiduous male courting would overcome her doubts.

The Female's Optimal Courting Age ("FOCA" for short, from now on) was a phase, not a finish line. As each child aged, her face and muzzle continually grew toward adult size, and her brain-growth slowed, making her forehead recede. Men's instinct is based not on her age in years, but on the childish form of face and skull. Targeting physique was men's only way to bypass personality. Males seek the skull shape where the most-gullible mind dwells in an accessible female body.

Individual females gain by attracting male attention as long as possible; if a girl was eight but looked six, she had two more years of peak attention. So the critical female competition was between a girl of six, and one who looked it but was older. An individual gained by maturing later (delaying the point where brain growth slows), and therefore looking younger than she was, long before her puberty, before she could possibly have the first pregnancy men competed to give her.

Her gains at first were small. Few females would be entirely mateless except when sexual fears held them back. Male FOCA-based competition only controlled sexual access for a woman's *first* pregnancy, and only duller men gained by pursuing child-brides. So women's youthful evolution was initially very slow, with two benefits. First, her sexual inhibitions were more likely to be overcome where male interest in her lasted longer. Second, by gaining the better male (evoking more male competition) she might leave stronger progeny.

After perhaps ten thousand generations, these small successes permeated the female population. Girls who are eight all possessed the more-rounded forehead their ancestors once had at six years. Adult women were also slightly affected, for those eight-year-olds eventually grew up and retained fractionally-larger brains, but they still weren't close to the male caricature. The only effect was to blur the FOCA by a couple years.

But the FOCA is a degree of ingenuousness, an optimal point between protectiveness and anger. It is not a shape of face or roundness of skull.   As females began to age more slowly, it was still to an individual man's advantage to court them at six, not eight. So natural selection corrected the male's cartoon slightly, again leading him to focus on six-year-olds. That is, a dull male gained advantage if his perception of six-year-olds was more accurate than some other man's. And as soon as his adjusted image led him to ignore eight-year-olds, another small adjustment became adaptive for women.

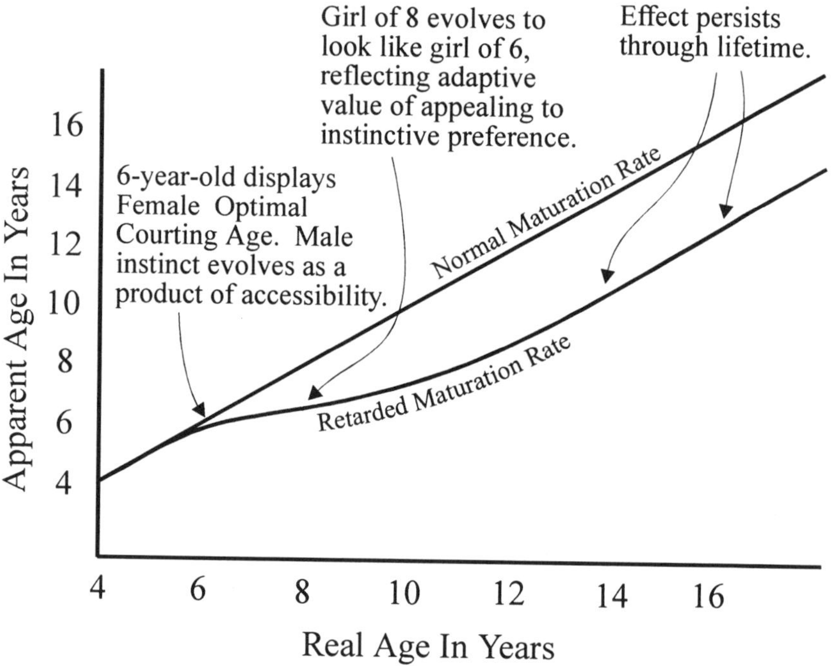

Figure 23: Leverage of FOCA

Time and again this repeated, less a cycle than a steady process, successive blurrings of the line between six and eight. It gripped our species over perhaps two million years, spurred by only the dullest men, inherited only when a child's form helped a girl to win one of those men, the new preference penetrating the gene pool more slowly than flowing water cuts rock. Each small change created an eight-year-old with the fractionally-rounder head of her younger sister, compelling in turn a sharper focus in men. This was not the net growth of infants' heads, but retention of the infant's rounded head further and further into adulthood.

Leverage of this narrow competition slowly flattened the entire course of women's maturation. Larger-headed eight-year-olds grew to a larger-headed ten, then twelve, before finally becoming adult females still with receding foreheads. An occasional man, seeing a twelve-year-old who was close to his caricature, courted far beyond the FOCA. Bit by bit fetal brain growth extended, first in months and then in years. Eventually women kept an infant's domed head into early adulthood, and now we retain it for life.

As women's camouflage progressed, men were increasingly unable to tell them apart by age. Eventually a fully adult female resembled the caricature and could evoke courting by even a child-fixated male. In her small face, slight build and round head, the adult woman was not the *goal* of beauty's evolution. She was only the furthest end of a long maturation line, bent to a more-gradual slope by the intense competition by and for child brides.

The Image Becoming

Now include adult women again. By today's standards, ancient adult women were ugly ugly ugly, yet to their men they seemed pretty. It is not a woman's face or figure that makes beauty, but a man's awareness of sexual possibilities in a non-estrous female. Men found greater beauty in young girls because this male susceptibility guided men to evade the courting wall.

Two million years ago a man thought his mate about as beautiful as men do today. The child-cartoon hovered at the edge of his mind, and he gazed at girls who matched it. But the caricature was an artifact of his mind, not the governor of it, and barely strong enough to shunt duller men toward younger girls. When adult women were very unlike the cartoon, the cartoon itself was vague because even a weak guide was enough to distinguish girls. The weaker the caricature, the greater physical difference it took to make men see a difference in beauty. Three million years ago, to the men who were there, all women were reasonably pretty, none were gorgeous, but none were ugly.

Beauty is not a timeless value outside the human mind, but the mind's creation. Men can find feminine beauty only within women's range of appearance. If cultural rules limit women's display to ankles, men will drool over them. If women are entirely covered and shrouded, men will drool over what they imagine to be there. Men hold women responsible in any case, because instincts are invisible. Beauty is only a male susceptibility, identifying the most-auspicious women within the population.

Stop and remember again that these men and women loved each other. There is nothing bad about love's presence in men and every reason for it to persist. Men inherit it from their maternal side, and for women love was life's very reason. As sex became frightening, love became stronger. As men became sometimes hurtful love grew stronger still. Men inherited this engrossing passion, and though it did not evolve first in them, it surely overwhelmed them.

Men possessed an innate caricature leading the duller to chase immature women. Our caricature was only a rough sketch, but any woman a bit closer to it gained a competitive edge. This was no conscious decision; men merely responded to an asexual and meaningless image inherited from their fathers. This image grew and clarified over generations, while a different image might have vanished, because this one consistently led some men to the courting wall's weakest point. Few men *needed* the easiest path, but no man lost for taking it, and so it grew.

Within the limits of practical needs, women gain by resembling men's caricature, by having a large head and small face. This broad pattern is enough to evoke a courting response in the male. So adult women can possess a number of distinctive adult features (breasts, armpit and pubic hair, an hourglass silhouette) without clashing too much with men's innate image. Men learn to recognize these, but instinctively they still note little girls' head shape.

Men did not reject adult women. If men had instinctively declined to mate knowingly with an adult, then women would have shed adult features. Breasts clearly announce sexual maturity in women; if men sought girls exclusively, breasts would be a handicap to women. Armpit and crotch hair would be a problem for the same reason.

Men seek only the best match to their innate caricature from within the population of available women. Any feature on a woman which helps evoke that male response will attract his attention, will lead him to speak to her. The woman need not look like a child, for the caricature bears only a passing resemblance to a child. The woman gains if she just looks a bit closer to the male's embedded image than do the other women he sees.

At the start this female evolution was only a small increase in adult brain size, a small reduction in face and jaw. Men inherited their preference like a thousand coats of paint from male ancestors, thickened to a powerful liking now only because the process has been layering for so long. As these features attracted men, such women left more progeny than others with smaller brains, who looked older sooner. As the new head shape became common among women, it was still the younger-looking who won the contest. This was never more than a fringe behavior, matching the few dullest males with the few slowest-maturing females. Each single coat of neotenic paint was immeasurably thin. Over countless generations this consistent matching led all females to grow larger skulls and smaller faces.

When adult women finally resembled the caricature, the cartoon had become powerful and clear. By evolving a youthful camouflage, women forced dull men to sharpen their eyes, and accidentally narrowed the range of women men thought beautiful. The stronger the caricature, the smaller difference it took to make men see a difference in beauty, and the more important beauty became for all. Men's caricature and women's competition gave beauty to many women, but also gave ugliness to many others who otherwise would seem attractive enough, and we all inherit the ability to see it.

Counter Evolution

Men and women, each competing against their own sex for progeny, coalesced into opposing armies. Today they still trudge ponderously past each other, circling in the dark, not seeing their predicament. No longer is our sexual evolution an alternation of themes, but simultaneous attempts by each to counteract the innate preferences of the opposite sex.

Women became neotenic not to please men but to fool them. Each tiny extension in childhood brain-growth gave a fractionally larger adult skull. A closer fit to the new male instinct for young women, this gave young-looking adult women an advantage in procreation. As soon as men's inability to court gave younger women a reproductive advantage, delayed maturity began evolving in women.

We have long recognized that animals simultaneously counter-evolve. As predators cull the slowest and weakest, a prey population becomes faster and stronger. As it does so, the slower and duller predators have less luck, and evolve to be faster too. The relationship causes evolution of both; every stalking, fruitless or successful, microscopically readjusts the balance.

I painted men's and women's competitions in black and white steps for clarity, but they are shades of gray and simultaneous. Not all men focussed obsessively on young girls; few men courted *only* young women. Women take the courting initiative more often than we generally credit, skimming choice males from the pool. It has always been the few least-able male courters who gain by targeting the few most-gullible females.

Evolution happens at the boundaries where choices are possible. Men's obsessions *evolved* in only a small part of the male population, though the preference spread to all by inheritance. The obsession never won free rein because women opposed men by their own unknowing evolution. Men's obsession grew not because it brought great advantage to one generation but because it brought tiny advantages consistently, if only to a few.

Retaining a youthful face does not make a woman stupid. Adult women were still smarter than men and still selecting mates who would stay home. Courting adult women was no less difficult a project for men, and duller men who attempted it made fewer babies. Nothing in this feminine *physical* evolution made women easier to court, or decreased a woman's anger if her mate were unfaithful. It evolved only because those women who looked younger had more suitors and an increased chance of procreation over their sisters.

We evolved as predators and prey evolve. As soon as women began growing the larger brain and smaller face of a child, men honed a sharper awareness of young women. When a young-looking woman lured a man by fooling his innate caricature, she won her competition by stealing a mate from some other woman.   But he still had to cope with her personality. The time he lost doing so was used by other men in repeated matings with more-accepting child brides.

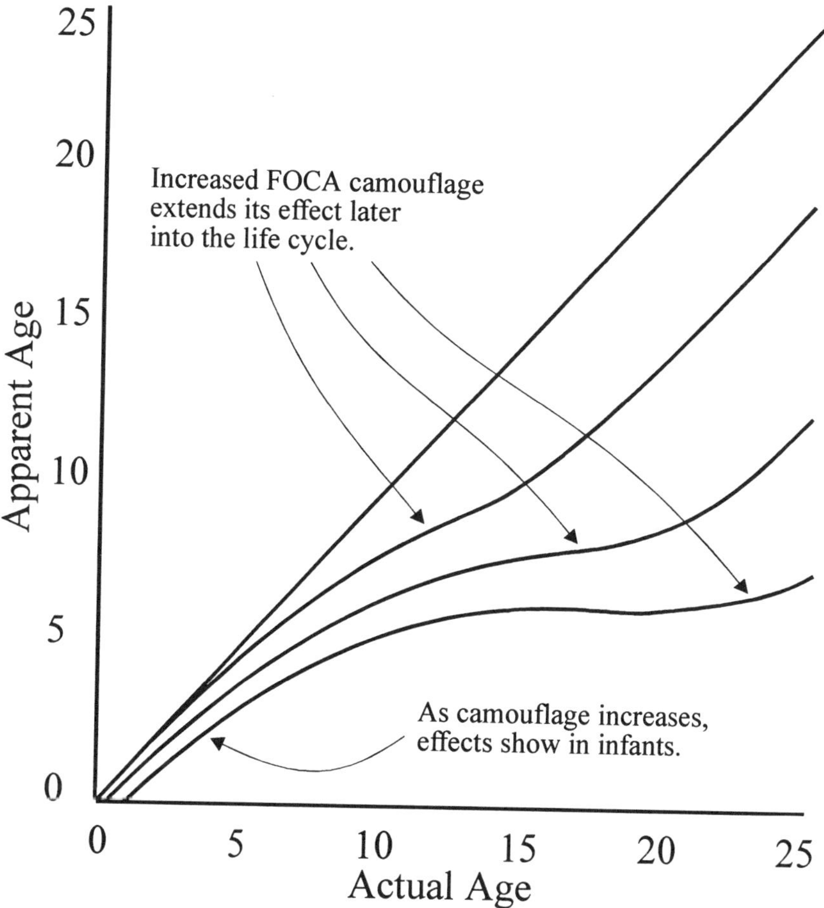

Figure 24: Camouflaged for Life

Men never wanted a younger-looking adult woman. Her appearance was only an outward indicator of the mind her body contained. Men were instinctively drawn to the female body where a gullible mind could be found, and "beauty" is only a name we give to this male focus. Men needed to target *young* women, not young-*looking* women. Regardless what women looked like, a man had an advantage if he targeted the Female Optimal Courting Age. His visual instinct, unknown to him, led him to the most-accepting of sexual partners.

Men evolved a visual cue to young girls, an image of large head and large eyes in a small face. A population of such men did no good for women as a group, it only records the successes of individual men who could bypass the courting wall. And as men became more discerning, individual women gained by again blurring the line between child and adult. A population of such women did no good for men. Women never evolved to be what men *needed*, but evolved to be what men were *looking for*.

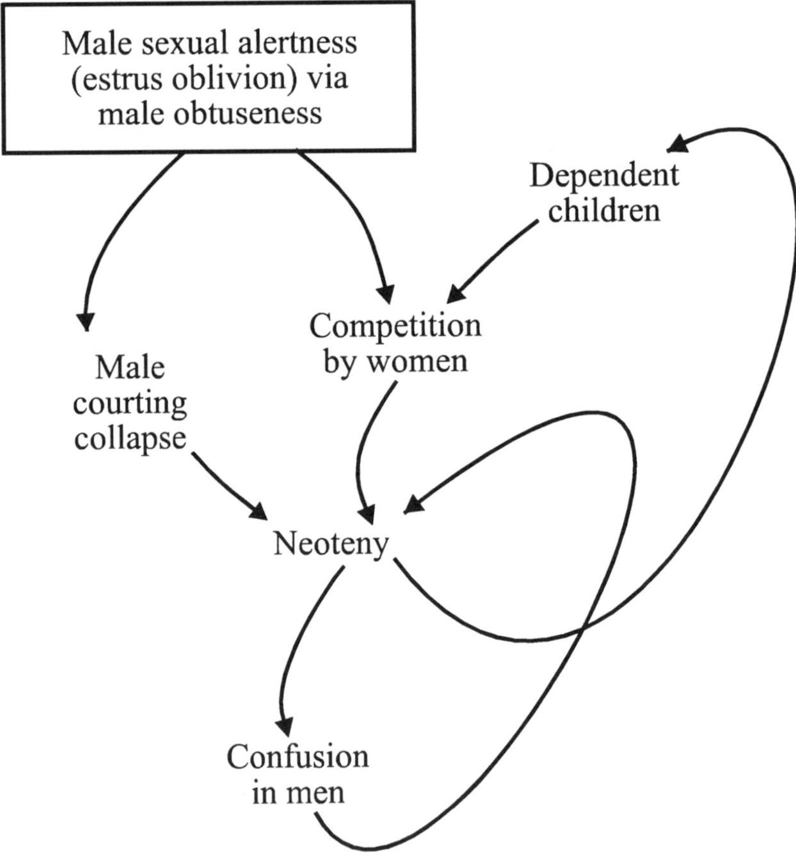

Figure 25: Neoteny Evolution and Feedback

Women were largely successful in their counter-evolution. By retaining childish features, they lured men into courting them long past the age of a woman's maximum ingenuousness. Women blurred age markers so well that a man now might court a woman well into her later life, though he labors under a thousand coats of paint inherited from incompetent grandfathers. We still can find beauty at any age. But because humans invented ugly along the way, some women can never find a mate simply because of their faces.

Menopause records the success of female camouflage, shutting down women's reproductive equipment to preserve their intellectual assets for the good of the tribe. Menopause could not have evolved if men's obsessions had free rein. If men were mating with women only up to age twenty, women would not have evolved an ovulation switch at forty. Menopause evolved because men can be fooled.

Women's counter-evolution brought other penalties. Growing larger brains makes infants dependent for longer times, increasing the mother's burdens and risks. Women see the risks, and need greater passion to accept them, making male promiscuity more intolerable. So obtuseness continued to be adaptive in men, and made them even less able to win mates. And as their glibness fell further, men were led even more to bypass the courting wall by sharpening their focus on young women.

This dual competition continues. Men evolved to be expert at detecting age in women, at targeting the youngest and most vulnerable. Women evolved to blur the line between the stages of life, to retain the large cranium and small face. Each sex's evolution records the successes of individual men and women, though their personal victories did nothing for their side as a whole.

This is the battlefield where we bloodied ourselves. Nowhere in the trenches was there any fighter who could say "Stop!" to the slaughter. Each fertile woman and courting man was the point of evolution's spear, thrown by the first frontal sex in water. We fight for advantage in the battle of the sexes, not knowing why we're fighting.

Growing Human

Our redefinition is long overdue. Hunter-gatherer myths are not wrong but irrelevant. Our nature owes more to coping with passions than flaking flint. Our saga pivots on women's misjudgments and men's obsessions, on delusions both useful and harmful. Women's counter-evolution shook us to our foundations and unlocked treasures and tragedy.

We're almost done exploring how physical evolution answered sexual problems. It's nearly time to see what the journey did to our hearts. The lessons I found on the way were not even close to my initial expectations. Much of this book's labor was in my shedding tenacious and unexamined presumptions, so conclusions could find their own light.

Evolution is not survival of the fittest but culling of the unfit. As Johanson (1989) points out, a species needs only to be adapted enough to produce another generation, nothing more. From bacterium to primate we all survived this culling, pushed by circumstance down different evolutionary paths. We are not the best of animals, merely the adequate.

Humans are different, dominant, arrogant, but not superior. We can define ourselves better by what we lost than what we gained. Language is not evidence of successful hunters but a sign of frustrated courters. Human monogamy is not a higher form of bond but the shell remaining when promiscuity collapsed. Human uniqueness lies less in what we alone seek than in other species' common behavior we alone shun.

We are not the hunting ape, nor the technological ape. We are not a killer ape or even a naked one. We are the passionate ape, the ape who falls in love. This was the minimal requirement for bringing frontal sex out of the water, and became the cornerstone of further evolution. We became the slowest-growing and infantile ape because human sexuality became fearsome. This is the course we followed, but it does not show all we gained.

I began this book intending to discuss beards and noses, bradycardia and trivia. I'm not a philosopher or an anthropologist, only an amateur. Human superiority was a convenient assumption, though hunter-gatherer myths left me unsatisfied. I planned only to restate our physical evolution in an aquatic framework. If we were once aquatic we should be able to imagine a sequence of reasonable (not inevitable) choices starting with a chimplike Miocene ape, including an aquatic phase, and ending in ourselves.

Imagining myself in these successive stages, I repeatedly noticed problems and fears an ape would face. I picked up clues and carried them as best I could, following many false trails and silly ideas, drifting far from well-traveled paths. When the pieces finally fit into place, I looked up and found myself in our species' emotional basement. At our foundation, rot and rust and treasure caught my eye, and I think you should see it.

I hinted at the treasure in describing our evolution. Women's self-repression measures passion's depth in their lives, making both a bond and a barrier. Men evolved to circumvent the barrier, creating their sexual obsession with girls. Men's obsession, for all its power, is only a symptom of women's emotional needs. Women's youthful competition is in turn only a symptom of men's visual obsession, as genetic shockwaves echo an echo.

Passion is part of our prize not because we like it or seek it but because it does not care what we want. Its value is not in making our pairings pleasant but in making separations painful. It happened to be just what we needed though it grew from our failures. The battle of the sexes saved us accidentally as if a soldier at the front escaped a plague at home.

I found a treasure standing as if a marker on a path. But I found it by coming through thick woods, while most have approached it on a more conventional trail. To show you what the marker means I must also show how it is misunderstood. Only my approach through the woods showed me why the conventional trail is so seductive and so wrong.

Growth And Maturity

Long before women extended infantile brain-growth to fool men, long before we were aquatic, we were already the slowest-growing primate. Our chimplike Miocene ancestor bequeathed infantile traits to all its descendants — gorillas, chimps, orangs and ourselves. Our slow-growing family implies that slow growth came from a common ancestor, but we don't know why it began.

Humans grow more slowly than other great apes; the great apes grow more slowly than any other primates. We are born abjectly dependent, with nerves half formed, unable even to lift our heads. We have the longest infancy of any ape and keep childish features for life. None seriously debate this; few agree on the cause.

Animals fight for their species' survival by evolving teeth, claws, and speed, and also by growing more quickly or more slowly. Primitive animals bear large litters of relatively undeveloped (altricial) young, naked and blind. Mammals in the last hundred million years have consistently moved to smaller litters of more-developed (precocial) young, and most large mammals bear only one or two infants. Large prey animals are born highly developed, mobile within minutes of birth, and development at birth is obviously part of a species' competition.

A species' lifespan is also competition. Small animals with high reproductive rates can expand into vacant habitats more quickly, can evolve faster to meet new challenges. By rapid turnover of generations they can surpass other animals who compete more slowly for the same food or space. It is mice, not tortoises, that infest our barns.

Where species have an important social structure, a *longer* lifespan might lead to success. Critical to an elephant's survival is the presence of adults whose example can pass on fifty or more years of valuable experience. The human child may similarly benefit from both parents and grandparents.

In 1918 Richard Goldschmidt proposed for the first time that genes might control rates of development. In retrospect it seems obvious that such clocks must evolve through natural selection, like other gene-linked features. By the early thirties, Goldschmidt expanded his initial theory to the concept of "heterochrony" — varying rates of different body parts' development, controlled by many timers.

We now recognize heterochrony in many species, including our own. Permanent teeth and crotch hair appear at different ages. The adult animal's form owes much to the complex interplay of timers, as well as to genes determining specific anatomy. Timing genes control the shape of a zebra's stripes, for example, by clocking their appearance in the fetus. Fetal stripes are identical in different species, but they appear at different fetal ages and this causes variations in the adult form.

Axolotls (Mexican salamanders) are classic examples of growth timers at work. The animal is a tadpole in early life, normally maturing into a terrestrial salamander in adulthood. But when conditions on land are poor it may breed and raise young while still in its aquatic tadpole state. A relative (the ammocoete) lives permanently as a tadpole. Both these animals breed as tadpoles by delaying or halting appearance of the adult body, while genital development continues on time.

We use genital maturity to define the adult state. Genital maturity isn't the only marker we might use, but it directly relates to a universally required function. We can measure other body-growth features by genital maturity, as if timing by a large clock that ticks once per lifetime.

"Paedomorphism" is the general term for juvenile features in a body possessing mature genitalia. The breeding axolotl tadpole is a paedomorphic adult because it can reproduce though it has juvenile features. This mix was recognized and labelled in the 1880's, leading to Goldschmidt's theory early in this century.

Paedomorphosis is not an inevitable result of heterochrony, and occurs only when juvenile forms *persist* into the adult state. Variations in growth rates are not paedomorphosis if all the bits eventually catch up. A puppy's foot bones grow quickly, for example, to be stronger than the greatest force the enthusiastic youngster can exert. Foot overdevelopment probably reflects a history of crippled puppies. As the dog matures, the over-growth is no longer needed, and the feet stop to wait for the rest of the body to grow to fit. The adult feet are not paedomorphic.

Stephen Gould (1977) proposes two ways for paedomorphism to evolve, the result of quite different selective pressures. In "progenesis" the genitals mature before the rest of the body has finished growing. "Neoteny" by contrast is *delayed* maturation in the rest of the body, while genitals mature on time. These are different phenomena, not different ways to say the same thing. In progenesis the animal rushes into sexual maturity; in neoteny the animal delays nonsexual development. The progenetic clock ticks early; the neotenic clock ticks on time, but the rest of the body runs late.

Accelerated genital maturation (progenesis) can be adaptive in three situations. When an animal colonizes new space, extremely rapid reproduction can give an advantage, as where an insect is the first to arrive at a new food source. Gould cites a mite which bears sexually mature young when food is abundant. Aphids similarly accelerate genital development when colonizing a new leaf.

In these cases, the reproduction rate is the only limit on population growth. Progenetic specimens overwhelm slower-maturing relatives, and as the resource is consumed it benefits the progenetic majority, who quickly pass their useful trait to succeeding generations. This strategy is suitable for highly unstable environments, and the degree of progenesis may adjust to the richness of the temporary resource.

Quick growth need not result in paedomorphosis; a rapidly growing generation may have perfectly formed adults who simply mature quickly, or they may be progenetic, with immature bodies and mature genitals. They must at least not be *inadaptive* for the available environment. Aphids produce progenetic wingless versions when colonizing a new leaf, and as food dwindles they begin producing normally winged adults. Wingless variants are perfect for exploiting the local resource but unable to find a new one.

A second progenetic benefit comes from small size. By reproducing before it reaches a normal adult size, an animal need not grow more. This is an easy route to dwarfism whenever an environment is hostile to larger sizes, and can persist as a stable strategy for a long time. High levels of progenesis are found in interstitial organisms — those living between sand grains on a beach, for example, where small size gives advantage in a turbulent environment. Progenetic dwarfism is also a useful strategy for males in species lacking male-to-male combat and where movement as an adult may be difficult. Dwarf males are common in the sea, where currents can disperse them to encounter the larger and perhaps sedentary females. These are not merely miniatures but are truly progenetic, retaining larval or juvenile characters and maturing more quickly than females.

The first two aim more at an individual's life strategy, but a final advantage of progenesis may be in the species' long-term evolution. When a progenetic organism halts normal adult growth, the resulting specimen may give a new mix of juvenile and mature features. The animal is as adapted to its environment as any juvenile, but not so specialized as the adult. Truncating adult development frees the animal to evolve along new paths, while genetic instructions for forming the adult body become superfluous and available for mutation and new evolution.

No blanket statement covers all reasons for accelerated maturation. Yet where a species is limited in its expansion only by the habitat, rather than by direct competition between its individuals, accelerated maturation frequently results.

Delaying Maturity

In crowded, stable environments, individuals may find advantage by *delaying* maturity. Continuing to grow in a juvenile shape while its contemporaries enter the fray, an individual gains strength, and experience. When the animal finally matures, it will have an edge over other individuals, who might be several years junior to it and smaller.

Delayed maturity is common in highly competitive environments. Parents use this strategy when they delay their child's entry into school to gain the advantage of another year's maturity. Gould mentions the royal albatross, with the longest period of immaturity known for any bird (nine years) and extremely scarce food in its breeding area. A bird that survives to adulthood is a skilled hunter, able to feed its young.

Larger body size frequently helps the individual, and delayed maturity often leads to sexual dimorphism. When males must compete for access to females, delayed sexual maturity gives each male more years of growth, a larger final size and better odds in sexual combat. In species with high polygyny (seals and sea lions, for example) combat is critical and male maturation may be profoundly delayed. Primates who keep harems tend to have larger males who mature later, while monogamous primates show little sexual dimorphism and more-contemporary maturation. Sarah Hrdy (1981) suggests that the delayed puberty and larger size of human males is physical evidence for past human polygyny.

Regardless of the benefit of larger bodies ("hypermorphosis"), delaying sexual maturity can be a convenient way to achieve it just as progenesis can give dwarfism. This may give an edge to the individual, but sometimes damages the species. Since the hypermorphic animal is descended from smaller adults who had been specialized for their environment, continuing growth can lead to those features' over-emphasis when the whole species evolves to be larger. Gould (1977) cites the Irish Elk as an example, and states that most hypermorphoses will ultimately extinguish the species.

Because growth depends not on one genetic timer but on many, when evolution starts tinkering they often get out of sync. When the entire animal matures slowly and synchronously the result is hypermorphism — a larger adult with classic features of the order. When one or more parts have failed to mature when the genitalia do, neoteny results.

"Neoteny" means retention of juvenile features in a sexually mature animal — paedomorphosis caused by delayed maturation. The genital clock ticks on time but the body runs late. Like progenesis, this can be an evolutionary path disconnecting the highly specialized adult form and freeing the animal to evolve along new lines.

By slowing other parts while the genitals mature at a normal rate, the animal retains a less-specialized juvenile form throughout life. In the axolotl this juvenile form is an aquatic tadpole, and the sexually mature tadpole can exploit aquatic resources through many generations without the risks of terrestrial life. Several salamander species use this strategy, based not on rapid reproduction but on delayed maturation favored by their particular environment.

Neoteny also has social consequences. By slowing the transformation from juvenile to adult, neoteny can make an individual's age readily apparent to other members of the group. Gould (1977) lists bighorn sheep as an example, where males retain a juvenile form (and look much like females) for five or six years after attaining sexual maturity. During this extended transition they don't attempt to challenge the fully mature males, nor do older males treat younger ones as threats. Most dominance issues in their complex societies can be settled by sight rather than by combat. African buffalo show the same pattern, where males are sexually mature at three years, but unable to compete for breeding rights until seven or eight.

Delayed physical maturity causes delayed adult behavior too, and a juvenile body leads us to at least suspect a juvenile mind. Bighorn sheep, with their thick horns, echo the juvenile's horn shape in mountain goats (a near relative). And adult bighorns use horn displays and kicking habits characteristic of other species' juveniles. The African buffalo similarly shows juvenile behaviors for a lifetime.

Juvenile behavior may evolve on its own or as an accident of physical evolution. Juveniles are typically curious and deferential; adults are phlegmatic and confident. Curiosity benefits the youngster who has much to learn, while adult confidence may benefit the group, and the combination may reduce conflict between the generations. If a whole species becomes more juvenile, and therefore less inclined to fight, populations can become denser before friction becomes a problem.

For any one species, we can't tell whether juvenile behavior or a neotenic body was natural selection's focus. But we can be confident a juvenilized animal *feels* like a youngster, though perhaps as mature as one gets in that species. The bighorn who declines a fight does so because he feels unsure of himself, not because he maintains a calendar telling him he has a few years to go. Animals react as their feelings direct them; neotenic animals act like juveniles because in their minds they are.

When we domesticated dogs we selected them for docility, trainability, and respect for us. We killed or drove away those we could not manage, as we still do to dangerous dogs. Traits making them manageable are all juvenile behaviors in canids, while adult traits of aggression and willfulness were culled by our ancestors. Under our artificial selection, dogs evolved lifelong juvenile behavior, retaining into adulthood the confused yapping of a puppy. Wild species' adults, which we did not juvenilize, do not bark.

Mentality and body tend to evolve together if allowed to. Domesticated dogs often retain into adulthood a puppy's floppy ears while wild canids have erect ears. Though any dog may attack, generally the floppier a dog's ears, the safer it is to be near. Timers controlling mental and physical maturity *can* be disconnected, but only where disconnection gives advantage. Floppy ears cause no harm in our juvenilized dogs, so they retain them except where we bred dogs for a different look. A juvenile mind and body will normally coincide, no matter which first gave the species some evolutionary edge.

As with progenesis, no blanket statement can cover all reasons for delayed maturity. Progenesis is common where the environment alone limits population; neoteny is common where members of the same species are the key to success (either by cooperation or competition) or where a juvenile body or brain helps in a particular environment.

Primate Neoteny

In the 1920's, we began to see how timing genes have affected primate evolution. Primates live longer childhoods than most mammals, maturing relatively late in life and at a later stage of physical growth. Primates typically reach puberty at 60% of final body weight while most mammals reach it at 30% of final weight, and other development milestones are similarly retarded compared to other orders. Gorillas, chimps, orangs and humans have the most retarded maturity of any primate, and humans are the most retarded of that group.

The reasons for hominoid (and human) neoteny are hotly debated and not necessarily resolvable. Neoteny need not be an obvious response to an evident force, but it wouldn't arise unless it conferred some advantage (or at least did not encumber) the species.

Even where neoteny's benefits *seem* clear, we can't necessarily reconstruct its evolution. Gould says the detailed sequence is less important than the interlocking system's total effect; we evolve in a retardation matrix with wide-ranging impact. Slow-growing young can appear only if the social structure tolerates burdened mothers. Extended learning time then becomes an important stage in forming the highly social and sophisticated primate adult. So the burdensome infant (whose survival this social structure allows) also is a *requirement* for the society itself. This broad pattern repeats in various primates in varying degrees — a mutually reinforcing set of advantages typically giving a complex and close society.

Neoteny is more evident in anthropoids than in the more-primitive prosimians. ("Anthropoids" are primates without tails, roughly speaking, and include the Great Apes as well as gibbons.) At birth all primates have mobile lips needed for nursing; prosimians lose mobility at adulthood but anthropoids retain some lip mobility for life. The infant's thick tongue becomes thin and wide in most adult mammals, but retains the infantile shape in anthropoids, and as a result the great apes (yourself included) can't lap water.

Primates 40 million years ago had larger brains than their forebears, and relatively larger than almost all other mammals. Brain growth is primarily a juvenile or infantile process, and a larger brain is easily evolved by delaying the time when adult brain stasis replaces juvenile brain growth. Big primate brains may be an accident of delayed maturity, or delayed maturity may have evolved because it gave bigger brains. No one knows.

Thirty million years ago, increased neoteny began separating anthropoids from other primates. This paedomorphism was caused by retardation rather than progenesis (by delayed adulthood, not accelerated genitals) and led to extended lifespans throughout the anthropoids. Interdependence, coming from slow-growing young and smart adults, created the distinctive anthropoid social matrix. Our ancestors were firmly rooted in this pattern.

The hominoids alive today (chimps, gorillas, orangs and humans) all descend from a single anthropoid species living fifteen to twenty million years ago. ("Hominoid" covers humans and other great apes, "hominid" means the human lineage.) There may have been many similar anthropoids at the time, and one of those divided just as it and others arose from parents. Four of its descendant species survive today, and my reconstruction of our terrestrial precursor draws on this existing selection.

The hominoids (gorillas, chimps, orangs and humans) are the slowest-growing anthropoids. Our cousin gorilla's lifespan is about 35 years and the chimp's is 45 to 50. A chimp attains puberty at eight or nine, but may not be mature until fifteen. In our Miocene-ape rootstock, an individual might spend a third of its life learning to be an adult. With delayed maturity present in all its descendants, we can reasonably assume the Miocene ape grew slowly too.

This population was likely the most neotenous of the anthropoids, slower-growing and longer-lived. The six-month childhood of prosimian lemurs had slowed to eight or nine years in early hominoids. Paedomorphosis had moved primate brain size beyond the standard mammalian brain/body ratio, and hominoids carried it to an extreme. Hominoids also were large for primates, and Gould suggests that combining a relatively larger brain in an absolutely larger body could give a brain able to think in radically new ways.

Human Neoteny

Humans lives run at half the speed of a chimp's or gorilla's. Our permanent teeth erupt at six years to the gorilla's three; we live about 70 years to the gorilla's 35. We seem not so much on a course of our own, as further along the retarded-maturity path other anthropoids are following. Humans are the slowest-growing species of the slowest-growing family in a slow-growing order. We have recognized our slow growth from prehistoric times, each learning in childhood that a puppy grows old before its owner grows up.

Only in the last 150 years have we seen ourselves as paedomorphic — retaining childlike forms in adulthood. In our case the paedomorphism is neotenic (slowed nonsexual growth) instead of progenetic (accelerated sexual maturity). When the first primate specimens reached European zoos in the early 1800's, scientists noted the human resemblance to other apes' infants. Evolutionary theories of that time held neoteny to be inferior, so we viewed our own appearance as an anomaly. As our view of evolution changed, we started seeing neoteny as superiority, and thought our neoteny proved human supremacy.

Human adults have the thin and fragile skull structure of an infant chimp, gorilla or orang. Our small teeth are mounted on small jaws; we have the canine teeth of an anthropoid infant, with mobile lips to match. We have no heavy brow ridge or cranial crest, anchoring no adult-sized chewing muscles. The cranium is thin, enclosing a relatively large brain for our size. The brain size causes the face to seem flattened, and makes our face relatively smaller than our cousins'. All this is well-accepted.

Not only do we look younger, we *act* younger, spending much of our lives as children. Humans are playful well into adulthood, as are other anthropoids. Younger chimps and humans spend much of their time in play, and even adults remain playful though it decreases with age. Both chimps and humans play with focussed intensity, including miming and make-believe. Given the wealth and freedom, many adult humans devote their lives to play. Less-affluent people may passionately watch the play of others.

Our curiosity also lasts for life, and we show more pride in this than in any other quality. Most mammals are curious when infants, and later grow into a less-probing mindset. Curiosity varies by order; primates are all more curious than ungulates, for example.   Yet even a calf is inquisitive when young and phlegmatic when older. Perhaps humans are the most curious primate, but language probably reinforces curiosity and surely affects our view of it.

Curiosity and extended juvenile growth are inseparable in us. A slower-growing child spends more years learning; an extended learning time makes an inquisitive brain advantageous. As we became more curious we acquired more knowledge and control over our world. With more knowledge available curiosity became increasingly valuable. This forms part of the Mighty Hunter view, with interlocking advantages evolving together.

Finally, our rampant interest in sex strongly resembles the interest common in primate infants. In most mammalian copulations, only the male and female involved have any interest in the act; they might be invisible, for all other adults care. But juveniles watch, in many species, and primate children are irrepressibly meddlesome. When primate adults mate, youngsters may surround them and even climb on top of the busy couple. Goodall reports great restraint by adults even with three or four youngsters clambering on them. In addition to kibitzing, young primates may actively interfere. Goodall mentions a daughter who repeatedly attempted to halt her mother's copulating but was ignored by both adults.

(I separate the coupling from the *right* to do so. In many species, other adult males may be powerfully interested in whether a particular male *should* copulate, and with which female. Harem keeping and ritual combat grew from this. But once that is settled, how it happens is of no interest at all.)

Attempts to interfere with sexual acts of others is (as far as I know) unique to primate children. Human adult interest also has valid social ends; protecting the young matters, because sex in humans can cause trauma. But when people try to limit *how* the act is done, having conceded that the pair have the *right* to couple, they are behaving exactly like infant primates.

Arguing Human Neoteny

Secure in a primate context of slowed growth, clearly the slowest-growing of them all, we can argue that even-slower growth gave the original human advantage. This is the beaten path through the woods to the marker I found by another route. If we can identify some specific benefits of neotenic body or behavior, we can conclude that natural selection for those benefits led to our unique state. It's a seductive theory because it draws on a real trend; it's wrong because the details fail to hold up.

We might group our own paedomorphic benefits into three areas. A paedomorphic anthropoid will (some argue) be more intelligent, be better-suited to an upright stance, and have a more-cohesive social structure. These all depend ultimately on brain size.

A slow-growing infant retains curiosity into adulthood, while slow growth gives a child a long period of learning. Longer fetal brain growth gives a larger adult brain, capable of holding more information and making more-complex judgments. As brains grew larger, our heavy-headed children became more dependent. The hunter's skill became critical, with knowledge his foremost weapon. And as knowledge increased in value, even larger brains would have helped.

Neoteny gave the shortest path to larger brains. A monkey brain reaches 70% of its adult size before birth, with the remaining growth happening in the first six months of life. A human brain at birth is only 23% of its final size, growing rapidly for the next six years, and not reaching final size until past age 20.

Forty million years ago, primates already had larger brains than their ancestors, and relatively larger than almost all other mammals. Fifteen million years ago, the common parent of modern hominoids was more-neotenous than other primates. Evolving a smarter human (the theory has it) took only another extension of fetal growth. If we needed larger brains, natural selection would enhance neoteny, further delaying the shut-down of fetal brain-growth speed, reinforcing our pre-existing trend.

Language arose with knowledge. The infant's mobile lips and thick tongue persisted into adulthood, giving agile mouths and then speech. Conversation let hunters share skills; women discussed food sources.

Language linked the minds and knowledge of the whole tribe into a vast information-storage system. Community brainpower became a defining human property, and this school of thought suggests that big brains were its essential foundation.

Using big brains we learned to fashion weapons. Chimps use crude tools and weapons when convenient, but only humans regularly create tools or weapons by shaping and assembling parts. With brains and language (both born of neoteny) we invented and then passed on new designs, creating the first hunting cultures.

Second, neoteny might help our upright stance. When we left the trees and became upright to carry weapons, natural selection encouraged any body change that helped us be vertical. Longer fetal brain growth extends the skull to the rear, placing the spine more-directly under the brain case. This tilts the face forward instead of aligned with our spine (looking at the sky), and makes a vertical posture easier to maintain. So large brains (the theory goes) not only helped create the weapons making bipedalism necessary, they also formed the skull shape making bipedalism convenient.

A more-juvenile skull has a smaller face and jaw, and never develops the jutting canine teeth of an adult chimp. Weapons made canines obsolete; losing them cost us little. Further, the new cooperative hunting life demanded minimal conflicts between males. So losing canine teeth *helped* us, reducing the chance that men would gash each other in a fit of pique while sitting around the campfire. Paedomorphism not only made hunters of us, it helped make *chums* of us.

Third, these physical features urged the distinctive human social structure. Long learning times meant that children needed long and intense parenting. Before a little boy first hunted, he heard years of stories and absorbed a body of knowledge. Children's needs demanded a stable bond by the parents, bringing the hunter faithfully home after every hunt. Love evolved in men and women to ensure this bond, reflecting the reproductive success of firm families.

Bonds of men and women created the core of stable society. Grandparents stored information, becoming a treasure spanning generations. They also baby-sat, helping their offspring by caring for their grandchildren. Menopause evolved to ensure long lives for women, affirming the value of the spoken word and the pool of wisdom. No longer isolated competitive individuals, we had become tribes of many cooperative families.

Children were like weighted boots, training the human species to work harder. The more childhoods slowed, the more our societies strengthened to cope. Stronger societies were more cooperative, out-reproduced less-cohesive bands, and licensed the next generation's children to be even more dependent. In 1733 Alexander Pope used this theme in his Essay On Man, finding that burdensome children create a superior society from necessity. John Locke expressed the same view in his Second Treatise of Government in 1689. Slow growth made females dependent, made male support adaptive, and so led to pairing.

An argument for human superiority based on our slow growth has solidified over several hundred years. It gathered strength since the 1920's based on Goldschmidt's heterochrony theory, and on our new view of paedomorphism as advantageous instead of anomalous. The concept draws on a well-understood trend and rolls over opposing views by sheer inertia. It seduces us because the pieces nearly fit.

Retardation Matrix

We can easily find benefits in our neoteny. The question is: Did these benefits actually separate us from other apes, or are they accidental results of retardation that evolved for other reasons? That is, did natural selection create our matrix of retardation because slow-growing and burdensome infants brought advantage through social ties or intellect, or do children simply bring burdens to endure? I think we only endure them, and have misunderstood our most-vital evolutionary asset.

Solving the puzzle requires that we explain the first tiny separation between humans and our cousins. The retardation matrix is the easy route to our marker in the woods, easy because it assumes that the benefits we find today must have initiated the process. Imagining later benefits of great retardation doesn't explain why small *differences* in maturation first appeared. If the same forces shaped all apes, something made the retardation far stronger in humans. If different forces shaped us, something caused us to land much farther along the very same path that other apes are on.

Their life speed, already slower than the primate norm, halved in us. Their infants' development, nearly normal, retarded so far in us that human newborns don't catch up to chimps until our babies reach eighteen months or more. Their children are capable but their species dwindle near extinction. Our children are helpless lumps, yet we thrive. As adults we have gone farther along the same path; as infants we have retreated.

The answer will lie in whatever made our infants first start going backwards. Mammals have consistently evolved toward more-developed fetuses; primates typically bear a precocial infant that can cling to its mother from birth. The non-human apes (gorillas, chimps and orangs) grow more slowly than other primates, yet their newborns are also well developed. Our infants reversed a trend of forty million years, and it happened for some reason.

If burdens bring any benefit, it's hard to imagine retarded growth being the path. Wolf packs and lion prides are close and cooperative units where infants have much to learn and adults bring food home to them. Yet none of them has evolved retarded development to seemingly force better cooperation. Instead they bear litters as large as they can ordinarily feed, and starvation culls excess young.

The simplest source of parental burdens is to have more infants, not that burdens are the point. Litter sizes are an evolved female trait, balancing the odds of infant survival against the nutrients required for a live birth. No wolf bears thirty pups because rarely do more than half a dozen survive; natural selection cannot form a thirty-pup trait until twenty are sustainable, cannot select for twenty until ten can often live. A whale gives birth to one because she can't keep two alive.

Twinning occasionally happens in mammals that have predominantly single births, and twinning is an inheritable trait. If a mother can handle one infant with ease, her descendants will begin bearing more. Humans twin once in 86 births, and twins are clearly a burden. If burdens were somehow good for us we could bear twins more often, increasing our reproductive rate and forcing our societies to cooperate, too.

Instead of becoming more burdensome, our infants are becoming *less* so. About four percent of chimp births are twins, and high mortality in multiple births suggests chimps are probing the limits of mothering capacity. Our babies are individually more of a burden, but we have twins only a third as often as do chimps. This suggests that humans are penalized by burdensome children, and natural selection has culled twin-prone lineages because our babies are harder to raise. Natural selection is (in the vernacular) trying to reduce the impact of our babies, not increase it.

The answer must lie in the first tiny separation; evolution happens when a new mutation brings benefit to a lineage carrying it, if not to the individuals. The ant dies attacking an intruder, but the lineage gains and the combative genes are selected for. An animal grows neotenic because something in its life (or its progeny's) gains (or does not lose) from the change. Slowed growth would first cause higher infant mortality, then require stronger parents, who after generations might evolve social structures to aid them, and then could raise those infants. But the infant carrying the genes died a century earlier because its mother couldn't manage, the neotenic lineage was culled, and the causation chain was cut.

Perhaps in a couple thousand years weighted boots could begin to train us. In those hundred generations, slow-growing infants might consistently need more, leading to supportive males and closer bonds. But retarded infants and struggling parents had a poorer chance of surviving each generation, not a better, and closer bonds could not begin evolving until slow-growing infants were a common phenomenon. What benefit did neotenic infants gain from the first 99 generations, each maturation delay with minuscule effect, such that the trait was passed on and not culled?

The weighted boots scenario gives no reason for humanity's initial separation from other apes. It does explain how retardation in a social animal might make a cohesive group after many generations. Benefits seem clear enough in their full expression and reasons for their selection then seem plausible, but this doesn't explain how they first began. The weighted boots concept can't explain why the first feathery weight was added. The retardation matrix cannot launch itself from a standing start.

Neoteny came not from society's indirect advantage in bearing burdensome infants but from individual advantage in each change. Neoteny in water caused our brains to grow for flotation, our lips to remain supple for sucking shellfish, our canine teeth to reduce for dental dexterity.   Already part of our hominoid heritage, neoteny gave the shortest evolutionary path to each of these.   Back on land women needed an infantile form to fool dull men, and neoteny gave individual women an advantage in their quest.   Neoteny evolved in us by classic natural selection, giving advantage to the individual who carries a new mutation, not to his society by making it work harder.

And in my scenario neoteny easily grows, for selection works on even the tiniest variation.   We need not argue for an evolutionary trend lasting hundreds of generations until an aggregate change conferred a benefit.   When drowning became an infant danger, the smallest increase in brain size carried some survival advantage.   When mobile lips were needed for diet, the briefest retention of infant mobility brought some nutritional gain.   Neoteny grew stronger still with estrus' death, when each tiny individual variation affected a female's ability to camouflage her age.

In the aquatic scenario, hyper-retarded growth can start from nothing, with each small change helping the individual who carries the gene.   In the classic view of retarded human maturity, neoteny brings no benefit to the group until individuals have already evolved it for hundreds of generations.   The retardation matrix demands an individual benefit to jump-start the process, and the only candidate explanation seems to be that we needed larger brains for increased intelligence.

<u>What Brains Are For</u>

The brain is our costliest organ; your mother gambled her life on yours.   Any human-evolution scenario collapses if it can't explain why large brains are worth occasionally killing mothers.   Even in the retardation matrix, the brain demands independent explanation.   Yet our brains are larger than cognitive functions demand.

In the weighted-boots view, brains might form part of the entire neotenic bundle without regard to the brain's intellectual contribution. That is, if a burdensome baby has value, then its over-large head may be a useful part of that burden.   Alternatively, bigger brains might give useful functions, and by their value carry the *rest* of the body further toward neoteny.   If a larger brain makes us brilliant, mothers' deaths might be a small price to pay.

Our paedomorphic evolution clearly focusses on brains. We do not merely *continue* fetal brain growth after birth but *accelerate* it during the entire gestation. Before birth the human skull is already hypermorphic and the largest object traversing the birth canal. Its size routinely kills women in childbirth and makes us the only species where parturition is demonstrably painful.

Growing heads became so dangerous that we began bearing our young prematurely. Our primate context and life speed led Portmann to calculate that our gestation should be 21 months. But an infant that age could not pass through the birth canal, so our infants must leave the uterus at only nine months. A human baby's helplessness comes partly from this premature birth, with nervous system incomplete.

Women's pelves expanded as our skull size grew. Sagan (1977) calls this a nice coincidence; we needed a bigger brain for intellect, and a larger pelvis lets a woman bear the larger-headed baby. But this was no pleasant parallel, it was the bloody record of millions of dead women. If your pelvis is too small you don't just happen to bear smaller-headed babies, you die in childbirth, leaving larger-hipped women to reproduce. Women's hips reflect a brutal culling in agony, exhaustion, and death.

If infants were useful by being burdensome, women could have evolved early births and nothing more. But early births were forced on us by infant brains and dead mothers. Was a weighted boot so valuable that women's deaths were a small price to pay? There are other ways to form social glue, and monogamous primates bear small-brained infants. There is too much variation in infant skull size, ready for natural selection to cull the big-headed, for big brains to persist without specific need.

Brains had some value, to grow so persistently. Perhaps this value initiated the entire neotenic mechanism, brain growth carrying the rest of the body along, if the first increased size brought some small gain. Or perhaps we merely continued the apes' slow-growth trend, with our brains growing at first as idle passengers, and a new brain function didn't even appear until well into the expansion. Gasoline engines grew slowly in power for many years before the first car peeled rubber — a wholly new phenomenon not the original purpose of a big engine but impossible without it. Perhaps our brains expanded over a million years of neotenic shift, still too small to harm the mother, before new functions arose. Once brains gained real power, perhaps mothers' deaths became a trivial expense.

Form follows function. Whatever an animal uses its body for becomes the focus of natural selection. If an animal runs to escape, the species evolves to be faster because the slowest are caught and killed. If an animal hides, the least-camouflaged is discovered and fails to survive. These courses appear first by chance, then by habit, and finally by instinct and anatomy. Evolution does not lead behavior, but records it.

Form follows function. This correlation is so intimate that we can judge the function by form alone. A sharp tooth denotes a carnivore; a long, slim leg bone distinguishes a runner. Any hyper-developed body part tells us that part was particularly valuable to its owner. Even if we knew nothing of courting, we would recognize that a peacock's tail must be important to the peacock.

Brain size alone does not matter. The whale's 4000cc brain does not prove that intelligence means more to the whale than to the mouse. And *relative* brain size does not matter. That a mouse's brain is relatively larger than a whale's does not mean that intelligence is worth more to the mouse. We might compare brains to bathrooms. Large houses have large bathrooms because a big bathroom is a trivial added cost. But even the smallest house has a minimal size for the water closet because we all need its basic functions.

The only useful measure of a brain's importance is the burden an animal endures to grow and maintain it. A whale can easily have a 4000cc brain because it costs a whale little to grow the thing; it kills no fcmalc in parturition. The human brain must have been critical to our success, not because it is absolutely larger or relatively larger, but because it routinely kills mothers in childbirth. Whatever function the human brain performs, regardless whether we have yet recognized it, it must be vital to have grown so relentlessly in the face of so many mothers' deaths.

Intellectual Muscle

In the classic view brains are cognitive muscle, grown to bear life's intellectual strains. When we left the trees and became bipedal, our brains had new tasks. We learned to make weapons, to hunt cooperatively, to remember lessons. But (unlike muscle) a brain does not grow with use.

So success came to those born with bigger brains, who bequeathed the trait to their children. The theory is intuitively obvious but wrong.

The theory fails partly because we are not significantly more intelligent than chimps. Until a child begins to speak, a chimp infant of the same age consistently scores higher on our own intelligence tests. An adult chimp can score 80 IQ to a human's average 100, and one gorilla (Koko) scored 85 to 95 on Stanford-Binet. Cultural bias in our tests makes this small difference insignificant. (Testers subtracted points when a chimp signed that it would run to a tree during a storm, for example, since any intelligent person knows you run to your house.) Chimps, gorillas and humans all make tools, form hypotheses, and plan for the future. There is no meaningful difference, barring language, between their problem-solving skills and our own.

The more we study other primates, the more we respect their mental powers. Chimps even surpass the problem-solving abilities of scientists testing them. In one famous case, a chimp with a pole saw a bunch of bananas fastened above him; researchers wanted to see if he was smart enough to knock down the fruit. Instead he held the pole vertically on the floor and raced up it before it could fall over, plucking the fruit unbruised, in a strategy the researchers had never thought of.

In another test, Robert Yerkes presented Julius (an orang) with two stout boxes and some hanging bananas. Instead of stacking the boxes, Julius led Yerkes under the bananas, raised Yerkes' arm, and attempted to ascend him! Yerkes declined to be scaled, but said he was amazed by the attempt. If we can forgive Julius for failing to correctly predict Yerkes' reaction, we can also credit him with an imagination more flexible than his examiner's.

Granting intellectual credit to our cousins places us in a dilemma. If they are nearly as smart as we, why do we have vastly larger brains?. Whatever our intellect's precise nature, it's difficult to argue that we need a certain size brain to support it. Nanocephalic dwarves are humans with smaller brains and fewer brain cells than chimps, yet they can learn languages. Small-headed humans can be high achievers, as Anatole France showed with his diminutive 1017g organ (against an average 1400). That kind of thing dismays a craniometrician.

Even severe brain damage need not be a handicap, especially if suffered by young victims. Humans have had over half their brain displaced by cranial fluid without apparent loss of intellect. Young patients can surmount even radical brain surgery if their brain cells are still growing. Brain functions seem only loosely linked to brain mass.

For the intellectual muscle theory, the stickiest problem is that the brain's expansion doesn't coincide with the time we apparently needed it. Though tools and weapons were our brain's creations, we invented them long before our brain evolved to its current size. Three million years ago, several bipedal hominids left tools in eastern Africa, none with brains larger than half our modern size. Large brains were clearly superfluous for a successful tool-making and hunting ape.

Even worse, humans who don't appear to *need* big brains have them anyway. Alfred Wallace (codiscoverer of natural selection with Charles Darwin) realized that all peoples had innately equal mental abilities. If form follows function, natural selection could not evolve anything until it was useful, yet brains apparently evolved in advance of need. This puzzle so plagued Wallace that he gave up on natural selection and returned to divine creation as the answer for our brains. Gould (1981) says this disgusted Darwin, who knew he didn't have the answer but still trusted natural selection as the cause.

So brains grew later in our whole species (but earlier in some populations) than our apparent need of them. Their growth also fails to match the life complexities that are proposed as the impetus. Instead of expanding continually over time, brains grew rapidly over a half-million-year span, and have *shrunk* since Cro-Magnon times. We have brains three times the size of our cousins', and can show virtually nothing for it except language.

Muscle Of Speech

This leads us to see large brains as an intellectual muscle for language alone, with brain growth reflecting language use. By specifically linking brains to speech, both uniquely human, we excuse ourselves from justifying brains on their own. We can point to fossil skull growth, shrug, and say that's when language must have started.

Speech fills several roles for hunter-gatherers. As we became slower-growing and more interdependent, any social activity would help smooth personal contact. As brains grew and could handle more complex conversations, our relationships became closer and more supportive. This could then permit even more burdensome children in the next generations, with larger heads (and better speech).

As our hunting lives gained complexity, conversation might have become vital. Hands could not gesture when filled with weapons or tools or food. The flint-knapper training his young son could *explain* what he was doing while he did it. A hunter in the field could signal the concerted rush at the prey, or explain failures to his mate. Language also let us plan the next day's activities. We didn't create speech from scratch just for planning, but evolved it through the advantages a more expressive grunter would gain.

There are, however, many cooperative hunting species, and none has a language in our sense. Group hunting does not imply complex communication — Bartholomew and Birdsell note group hunting in many canids, fish-eating birds, and killer whales. Nor does communication necessarily imply language. Wolves apparently use at least twenty-one signals without a language. Morgan notes that visual signals help a hunter far more than does speech, and complete silence is often the key to success.

Any species might find value in communication, but only we among the terrestrials seem to have a language. Any group-hunting species could well use more-precise signals (many lion attacks fail due to bad timing) but the gift seems ours alone. Though evolving a verbal language to communicate is an appealing idea, we can't find a terrestrial parallel.

Physical and mental attributes of other apes are no barrier. Chimps can manage vocabularies of several hundred words, and can communicate with humans via a keyboard or other non-vocal device. Physiological studies of brain lesions in humans and chimps indicate both have a latent linguistic ability; chimp vocal cords are essentially the same as ours. There seems nothing to stop a chimp from speaking. We can easily find a reason to keep talking once we began, but it's hard to see why suddenly we alone started.

Morgan (1972) depicts an aquatic ape with its own need to communicate. Hand gestures don't work when treading water, and one can't clearly direct another's attention to a submerged object of interest. Starting from crude beginnings, some pair might work out the first verbal symbols, become a more efficient team, and others would follow their lead. Those who can learn this communication technique would succeed, and the predisposition to learn language would be a valuable and inheritable trait.

Visual difficulties consistently foster vocalization. Jungle-dwelling animals are often noisier than plains' dwellers, and aquatic mammals have the most complex vocal patterns known. It's easy enough to imagine that a primate might evolve language, as Morgan depicts, when troops and their interlocking relationships moved into the water. But both the terrestrial and aquatic approaches fail to explain why we first differed from other apes. They show why a language would be *handy*; none explains how the first message was communicated.

Instead of looking for value in the information communicated, look for value in the behavior itself. Language evolved for competing through audible innovation. Such competition easily led to babble and finally to language, and did so precisely because there was *no* need to communicate information. Audible play could flower because there was at first no requirement that it be understood.

Language in the classic view, like intelligence, fails to fit the fossil record. In the human brain, Broca's area controls speech and Wernicke's area deciphers it, very simply put. Broca's area leaves an indentation on the inside of a skull, and Ralph Holloway found it in two-million-year-old *Homo habilis*. This bipedal ape, with a brain half the size of a modern human's, had been talking long enough for specialized brain areas to evolve in support.

Hence the first problem. If speech is so valuable, how could we have had it for two million years and make so little progress? Alternatively, if speech had little practical effect, how could it be worth killing women in childbirth?

Language has the most value in the first few words. It's vital that we can shout "Look out for the hole!" It adds little to our warning that we can choose from "gully," "arroyo," "canyon," "ravine" or "excavation." For a creature without language, the first thousand words are an incredible treasure; a child's first hundred words transform it forever. The next hundred thousand words are trivial, marking the change from plain-spoken to florid.

And flowery it is — the second problem it poses. No primitive population speaks a language less complex than we find in highly technical societies. This means that primitive tribes have verbal abilities far beyond their obvious needs, evolved long before technology appeared. What happened two million years ago, that our parents evolved brains to hold a hundred thousand words?

We don't merely *cope* with language; words to us are playthings. We enjoy language and singing as art forms, in literature and symphonies. We enjoy them for their own sake, and admire the skills of composer, writer and performer. This is all pleasant enough, but if the brain grew to support language it also began to kill mothers. I can't help but think that the value of a well-phrased campfire saga is less than the value of a living mother.

Singing and language evolved under the same pressure and for the same purpose. With no other courting competition available, men used these to impress and win mates. Spurred by the need to compete, men faced powerful selective pressure to evolve skills far beyond mere communication needs. We evolved elaborate discourse as male plumage, because nothing less could win a heart. And I explicitly deny any connection to brain size.

We evolved language in a small brain and we still find language in small brains. Nanocephalic dwarves learn language, with a brain a quarter of the human average. Children master it long before their brains are grown, learning thousands of rules and tens of thousands of words in their first few years. If no adults teach them, children invent language spontaneously, with syntax prewired in our brain. Verbal fluency requires a certain brain structure, but not a specific size.

Our touchstone for brain growth must always be dead mothers. We can *imagine* many benefits of big brains, but each increase makes childbirth more hazardous. If brain growth gave us language, then the first small increase transformed us with a thousand words and killed few; the next hundred thousand words merely made our speech fancy and killed millions. Language could not have caused our brains to grow, because elaborate speech grows well enough in small brains and is not worth the awful price we pay for big ones.

Fat Has A Function

Brain cells are fat cells. Human brains grew large to build visible fat deposits without regard to *intellectual* function. When we see fat on a seal, we presume that it gives buoyancy, adds streamlining, stores nutrients, and insulates against the cold. We don't think it makes the seal more powerful, for fat is not muscle. We need to stop thinking of our brain as intellectual muscle and see it as fat.

Nerve cells evolved from fat cells, free to serve as neural transmitters because they weren't busy as muscles. Brains are still fat cells, evolved into networks of awesome complexity. Brain fat is part of the body's nutrient store; when you lose weight your brain loses mass. The brain's ability to grow and shrink along with other body fat is one reason scientists largely gave up weighing it to compare brain sizes.

Form following function applies to the brain as to any other organ, though we err in defining "function" only as cognitive musclepower. Fat has its own function, and we have two reasons to think that brains evolved to simply take up space and expand the skull. First, my theory of brain growth matches the evidence from both archeology and current cultures, and within my theory brains grew to fill space. If women's brains grew as useful camouflage then any growth is valuable without regard to intellectual power, or evidence of language or toolmaking.

Second, on a cellular level, human brains show that taking up space is what they do best. We have brains three times the size of a chimp's, but ours contain only 25% more cells. Each of our brain cells is two and a half times the size of a chimp's cell. When your belly gets fat, you mostly just expand the cells you have, each filling up like a little pocket. This is exactly what the human brain has done over thousands of generations. Brains make up twenty percent of a human's metabolic requirements. If you need a big brain (but not a particularly smart one), it's more efficient to make brain cells fatter, since they were fat cells to start with, rather than adding more cells to be fed.

More than any other organ, the human brain demands efficient design. Internal organs can compress to pass through the birth canal. The skeleton bends; the shoulders pass one at a time. But the skull passes through intact, though it flexes and cranial bones partly overlap. Female pelvic size commemorates wholesale deaths from big-headed babies.

More than any other part, a baby's brain needs to be as small as possible, yet it has evolved to make childbearing the greatest hazard a woman faces.

I was unable to find any report of chimps dying in childbirth, suggesting that their infant brain size exerts no evolutionary pressure. Smaller-celled chimp brains would give no advantage, but smaller-celled human brains (and therefore smaller brains) would have great value. Yet we kill our mothers with a brain less than half as *efficient*, though very much larger and therefore (perhaps) slightly smarter. A large brain clearly has great value apart from its intellectual power.

If we needed more cognitive muscle, we could have increased brain cells' number instead of their size. Any infant with smaller brain cells could spare its mother's life and still be brilliant, if brilliance were the point. Only when cellular-compacting had reached its limits should we see the cranium expanding, if mothers' deaths are the toll we pay. Any increased brain size should correlate to increased intellect, the price of growth being so high. Our unusually fat brain cells are precisely the wrong thing to find if brains grew large to make us smart.

So I conclude that our brain's function is partly in its bulk. When we evolved larger skulls for flotation, and later for sexual attraction, we accidentally acquired extra brain space. Some other fat deposit might have grown, but the skull had no other fat for natural selection to work with. The bigger brain that made our foreheads childishly bulge was the raw material for a fortuitous intellectual increase. Though cognitive skills were not the *focus* of growth and had no great value, when the brain grew it did no *harm* if cognitive skills rose. We became smarter by accident, not from necessity, accepting (we might say) a mental engine more powerful than we requested.

I claim no unique evolution for brains. Most anthropologists will tell you that breasts evolved for sexual attraction by mimicking buttocks, and can present ingenious arguments to support their view. Breasts are fat deposits built around existing mammary glands whose function is undeniably important. Similarly, brains are fatty enhancements of pre-existing and necessary organs; though not *merely* fat, brains are *also* fat.

Complex animals require the brain's neural processes. Shot in the brain, a human and a rabbit are equally dead. Though we can do wonderful things with our brains, we did equally wonderful things (but left no archeological record) over four million years ago with much smaller ones. Like a diamond in a bucket, our mind rattles around in a container larger than it needs. I don't say that the contents are trivial.

Hypermorphic Brain

Growing large heads as a sexual lure seemed laughably improbable when it first occurred to me. As I picked my way along my theory's path, looking ahead and behind to keep my bearings, I passed far beyond our brain growth before I saw its significance, which goes far beyond mere intellect or mating competition.

We stand on bloody ground. More than just a juvenile feature, our overgrown skull records women's ruthless competition for male attention, relentlessly pushing infant heads to larger sizes and pushing all humans to neoteny. We gained large skulls not from paedomorphic accident, but instead became paedomorphic expressly to gain large skulls.

Female visual competition reflects male dullness. When men ducked under the courting wall, a juvenile appearance became adaptive for adult women. By retaining a youthful head (a small face and large brain) older women competed against the girls men were pursuing. Women camouflaged themselves by flattening maturity's slope, costing men little but gaining much for each young-looking woman.

Stimulus signals evolve in both the observer and the observed. Peacock tails grew large because peahens select mates by their tails. The male's tail "releases" the female sexual interest; his tail and her instinct grew together. Estrous odors and recognition similarly evolve in lock-step from barely perceptible beginnings, simultaneous in both sexes. But any attractive anatomy must also cope with the real world. A peacock with twenty-foot tail feathers might impress a female but couldn't move or feed himself. Any signal is a compromise between its attractive value and its practical drawbacks.

Other birds that attract females by male tail-feather length have illustrated the mechanism. If a male's tail feathers are shortened, his courting success drops precipitously; when longer feathers are glued on, he may enjoy an abundance of female attention. When researchers glue longer tail feathers on a male, they allow him to bypass evolution and life's practical limits. His antecedents, too, might have grown longer feathers, except that long tails' burdens weeded them from the gene pool. The surgically enhanced male then presents females with a super-releaser, an over-sized version of the stimulus females instinctively seek. Female response may be correspondingly overblown; their sudden interest tells us that tails were what they'd been looking at all along.

We humans show the same response in our frequent emphasis on breast size. Breasts are not inherently sexual, but women perceive them (and present them) as sensual, and men learn to appreciate them. Once a male learns this link, overlarge breasts can evoke a powerful male response, to the dismay of small-breasted women. Any feature that evokes a response, whether learned or innate, can be a super-stimulus when overly emphasized. Tinbergen (1951) showed how a female bird will ignore her own egg in favor of a larger fake having the same pattern.

When men began to target young women, a juvenile skull became men's instinctive sexual trigger. Men can learn to appreciate many feminine lures, but only skull shape led them reliably to court prepubescent (gullible, available) girls, and so created an instinct. Women's retaining this skull shape through later life records female competition for male attention.

Human maturation's flattened slope reflects sexual competition between women of *different* ages. The eight-year-old, selectively favored for looking more like six, caused maturation to slow. Repeated over endless cycles, this pushed childlike faces later into women's lives, eventually reaching their adulthood. This female change catered to the small but steady influence of duller men who bequeathed their preference to all. Male preference and female camouflage, in slow and constant opposition, led to a clear and instinctive male visual focus on younger women.

When skull shape became a releaser, larger skulls became super-releasers. Each female child unknowingly competed for males against her contemporaries as well as against older women. A larger head emphasized the childlike cue men sought; a larger- headed girl had more men competing for her and was more likely to have children. So girl children began to evolve as caricatures of themselves, growing larger skulls as children, not merely continuing brain growth past childhood. Retaining a large skull into adulthood reflects competition between *different* ages; a large skull at birth reflects competition between girls of the *same* age.

The easiest route to larger infant skulls was the paedomorphic path we were already on. Retarding maturity had already flattened the growth curve; it gave us lip mobility, reduced our canine teeth, and gave big-brained adults. These all came from delays in maturity timers that normally clock our growth phases.

Our brains grow most rapidly early in gestation, and slow as we approach term. To grow a larger-headed infant we needed only to continue the faster (early-fetal) growth speed, just as to retain a childish skull shape we had continued early-childhood brain growth. Early-fetal development, like other timed growth, showed immediate morphological change from the smallest adjustments. So any tiny continuation in early-fetal growth gave a larger-headed child and a more effective releaser of male interest.

None of this happened in water. Males were capable courters in the wet; children were not sexual targets though active sexual participants. Larger brains had value only as flotation, and they had to grow only a little to elevate nostrils into the air. The fossil record shows brains growing slowly just a bit, remaining stable for millions of years, and then growing explosively. Slow growth in a frail skeleton might mark our aquatic time.

When aquatic-infant brains became large enough to conflict with birth canal limits, they caused a few childbirth deaths, and expansion stopped. In the sea, brain size reached equilibrium when childbirth became as great a danger as the risk of a child's drowning. And if a living mother was worth more than an occasional drowned baby, she is surely worth more than a slightly smarter or more-eloquent one.

On land it all changed when women's rage bred dull men. Men's poor courting led them to target young girls, and men's obsession eventually barred women who looked older. Slow brain evolution marks our early return, as the few duller men gained from chasing girl children, and as girls very slowly evolved to match. Over countless generations, dull men's choices accumulated to build a narrow eye for beauty in their sons, a caricature of a female child rendered in thousands of paint layers.

When our caricature became strong enough, when the paint had grown thick enough, all men began to see beauty not in all females but in young females. Instead of a youthful face gaining a dull mate for a woman, from that point a mature face might cost her a *normal* man's interest. Women had to compete for *all* men's eyes, and the brain began growing explosively, killing narrow-hipped mothers in childbirth, leaving wider pelves in the remainder. Brains reached a new equilibrium when a woman's burdens (from slower walking and uterine prolapse) balanced her daughter's risk of finding no mate.

Some say we use half our brain or less, while others think we might use it all in subtle and unrecognized ways. I believe as cognitive muscle our brain is largely flab. We had little need for added intellect and much need for a little male stupidity. Our brains grew as camouflage when men sought gullible females, to make a man think a woman was a child. When a male took the female visual bait, his *perception* of her as a girl was *her* path to procreation with a *gullible man.* His error validated every bit of brain she had; when we are *looked at* all of our brain is used.

We don't consciously think of child-shaped heads as sexual because we all have them. Men inherit body-structure from their mothers, and male babies are born as large-headed as females. (Suppressing the skull size of males gains nothing if your birth canal already can pass a female head.) The lack of infant sexual dimorphism clouds the initial impetus for brain growth, forcing me to build my case from such small (but tantalizing) clues.

The brain demands efficiency more than any other organ, balancing function against size. We don't pack brain cells densely because intelligence is irrelevant to the function being weighed. The brain's function and danger are both in its size; brains grew until mothers' deaths stopped them. There they hover, always killing a few, keeping brains near the limit that mothers can bear.

Our present brain size records a balance between danger to the mother and failure to find a mate. If a girl can attract no male she can pass on no genes. Bearing a small-headed girl baby was little better than bearing no girl baby at all, once the caricature gripped us. So powerful was the competitive pressure for child brides that the death of a mother became a small price to pay. No one intended to hurt women. Men's preference was (through several cycles of cause and effect) only the result of women's emotional needs, and I place no blame either way.

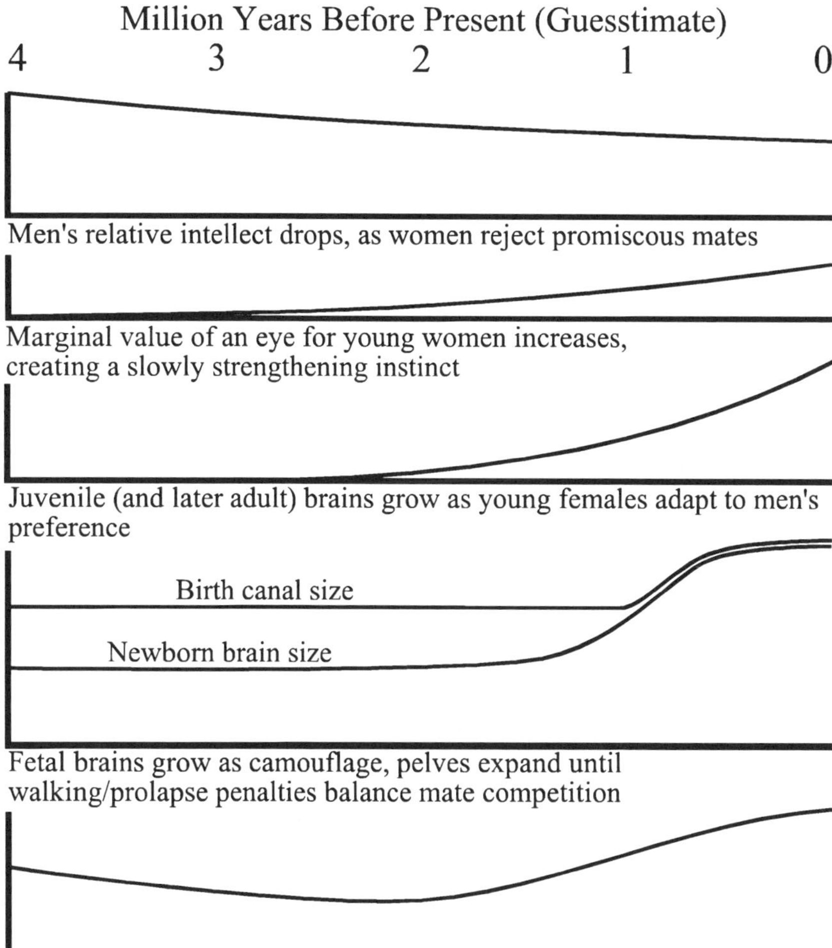

Figure 26: Hypermorphic Brain Growth

Small-hipped women died away, and as survivors' pelves expanded over generations, they became awkward and hips began to swing. Walking took greater effort as more weight concentrated low in the torso; with broadened pelves came the danger of uterine prolapse. Practical impediments now limit further pelvic expansion, like movement limits bird tails, with the opposing forces worked out lineage by lineage, not in an individual life. Pelvic expansion stops when a woman's *personal* disadvantages balance competitive pressure from females of other lineages, weighed though their *daughters'* successes at finding mates.

Ironically, brain expansion need not have stopped in water. Women there faced no danger of prolapse and no problem in locomotion. Their pelvis could widen at need, and fossil remains of *Australopithecus* show a very wide and delicate pelvis. But in water only moderate brain growth was enough for flotation, and the much-larger skull did not evolve when it cost little. Once the need for brain expansion became intense on land, new pelvic limits arose to stop larger heads from evolving.

All cultures have large-headed people regardless of apparent sophistication, because girls' brain/face ratios are a universal human courting cue. Women everywhere die in childbirth because small-headed girls don't trigger male sexual interest. None of this concerns intellect, for intellect is a trivial asset.

Human paedomorphism, and specifically skull paedomorphism, is the focus of retarded development, not its byproduct. Our brains evolved to match the caricature that grew in the minds of dull men. Our slow growth and big brain commemorate a battlefield of the sexes, a place where men and women for millions of years have tried blindly to outmaneuver each other on a plain without leaders, unable either to stop the fight or to break their bloody embrace. We delayed our maturity not to stay inquisitive or to form closer bonds, not to extend our learning period or force men and women to pair. Useful as we may think them, those effects are merely adjustments we made to endure the immature form inflicted on us by our sexual battle.

Chapter 9 - Invested Visions

Men's female-child caricature is the largest artifact humankind has created; it defines how we are evolving today. It spans a universe of the imagination; the pyramids will crumble before this image collects its first dust. We shaped it on our way to procreate, bent under sexual burdens, unaware of what we did. Once shaped by us, our caricature shaped and is shaping us in response. The molten vision, when set, became harder than the mold into which it was poured.

You need not believe me yet. I have no compelling evidence for my thesis, but can only show clues that hint at the pattern. With clearer evidence, we'd have seen it long ago. Having stumbled upon it while on other business, I feel obliged to report it.

It's not our weaponry or social structure that now saves us from extinction, but the products of sexual failures. When our cousin apes began dying from a melancholic plague at home, we were off fighting the battle of the sexes, creating a treasure. Our cousins will shortly be extinct because they avoided the battle and the scars and so lack precious assets you didn't know you had, formed like diamond from deep pressure over long time.

A geologist detects oil deposits through surface evidence. Hills and minerals don't point to oil, but betray a hidden structure reaching miles down into the earth. Understanding the structure, a geologist can predict where inside it oil might be. The deep system gives meaning to the surface features; we believe the whole configuration exists because its presence makes sense of the small part we can see. Often finding oil validates our comprehension, but we shouldn't take our awareness for granted. We lived with mountains for millions of years before we saw them as results of crustal movements.

Too often we seek surface phenomena to explain surface events. We view the human state as a smooth plain with rocky outcrops that each have an individual reason for existence. We explain slow growth by dependency's value. We explain love as a hunter's way to evolve fidelity. We argue that bipedalism came from a hunter's need to carry and throw. We explain rape and child abuse as acts of bad men. Murder flows, some suggest, from excess aggression our forebears once needed. We explain each in isolation, and think then we can leave it be.

I propose a deep structure of the human mind, an artifact of tectonic human emotions. These seemingly disconnected features are only surface manifestations of hidden selves. The deep structure takes its shape from our primate context and the courting difficulties of frontal sex. Fear and doubt gave us singing, language and love. Rage and pain led to obtuse men and child-caricatures. And this image, cast in a mold of soft and fatty brain tissue, became the hardest and most intractable piece of our sexual baggage. We now must plan all our travels with this luggage in mind, but it is also our ticket to continue.

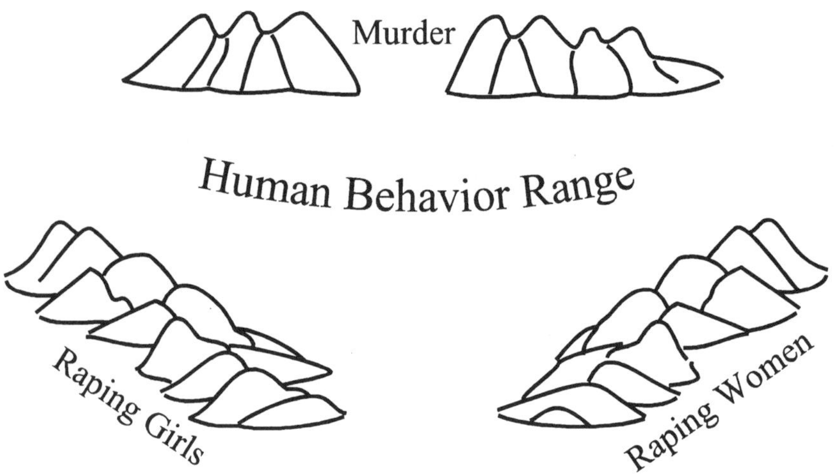

Figure 27: Behavioral Outcrops

I hope to persuade you that the deep structure exists, and also that I've described it fairly. All the outcroppings on our smooth plain you have seen before. In this chapter I discuss murder and two kinds of rape, none of them hot news. My point is not that these *exist*, behavioral outcrops on a level plain, but the idea that they *align*. If I can show you their grain, we can see them as tops of a mountain range largely hidden under sand. And if we see the mountains' alignment we can see where the crustal plates move and how they inexorably shape our course.

Our subterranean mental structure displays its depth and power in how we treat each other. We measure others by our innate caricature. We handle others clumsily, ourselves half-crippled from wounds that our species suffered breaking down courting walls. We murder and rape not because we are bad but because we are broken. Careful training can patch over breaks, but our mental fault line is instinctive and defies repair.

Miscommunication contributes to misbehavior. Female camouflage and male obtuseness leave us deaf to basic messages of age, sex and intent. The same evolutionary line that gave us love also burdened us with child abuse, rape, and murder. Men rape children partly because men cannot recognize sexual maturity. Men rape adult women partly because men cannot understand a spoken refusal. We kill partly because we don't recognize who we murder.

In conventional understanding these are unrelated. But when we hold the *idea* that some hidden structure guides our lives, then these can not only outline its contours but also take new individual significance. Murder, and rape of women and children, all are more than we have thought.

These three are primarily male behaviors. Few males do any of these, and I've done none of them. It is the fault of neither sex that men are the common instigators. It is the task of both sexes to see the problem as best they can; only understanding gives a hope of prevention. Each person remains responsible for what they as individuals do.

<u>Sex With Little Girls</u>

Forcing sex on a child seems insane. No other animal does what men do to little girls. When we explain it away as merely the act of a bad man, we conceal its true nature. No random outcrop, what we call child abuse is a predictable effect of evading the courting wall. Every human carries the instinct.

If you weren't already aware of child abuse, I'd be hard-pressed to persuade you that it exists. Imagine an explorer returning from remote lands, describing a species whose adult males raped female infants. It would be incomprehensible. Imagine a bull elephant in a zoo trying to couple with a calf. Flabbergasted zookeepers would isolate or shoot the male. Yet we are so accustomed to it in ourselves that we routinely warn children to avoid strangers. A behavior no other species ever displays makes a constant background in human lives.

Raping a girl hurts the gene pool and the girl. If she is not yet fertile, the man can cause no pregnancy. If she is barely pubescent, she may lack the resources to raise her own infant. In either case this male preference should quickly disappear from the breeding population, the prospect of healthy offspring being so slim. Men who pursue genetic dead-ends should lose out to men who court adult and fertile women. The hunter-gatherer view of ourselves gives no reason why such an unlikely act would evolve, and good reason to think the preference would be culled.

Some cultures allow marriage of child brides, but most discourage sexual contact with children. Yet many men seek this contact, even after punishment. Child-rape's widespread presence and intractability both imply a powerful predisposition to behavior that should never appear. This seems insane. That the act persists (in spite of social and probable evolutionary pressure against it) suggests that the *preference* has value though the *act itself* may harm.

What do child-rapists want? What do men generally want? In all human populations, we can identify women by their smaller and slimmer build, smoother skin and higher voices. A child's body resembles in many ways an adult female's.

We *might* explain away child abuse, then, as an outcrop on a level plain. If a poorly socialized male cannot court adult females, he might target a more-vulnerable sexual object. Girls have feminine-enough appearance to evoke misplaced erotic interest. We might say he wants a woman but settles for a little girl. This view says child abuse is aberrant behavior from an aberrant mind, out of the question for any healthy person.

We can then dismiss the abuser as no more than a bad man. His personal inadequacy, perhaps in more than just his sex life, leads him to carelessly inflict pain on his way to orgasm. For a callous and horny man, a child is a pliable sexual object, easily controlled and easily hurt. Because the family is so intimate, and so many of its relationships carry sexual tones, mistreatment in a family is often sexual mistreatment. A damaged child grows into a damaged adult, later inflicting pain to echo the pain of his or her childhood. This leads us to see child rape as a cycle of sexual violence that we can fix by curing one generation.

I disagree. We do not learn to rape children, though we do learn to cause pain. Trauma can surely foster trauma in a cycle, but child abuse would not carry such power unless the act tapped something deeper. Sex with the child is the constant underlying preference, unleashed by failure of training or unhealed personal wounds. Child rape cannot be cured by fixing only *one* generation; it requires us to fix *every* generation.

Men And Little Girls

Two distinct forces acting at separate times caused men to become sexually enmeshed with girls. Child-coitus is an act we did not seek but were led to, and we must adjust our view of a man who rapes children.

Sex with children is a predictable (not inevitable or desirable) result of life as an aquatic chimp. Non-primate aquatics didn't evolve it because only *primate* clinging instincts lead to an erotically-useful posture with legs wrapped around another. Girls clinging to a man's body initially excited him no more than a sack of rice. The pose became erotic because girls needed something to masturbate against.

Erotic demands of orgasmically insatiable primate females became useful to both males and females in the water. A male had no pre-existing sexual interest in the girl, but learned early that her masturbation was fun for him too. Females grew to like a normally frightening pose; males became hyper-sensitized to non-estrous females. The entire process worked because primate females need to masturbate and found that males were convenient aids.

This was an entirely normal part of growing up and had no moral taint. Men made no attack and used no force; a girl could outswim men and approached only those she trusted. Frontal sex caused no strong fear, only mild hesitations that this habit could allay. It was the child's *lack* of fear that led her to instigate the act, teaching her that men provided pleasure, preventing her learning the fear she otherwise would have felt as an adult.

The behavior vanished immediately on our return to land. No terrestrial circumstance made hanging from a man's beard useful for a girl. Estrus reappeared when water stopped drowning the odors, and men had no more confusion about their sexual target. Men had never evolved an instinctive *preference* for girls in the water, for men were (at first) competent to court adult women.

But feminine emotional needs demanded faithful males on land, and women reproduced better with duller men who stayed home, who couldn't detect other women's cycles. Men's estrus awareness vanished along with a measure of male intellect. Unable to sense estrus odors, men were unable *instinctively* to detect puberty and had to *learn* to recognize adult femininity.

When dullness made courting tougher, men evolved to bypass the courting wall. Men no longer had an instinctive estrus perception, which might otherwise have stopped their search at girls' puberty. Favored by a growing instinct leading them to *courtable* females, they now lacked the instinct to detect *fertile* ones, and men crossed the line from adult to immature female. Once sexual focus led them to prepubescent girls, there was no other female anatomical marker to stop the impregnation race, and only mothers' protectiveness set a lower boundary. Men evolved an innate caricature of a girl and then sharpened it to counter women's camouflage.

Coitus with a child of six or eight never had value for the species. The man gained a genetic victory by pre-empting her *emotional* allegiance before another man could get her. Her devotion to him, easily won at her younger age, held her heart until she became fertile. She gained nothing from prepubertal copulation.

But the man wanted sex. His obsession was not love's emotional susceptibility but a specifically sexual interest, a hunger to feed his evolving instinct for beauty. Orgasms were his short-term goal; he cared little that reproduction was his long-term benefit. Unable instinctively to detect maturity, instinctively focussed on prepubescent girls, he thought she was sexy.

Remember again, not all men took this route. Duller men gained more by targeting the most vulnerable, and competed for the pool of young women. But all men had a few dull fathers and inherited from them an eye for young women, and no man suffered for courting girls. Though many older women had suitors, all the younger ones did.

So sex on land was not at the girl's instigation as in water. Copulation came from relentless persuasion and perhaps some force. Sex had already become threatening on land, and love had evolved as both women's response and men's barrier. If a woman felt danger for herself she felt it even more for her daughter. Motherly vigilance became a courting obstacle for men, relaxing as her daughter grew and so defining the man's lower courting boundary.

Pedophilia began here, in our species' youth, when incompetent males (lacking innate age-preference) courted the least-critical female age. Each successful mating injected another miniscule taint into our common instinctive pool, accidentally making girls attractive to normal men. Duller ancestors skewed all men's eyes, pulling us toward the young and vulnerable. When a man beholds beauty, he sees the ghost of a girl his grandfathers courted.

Child rape is not the helpless act of a sick man but outrageous behavior by a normal one. Sexual violence against children is not a single act spread virus-like down generations and across cultural lines. It is the surface manifestation of a subterranean instinct common to all cultures, the consistent response of an average male unless he has learned better.

Fathers And Daughters

Sexual prohibitions fracture most frequently between fathers and daughters. Fathers remain with their mates long enough for daughters to attain the age men evolved to seek, while girls offer less resistance to familiar and dominant males.

Daughters become their fathers' sexual targets because fathers stay around too long. Aquatic females hesitated only briefly at frontal sex, and men's promiscuity had fewer restrictions. Aquatic life reduced combat's hazards and men courted well enough, so fears were small and love evolved as only a tepid emotion to overcome minor fears. Though aquatic sex focussed on emotions, passions were mild, and pairings were loose and brief.

Long before an aquatic girl reached five or six, her father left her mother. Their pleasant but light affair held him for a time, but then he rejoined other men flowing past matrilineal islands. Only rarely would a woman couple with her father later in life as occasional humans do today, when neither knows of their blood link. Such unusual events have negligible inbreeding effect.

On land women's fears led to passion; only deep emotion could enable inorgasmic frontal sex. Powerful emotions demanded stable pairing, leading to men's courting collapse and long-term bonds. For the first time men stayed near their mates long enough for daughters to grow up. Even though childhoods lengthened, much-longer bonds kept a man at home as his daughter reached courting age.

In daughters, parents see their innate caricature made real. Not a sexual image, this only focusses interest by giving pleasure in contemplation. But when the man begins to see his daughter as a beauty he may easily see her as sexual. Sex brings danger, daughters are to be protected, and most men can ignore their brain's sexual nudging. Even so, perhaps unknowingly and seldom intentionally, fathers court their daughters.

Girls approach incest by another path. A girl lacks the strength to prevent a father's acts, and the objectivity to know her own rights. But girls are also sexual creatures whose erotic exploration begins with their fathers, and girls have mental traits making them the least-resistant victims.

Promiscuity has been useful in many primates, giving paternal uncertainty and protecting children from infanticide. High sexual interest, necessary for inducing promiscuity, becomes apparent first in childhood. Goodall reports constant sexual solicitations by infant females, all ignored by the males. Many females are undoubtedly soliciting their own fathers, both unaware.

Bodies may be orgasmic shortly after birth, but female infants must resort to masturbation because they have no estrus to attract a male. So the primate vagina evolved insensitivity in its lower third to discourage digital exploration, with a clitoris for masturbation. Masturbation is vital for primate infant females, and they evolved genital anatomy to make masturbation both easy and safe.

An aquatic female grew up with adult men for company. She used them as mooring-posts and masturbatory devices. She cared nothing for family lineage, nor did she need to. Promiscuity had always removed her father before she became fertile; there was no evolutionary pressure for feminine paternal-awareness. She followed her sexual urges, and her erotic readiness led to reliable impregnation as an adult. There was no reason why a girl should not play sexually with any man she found, so girls evolved nothing preventing such a response.

On land women evolved sexual self-repression. Not an absolute bar to self-gratification, it counters the decreasing rewards of coitus with a growing disinclination to masturbate. Self-repression shows female masturbators' subtraction from the gene pool. Women now instinctively turn to men for sexual relief precisely because men are such a poor source of satisfaction.

A terrestrial girl child, in sum, prefers a safe adult male sexual partner to masturbation. Lacking an instinct for eliminating relatives from the male population, a girl will sexually target any nearby man. Since the average woman now dislikes promiscuity, the girl's nearest man is usually her father.

In fathers most women find their first sexual relationship, though only rarely does it lead to coitus. A woman may spend the rest of her life looking for a man with whom she can recreate this bond, though she may be unaware of his impact. A father is a mythic figure to his daughter: Dominant, sexual, unreachable and mysterious. Her concept of femininity and of her own self-worth comes first though his response to her, and no other man can ever fully replace him.

Some treat this bond as uniquely evolved for a special human need, but it is a fluke born of converging evolutionary paths and accidental proximity. Normal responses of adult males and young girls will always lead to powerful sexual currents. The family is a social structure newly invented in humans, necessary for female emotional needs and vital for our species. Unfortunately, this brings together men and girls who are highly responsive to each other but inappropriate as partners.

Within families we learn to suppress troublesome urges, and any weakness in our training raises the risk of incest. Women didn't coerce men into monogamy; they bred men for it by mating more often with men who stayed home, men who found courting more daunting. Love aside, the more faithful a male the more likely he finds courting a harrowing prospect. Care aside, only cultural rules stand between him and his daughter. His daughter may pay the price if his training is faulty.

Relationship aside, a daughter is exactly the gullible and compliant female men evolved to seek. Constant contact lets a father understand her better than he can know any other female. For some men a daughter is the only female who will do as he wants, whether by force or persuasion. She in turn tends to see her father as a trusted authority, holding the dominance she seeks in a mate. The daughter is the easiest of all females for a man to court, in a species where this function is specifically degraded. And incest happens most often when the girl is prepubertal, matching the age my scenario predicts.

Some men marry with this in mind, acquiring stepdaughters too young to defend themselves. Blood ties or not, the damage can roll over generations as if it were merely learned and then spread like a virus, but that is not its source. Incest rests in all of us barely beneath the surface, where small disturbances can bring it up. I can't bring myself to call it normal, but it is undeniably typical.

Fathers are rarely dangers and often treasures; sharing a flaw does not make them all monsters. From the time women opted for monogamy, men have been fathers. Humans could evolve slow growth only because two parents wrapped infants in parental care. We can afford to be born savages because we've always had two to train us. From our return to land until now fathers have been vital models and teachers. A fatherless child is forever incomplete.

Fathers and daughters share true love and unhappy circumstance. Most girls can use their father as a safe sexual testing ground. Most fathers can love a daughter like no other female, without doubt or sexual bargaining. But they are also (were they not parent and child) the mates each evolved to seek. Only a man's training can protect the daughter, and for most men most of the time, this will do.

Men's interest in little girls is the first of three outcrops on a level plain. We cannot dismiss it as evidence of a few aberrant minds. It is instead only a surface phenomenon helping us sketch the underlying structure, shaped by and shaping men's courting incompetence.

Raping Women

Rape is the act of a broken animal, a bad thing any man might do. Its existence confirms the courting wall, where a woman's preference bars a man. Child rape commemorates men's age-long ducking-under-the-wall; raping women came from men's collision with it. Rape is another sign of our hidden structure, not an isolated traumatic outcrop on a level plain.

Defining rape eludes me. As a moral litmus test, we ask each other whether rape is an act of violence or of sex. Only the combination makes it rape, of course; one or the other would be coupling or battery. But I cannot tell what elements in what amount make up its essential components.

Many species suffer painful sex. Snails routinely wound their partners during sex (sometimes fatally) using chalky daggers evolved for the purpose. Many reptiles, birds, and mammals bite or otherwise inflict pain during sex, and Kinsey reports half of all men and women respond positively to such pain. But more than just pain in coitus, rape implies a violent disagreement whether coupling should occur at all.

A man can range from loving insistence to brutal indifference. He may delight in a woman's pain and subjugation or be oblivious to it. A rapist may be a sociopath, or a normally gentle man now caught up in sexual tension and circumstance. For any man, some combination of sexual arousal and frustration, of power and opportunity, will lead to rape. Though few men rape, every man is a potential rapist.

A woman's response varies too. She may seek dominance as a sexual trigger. She may enjoy rape fantasies, ashamed to admit she has them. Perhaps her training has left her not knowing that she can refuse. Hite cites women who thought themselves willing at the time, and only after reflection concluded that they'd been raped.

Perhaps no sexual act can be totally free, if we dig through enough layers of need and expectation. I depicted passion as hidden bondage, compelling a woman to do what she would otherwise avoid. That women seek an addictive delusion does not make it any less their captor. Women and men stripped of choice by passion are no more free than those stripped of choice by force, though they at least feel less terror.

Variations in men and women work across each other, making a plaid landscape of infinite colors. Rapes occur all through it where the fabric of male and female rubs wrong. Brutal serial rapists may attack any woman; a sufficiently timid woman might feel threatened by the meekest man. A man might endanger a weak woman but not a strong one. A woman can feel overpowered by a man who intends no such thing.

I can't predict how any one pairing would work out, or whether any friendly episode might, on reflection, be remembered with pain. Long after our companionable walk by a river at night, as it seemed to me, a girl stunned me by saying that she had been afraid at the time. I blame no one for being raped, and forgive no rapist. Though I cannot define rape it surely exists all around us, and though I have never seen one I take its presence on faith. Yet I can find no one point where man's behavior crosses woman's such that the intersection reliably denotes rape. Many claim to know the point, but never seem to agree on its location. Our courts show the same confusion.

Fear of rape is baggage of a broken animal. Without regard to what a court might say in an individual case, if women fear sexual attack then something has gone badly wrong in our species. If a man can use sex as a threat, then a deep chasm has opened between ourselves and the cheerful promiscuous picnickers from whom we descend. Whether a woman hides or shows her hesitation, her fear alone proves our sexual failure.

Rape is virtually unknown in other species but common to nearly all human cultures. Only one other primate (the orang) ever rapes, and then rarely. (The orang is also the only other primate who regularly copulates face-to-face.) No monkey or other mammal rapes in the wild. Yet one in three women (by some estimates) suffers violent sexual attack, and all women fear it.

Brownmiller thinks rape led us to marriage; Morgan believes it was the path to frontal sex. Both think rape persists because its value outweighs its trauma. I suggest it is neither helpful nor necessary but a symptom of deeper forces. Men and women met in tectonic conflict, a deep collision of emotional continents, building mountains where they hit. Brutal rape is a delicate ripple on the surface compared to the pressures below. We need to set aside individual cases to see our continents from a distance. What did mankind and womankind do to each other, that rape so often results?

Men And Rape

Women's sexual choices helped form men by creating the courting wall and stripping men of the ability to surmount it. Women evolved powerful emotional hungers as they returned to land. Terrestrial frontal sex became threatening; strong love kept it tolerable but made men's promiscuity infuriating. By selecting men who stayed closer, women consistently reproduced better with good husbands but bad dates, and made men dull.

Males who can't detect estrus don't get subtle signals. A woman may attempt to warn off a man in several subtle ways before in desperation she yells "Leave me alone!" If men caught subtle signs, women wouldn't have to raise their voices.

Men's acts are not women's fault, but women partly formed men as gazelles formed cheetahs. Each gazelle seeking to survive by speed helped evolve cheetahs into formidable gazelle-killing machines. Each woman seeking the man most tolerable in her view helped create a population of rapists. Neither gazelle nor woman bears any guilt for trying to survive.

Women speak to men as they speak to children, in short sentences of simple words. "If she said 'NO', it was rape." Questions of subtle shading become black-and-white for men's benefit. Between themselves women discuss the complexities. They admit to each other their rape fantasies and mixed feelings, the ambiguity of victims' actions, their sympathy for men's confusion. But when facing men, women link arms and present a solid front. "If she said 'NO', it was rape." They can hardly do less.

Rape for men (like war for nations) is the simpler alternative when negotiations go badly. In most mammals, it is *other males* who prevent a male's mating; ritual combat between approximate equals decides the toss.

A female usually accepts any male who has won the right to strut, though she occasionally evaluates territory he holds. Male combat does no sexual damage to other males; rare deaths break no female heart.

In humans it is the smaller and weaker woman who directly opposes the man's interests. No competitor daunts him like her personal preference; no ritual victory ensures successful courting. A man with dulled courting skills wages a desperate campaign for her heart, seeking the line, the persona, or the prowess that will sway her. Her unresponsiveness proclaims him a failure before all other men, as he competes to win an allegiance he does not quite understand.

Obsession increases frustration. Obsessive focussing was a female requirement for mate selection, and all men inherit it from their mothers. A man focussing on a reluctant woman may convince himself that he can't live without her, as a woman in love convinces herself about a man. Violence and terror may spring from his frustration and her unwillingness, but a stalker feels the same obsession as a lover.

To this mental state we add a physique that could hardly have evolved better for rape. Like all primates we can grasp with our hands, and (as in many species) the human male is larger and stronger. But in most species the female has her feet planted on the ground and can run out from under an unwanted male. Human copulation by contrast often puts the female on her back. Add to this her weaker upper body, and the stage is set. Men know well their physical advantage.

In rape, a man may find immediate rewards and distant punishments. As in raping little girls, a woman's presence and vulnerability can free acts held just barely in check. Any small failure of training can unlock rape, as can a man's inability to empathize. Small comfort, but it's remarkable we don't rape more often. Violent disagreement between men and women rests always near the surface.

Women And Competition

Into an arena of male frustration and incapacity, women brought exactly the wrong tools. Rape was not an evolutionary goal for women, but they, like men, evolved for their own gain on their own path with rape waiting at the end of it.

Women discarded cyclical sexuality when they bred men who were numb to estrus. Making all men dull gave women no evolutionary gain; dull men prevailed only because as individuals they provoked women less. Our mothers did not, as it happens, then evolve heightened estrus scents to attract duller males. This may measure passion's importance to our forebears: A woman's need to attract a stranger had less evolutionary weight than her desire for a mate who stayed home. By preferring men numb to cycles, women doomed estrus.

Estrus originally evolved not to lure one male but to announce a female's availability to all. Her choice then follows one of two paths: If males fight each other she accepts any winner, and if not she judges them herself. In only a few primates do females seem to evaluate already-victorious males, probably with trivial effect. Furthering her matriline and cultivating useful friendships appear far more important than mate-choices in a female primate's long-term success.

Love burned women's competitive bridges. Women sacrificed matrilines by renouncing practical alliances with women in favor of passionate bonds with men. Constantly attracting all men in hope of getting an alpha became a woman's strategy. Being partly an addictive delusion, love let her imagine alpha-ness in any man and weakened the reliability of her choice. But whether or not a woman could objectively evaluate men, she still gained by luring them so the best, if she was lucky, would choose *her*. So as men's youth-fixation grew, she evolved to look always younger.

Once evolved for periodic hyper-attractiveness, women were now handicapped if they showed cyclical variations. Consistent attraction was vital in winning and holding men who were, after all, monogamous partly from obliviousness. Any cyclical signals would (by their absence) show non-availability, and cost her a man's attention perhaps at a critical time. A woman held her mate best by stable attraction, so he would daily be thinking this could be his night. He wouldn't learn otherwise until too late in the evening to wander.

Once women bred men who couldn't detect estrus at a distance, male dullness bred estrus' last traces out of women. When a man could no longer sense estrus at twenty paces, and stayed home for lack of sensing another female's state, his mate had to hide her own estrus at three feet. If she displayed any cyclical allure he would know it, and would leave on the evenings she seemed less attractive, to be snapped up by some other who might hold him better. Women suppressed their part of estrus not to make sex continually available but to conceal the times when it could cause no pregnancy, not to attract all men but to avoid losing the mate they had.

Lacking cycles then, women must compete by advertising constant sexual possibilities and choosing from men who court in return. Women don't always want sex and may feel crummy even when attempting to look alluring. But in the face of other women's efforts, a woman wanting to find or hold a mate can't afford to do less. Women do not, I think, compete to be the best that men can see, so much as they watch *other* women to avoid being left behind. Fashion runs not on men's likes but on women's fears. By selecting dull mates, women compelled their own endless competition and men's endlessly hopeful attention.

Wearing a short skirt does not announce readiness to be raped, any more than wearing a shirt with concentric circles announces willingness to be shot. I don't blame women for their rivalry, but women's competition is one force leading to rape. By always trying to look their best (something no female chimp apparently worries about), women took from men the chance to see them at their worst. Men needed to see that.

Without estrus as a signal to proceed, men sense no time for waiting. Male chimps are uninterested in swollen females lacking estrus scent because they know a ripe female will show up soon enough. Knowing stronger pulls will appear tomorrow lets them ignore weaker lures today. But if no estrus scents ever appeared, horny males would quickly find any female attractive. When rats cannot smell estrus, they rape.

When women evolved stable sexual signals, men had no cyclical releaser for their courting efforts. Seeing no peaks, men could see no valleys. Bred for lack of subtlety, men see no difference between a woman seeking a mate and one shunning all men. Faced with a male mind demanding black-and-white, women evolved a uniform shade of grey.

Rapists and their juries often believe that women asked for it, and more than any other criminals, rapists claim innocence. While some women walk into hazardous situations, none asks to be raped. But at least some of these men speak the truth as they see it. We still need to prohibit the act, and some men will suffer unjust punishment as women suffer rape. Explaining rape is not the same as permitting it.

Women didn't change their behavior, they only smoothed out what had once come in peaks and valleys. The woman who trolls for sexual interest among the men passing by, rejecting all until one impresses her, mirrors her female aquatic ancestor. Emotions became more powerful during our landward move, but the basic tactics of female pairing remained the same. Women's behavior endangered them only when men lost perceptiveness and women lost cycles.

Women did not evolve to be raped, and I hold any man responsible for his acts. I have been a victim, but not of rape. Some professing friendship have stalked me in the dark; I've faced crowds that did not care much for my welfare. I know something of fear, and I don't dismiss women's fears.

It is pleasant to imagine a world where husbands have no flaws and all rapists are bad. We want to kiss only princes though frogs lurk in the pond. The rapist a woman once thought was her friend must have been a monster all along, and she had only misjudged him. That women may think only bad men rape, or honest men may think a woman asked for it, shows the divergence in our minds. Such views let us deflect responsibility, but frontal sex and the courting wall together created dangerous pressures under which any man may fail.

Rape as Aggression

Many species have frustrated males, but no rapes. Killer-ape theorists suggest we became hyper-aggressive, and this necessary adaptation spilled over into our sexuality as rape. Woodland chimps had the luxury of shyness, but when we left the trees to become terrestrial only the most-aggressive could survive. Some people find enormous satisfaction in this image of dangerous humans.

Rape (in this view) is an outlet for heightened male aggression, broken eggs in a Mighty Hunter omelet. Only the smaller and vulnerable female poses an obstacle to male sexual interests. A male who forces a female into submission can both gain sexual access and re-enact his own daydreams of the conquering hunter. She wouldn't confess her rape fantasies, but the female's secret thoughts of being overpowered may mirror the male's wish to prevail.

Some would reverse this, arguing aggression follows from rape rather than rape from aggression. Morgan thinks the semi-aquatic ape, attempting terrestrial copulation, faced increasing difficulties from the female's deepening pelvis. In frustration a male might have flipped a female over, and despite her protests impregnated her in that pose. Though her objections should have stopped his attack, he ignored her. Over generations men evolved to ignore all surrenders, suppressing our ability to recognize it.

Women (many think) evolved to enjoy what they couldn't stop. If all sex is frontal, those women who protest the least will make the most babies. So women first evolved to tolerate and then enjoy the pose. Those who like sex more (or hate it less) compete harder for the Mighty Hunter who brings the provisions.

Female willing submission then becomes one of the sexual rewards for the hunter. Evolved for heightened aggression, he enjoys seeing surrender at home. Frontal sex is also personalized sex, with identity apparent during the act. So females who submit to the pose have a better chance of holding the attention and fidelity of the hunter whose aid helps raise the children.

Classic Freudian thought is close to this. The vagina, Freud thought, must be erotically awakened by forceful sex. To help them tolerate it, women evolved to enjoy pain. They also evolved a rudimentary transfer of eroticism from clitoris to vagina, with its unreliability explaining women's low orgasmic rate. Over generations women came to accept men's powerful embrace and violent penetration as enjoyment. Freud's compliant woman then might foster heightened aggression in men, encouraging rape to persist.

I cannot accept any of this.

We evolved in spite of rape, not for it. I will later suggest we evolved not hyper-aggression but hyper-bravery. Rape does not awaken a woman but wounds her, does not help form relationships but often destroys relationships that might otherwise have grown. We rape not because women's bodies constantly say "Go ahead" but because women's bodies say nothing at all, and any man some of the time (and some men all the time) may not understand "Stop".

When a woman's spoken refusal fails to dissuade a man she may think he planned on rape from the start. Perhaps he did. But spoken words are by nature vague and fluid, and female self-repression means a male must get past several preliminary refusals before he hears acceptance. Bred for lack of subtlety, verbally limited, and growing accustomed to refusal, men become deaf to "No" when aroused.

This may help explain the orang, the only other primate who rapes in the wild. Orangs hang from limbs by their hands, and copulate in that pose for long periods. This is necessarily frontal, and has accustomed them to frontal sex. Males are much larger than females and routinely frighten them. The combination favors a male who (as field observers report) ignores female half-hearted protests. But it also makes the male unable to distinguish a true objection from a sexual invitation, though the female knows her feelings well enough.

When a woman's pleas fail to deter an advancing male she may see him as hyper-aggressive, but this supposes that the male himself thought he was attacking. Worse, submissive gestures are also sexual offers for primates, while courting gestures are also aggressive poses. We have enough of those in us that men and women actually play-act dominance and submission as part of courting. If a frightened woman responds less than savagely, the man may get exactly the wrong message.

Courting is a dance with danger, where sexual messages are the loudest, and stress is high. Some can't take the strain without a couple drinks for self-medication. Dating is a cautious sexual approach, shutting down inhibitions and fears until we can achieve coitus. We do this repeatedly, with promiscuity deep in our heritage; we do it badly because both men and women have ill-made instincts.

Women want the same things ancient mothers wanted: A dominant male and sexual relief. But frontal sex and competition's breakdown in water meant women had to invent new rewards in intimate contact. Women evolved the fear-thrill and obsession, making them want men regardless of orgasms while disabling women's objective judgment. Sane people avoid the risks of mating; love-hunger pushes us to continue the species though sex has lost luster.

No longer sending signals men evolved to receive, women compete to evoke a courting response from the world's most dangerous animal — a frustrated male. Men lost the skills to court and the ability to detect female readiness. Men lost the subtlety to detect female intentions, and most signals women send are never received. No wonder dating sometimes makes us want to throw up.

A rapist is not necessarily a psychopath, though many psychopaths rape. Alfred Kinsey, Paul Gebhard, and other Kinsey staff found rapists to have an essentially healthy heterosexual adjustment. Menachim Amir finds rapists unextraordinary, if violence prone. To many experts the rapist seems an average man. Brownmiller (1975) says this shows that experts have questionable standards of heterosexual adjustment. To me this shows how hard it is to see a rapist as normal when his act is outrageous. Rapists are not inherently bad men any more than only bad women have rape fantasies.

We want to consider rapists monsters, because rape seems a monstrous act. Yet studies find only minor accidents of circumstance and aggression make one man a rapist and another not. Between any one man and woman, rape's absence demonstrates only that training and empathy work well enough, so far. It is not inevitable that any one man will rape or that one woman will be raped, but inevitably *some* will.

Only recently has sexual predation become a significant danger to women, because only recently have cities become large enough for anonymity. In small tribes a rape was never secret, a rapist could not hide. The !Kung don't even have secret lovers, for all members of the tribe know each individual's footprints. Any tryst leaves its participants' identities behind. Sexual predation comes from large populations and privacy more than from modern mentality.

Because rape in a small group is no secret, the rapist can't repeat it. The unlikely chance of procreation on his first attempt can't outweigh the certainty that all women will then shun him like the plague. Since he can't get away with it twice, rape is unlikely to be inherited behavior.

Rape is not a reproductive strategy but an orgasmic tactic; not a way to deal with women but a way to avoid dealing with them. It serves a man's short-term erotic interests but sabotages his long-term genetic gains. It did not evolve by its own value but splintered off a rough corner where men and women met misaligned. A cigar, Freud said, is sometimes just a cigar. A rape is *sometimes* an angry attack or a way to cause fear, but is often just sex done badly.

Rape dates back to the earliest male clumsiness. Raping an adult woman, like raping a child, is what a man might do if he cannot court a willing female. Not an outcrop on a level plain to be explained away on its own demerits, rape is the surface sign of a deep structure where continents collide.

Murder

Murder is mistaken infanticide, not hyper-aggression. While rapes reflect fringe men trying to bypass the courting wall, murder is a bruise in the earth showing that the courting wall has rolled by.

Infanticide is a special kind of killing. In many species males kill unrecognized infants, in a routine competitive tactic that erases other males' genes. Infanticide can also quickly bring females into season, increasing a newly-arrived male's reproductive odds. Not an act of rage but removal of a genetic irritant, infanticide is as passionless as swatting gnats.

Infanticide bypasses surrender instincts. Many species engage in ritual combat for breeding rights, and use instinctively recognized surrender signals to halt it. Once the appeasement signal is given, the aggressor is unable to attack further (though occasionally this response is flawed). Primates have such signals, including raised hands to show non-aggression, and bending down or lowering their bodies.

Mammalian infants typically grovel before adults as part of their socialization. Adult surrender poses often develop from infant fawnings; a subservient adult acts like a juvenile facing an elder. For infanticide to work, the attacking male (or female) must ignore submission signals which would stop the attack if the victim were older. An infanticidal male is not sociopathic; his surrender reflex works when it is an adult who yields. Infanticide requires a disconnected instinct for a special form of killing.

Infanticide is common in primates. Males are often brutally intolerant of competitors' young, though helpful to their consorts' infants. Males under attack by other males often pick up infants. Once this was thought to be protection for the attacked male, the infant defusing the other male's aggression. Study now suggests that males are picking up their consorts' young, to protect the infant from infanticide in the confusion of a fight.

Infant killings disturb our picture of helpful and social primate groups. Rapid attacks and fatal wounds are not necessarily obvious, but infanticide seems widespread. It is common in monkeys and great apes and may be the greatest danger any primate faces, accounting for up to a fourth of all gorilla deaths. In one langur troop, 46 infants disappeared after a new male arrived.

Our relatives' promiscuity owes much to dead infants. Sexually insatiable females have more consorts and lose fewer young to strangers' attacks; they reproduce better by having more friendly males, not more sex. Promiscuity comes from infanticide as the gazelle's speed comes from having cheetahs around. Women's multi-orgasmic capacity suggests our lineage has coped with infanticide as a mainstream behavior for millions of years.

Realizing that an individual is an infant is enough to trigger the act in most infanticidal species. (In social species, the male must also be covert. A grizzly might kill an infant on sight; a chimp will hide his act.) Along with a smaller body, two things distinguish all young mammals: They lack mature odors, and have reduced muzzles and relatively larger skulls. Both of these characterize humans, and both came from our sexual collapse.

Mature odors let a male detect sex and often status. Males rarely attack a strange adult female, and instead evaluate her as a sexual partner. Males size up strange adult males and attempt to establish their relative rank. But an unrecognized youngster without adult odor is not a full person, and without value to the male who meets it. A male can safely kill any odorless stranger, for it must be some other male's child. Odor's presence is a safety-catch, whose absence unlocks attack.

We evolved a suppressed smell sense to keep males oblivious to estrus. Humans have perhaps the least-sensitive nose of any primate, and this handicap alone makes it unlikely that we evolved as hunters. Only in the human species can an adult female pass as a juvenile male, a disguise we take for granted. Unable to detect maturity by odor as other animals can, we live in a scent-vacuum and have to *learn* what makes an adult. In us, one of infanticide's safety-catches is permanently disabled.

Infant skull shape sends a more-powerful signal. Small muzzles and bulging braincases characterize all young mammals, but female competition blurred human boundaries between infant and adult. Men inherit juvenile skulls from mothers, so we all now have the shape of other primates' infants. Every human matches the form we've been freely killing for forty million years, a form that disconnects appeasement gestures while it triggers an attack. We constantly see our innate caricature in the adult faces around us, and kill under the right circumstances because our victim resembles an infant.

Men kill women at astonishing rates, which we attribute to many sources of frustration and rage and cowardice. Though some think men attack as part of a campaign to maintain supremacy, I doubt that men are so organized. But women's deaths by mistaken infanticide exactly matches my thesis. Perhaps men attack women due to misrecognition.

Women unknowingly blurred maturity to keep men courting. Females keep childish voices into adulthood, remain small, and can easily disguise themselves as children. In at least some cultures, men often refer to women in infantile terms ("baby"), which many women reject as a belittling label from male-supremacist minds. But it may instead be a surface sign of men's subconscious recognition, verbal evidence that women's camouflage works.

Men evolved to court girls as a weakness in the courting wall, while women evolved to face dangers to gain love. When men and women court, they both seek precisely the combination of privacy, vulnerability, and opportunity where infanticide occurs in other primates. The courting couple risks an unexpected hazard beyond their comprehension. Sexual dimorphism that helps make courting deliciously dangerous for the woman also signals vulnerability to the man. Privacy they both seek may unleash a male response even *he* has no reason to expect. Under the right circumstances a man may kill his baby, and never really know why.

Whether men or women are victims, surrendering doesn't work well for humans. Morgan says we ignore surrender because we evolved as rapists; Lorenz says we haven't had time for new surrender instincts to match our new weapons. I suggest we can shoot a grovelling person because we are never far from an infanticidal mindset, where surrender-recognition disconnects. Killing each other does not demand hyper-aggressive killer apes; we are not more-aggressive, merely less-controlled.

Murder, like rapes of children and of women, suggests a hidden structure. Mistaken infanticide grew from neoteny, and neoteny came from male reproductive strategy. While rape recapitulates our impact with the courting wall, murder is a dent left where the courting wall rolled over.

The Sexual Mold

We are evolving now, oblivious to forces that change us over the generations. Men's female-child caricature is the vital product of our recent evolution, and we continue to focus it more precisely with each passing century. By shunning rapists we smooth men's rougher edges, nudge them toward the optimal female age, and show the direction we are evolving.

Men and women were not dropped from a clear sky onto a level plain and told to make the best of it. "Here you are. Work it out. Try pairing!" Each sex evolved along its own path, obeying selective pressures the other didn't always share. Each became part of the other's environment, forcing the opposite sex to evolve in turn. Men who sought accessible girls bred children who kill mothers in childbirth; women who spurned unreliable men bred rapists.

We cast men's sexual caricature in a mold of women's design, together making the largest artifact humans will ever build. Women's fears and self-repression led men to evolve an innate image of a young girl. Men's caricature then shaped how women evolved, but also became a key component that now saves us from extinction. Gorillas, chimps and orangs are dying partly from the lack of it.

The caricature I have proposed could evolve only under pressure. Treating all females as potential sex partners brings such clear advantage to all species that a narrower sexual focus could evolve only if it brought real benefit. If the caricature is real and my theory has it right, then rape and murder are reasonable products of the same force that created the caricature, and they align as a mountain range. Or see it the other way: If rape and murder come not from individual evil but from broken instincts and failed communication, then they suggest the caricature does exist.

We have been armies passing in the night, hearing the din of battle in the rustle of sheets and whisper of silk underwear. With smiling faces, uncertain hearts and gritted teeth, we search for love and sex. We know what we want but not why we want it. Women want emotional rewards without knowing their orgasmic sacrifice made emotion a fair trade. Women of all ages compete to draw men, not knowing why young women so often win. Men want young women without realizing how their grandfathers' dullness led the preference to evolve. Men compete for female attention lacking the competitive tools women evolved to appreciate, the tools women took away. Human sexuality is truly a mess.

Passing armies, as tectonic plates of emotion, shaped a deep formation in our mental earth. Opposing sexual pressures sliding against each other created a vast convolution in bedrock. Each sex was bent by the other, each was a force bending the other. Rape and murder are lumps on the plain above, giving surface evidence of a deep collision. Rape and murder, like the height of mountains, indicate the pressure far below where diamonds form.

We can map the battle as in a snapshot, where men confront a three-sided pen of copulatory possibilities. A man who wants sex with a female human has three choices: He can rape a woman, rape a child, or persuade. Rape is a refusal to play by the rules; cheating for orgasms. When a man playing the game courts young women, he encourages neoteny and accidentally fosters murder as mistaken infanticide. Only courting a contemporary causes no evolutionary change.

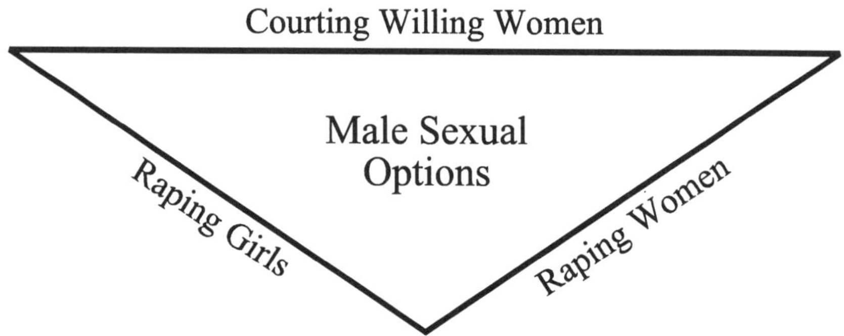

Figure 28: Male Sexual Options

This narrow three-fold choice defines man's sexual universe. Men face their choices today with the same desperation and hope as men a million years ago. Their options don't flex because they arose through deep pressure over long time. Men can have no new choices until women change drastically and women have no reason to change. We travel bound together, bearing sexual baggage, trudging an old evolutionary path.

The Moving Barrier

The following figure sacrifices subtlety for clarity to define a feminine psychological space. The top represents enthusiasm and the bottom reluctance. The time line defining a woman's lifespan stretches left to right. Within that space a diamond circumscribes the range of female sexual responses, the left half before puberty and the right half after. Lumpy boundaries reflect individual variation.

In our Miocene ancestors, male sexual interests focussed on the upper right, corresponding to the enthusiastic promiscuity of estrous adult females. Within this space, males had no preference for female youthfulness. Estrus cued sexual interest, and a female's odor and popularity, not her age, made her desirable.

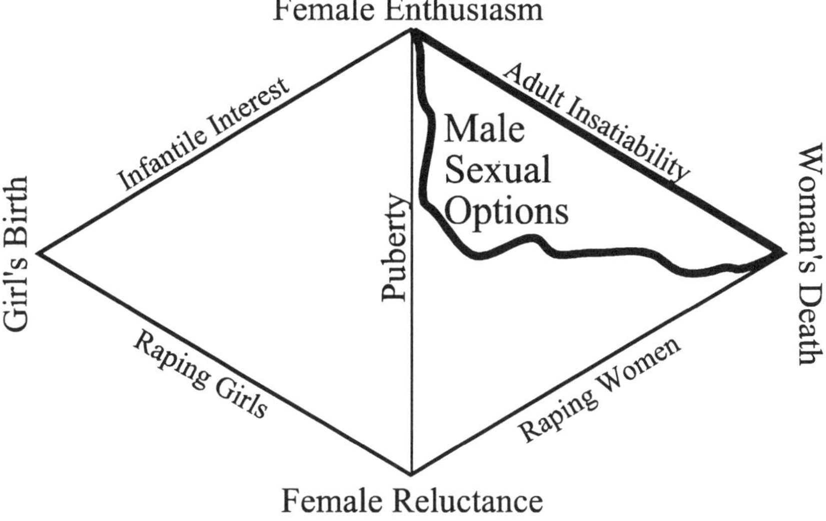

Figure 29: Ancient Sexuality

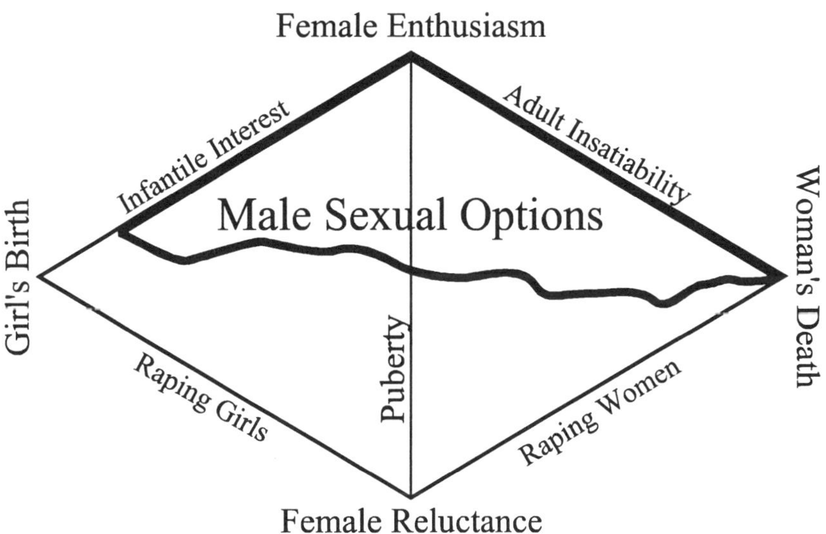

Figure 30: Aquatic Sex

Life in the water masked estrus, and male desensitizers evolved to ease frontal sex. Male sexual interest spread out to include all willing females because aquatic males couldn't recognize female maturity by odor. Prepubescent female primates are always interested, they merely lack odors that evoke a terrestrial male's response. So in water, a female's sexual contact with males spanned a larger part of her life, including before her puberty.

When we returned to land estrus became clear again and males refocussed on mature females. Only passionate women could tolerate scary frontal sex on land, and female hesitations pushed love's evolution.

Passion was incompatible with promiscuity, and estrus collapsed under the strain. Estrus' died when women bred terrestrial men who couldn't detect ovulation's cycles, as men had earlier been unable to in the wet. Male interest again spread out to include any willing female including girl children, but women had shifted toward reluctance and protectiveness. Feminine doubts and male dullness built the courting wall.

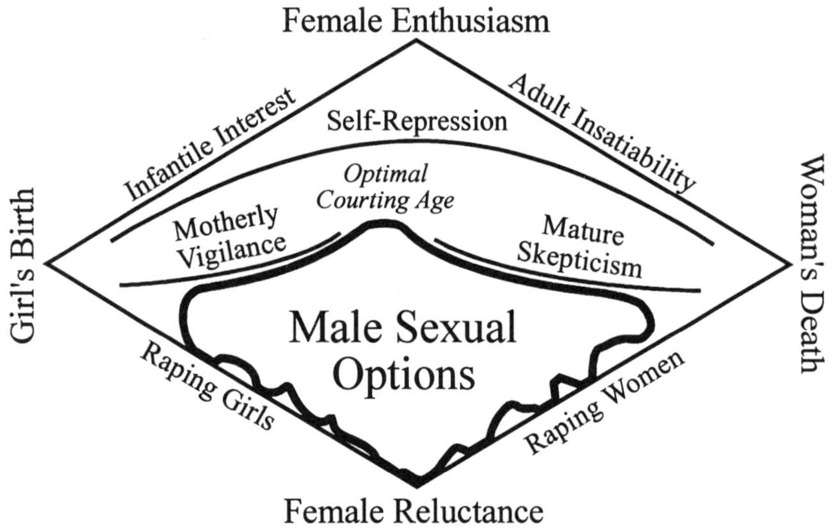

Figure 31: The Courting Wall

The wall men face is not a single feature, but a barrier composed of two separate feminine reactions. Because women see danger in sex, they protect daughters against male contact. And each woman's emotional bruises cause her to become increasingly difficult to court as she ages. At some point between these two is the lowest female resistance, and therefore the optimal female age for a man to court. The Female Optimal Courting Age is not an evolved feature but a weak interval between two barriers.

Men who targeted that age gained a competitive advantage, and evolved an instinctive caricature to help focus on it. Men's obsession with young women is a product of women's hesitations, a shadow cast by feminine needs. Once evolved, male obsession reduced reproductive prospects for older women who *were* willing, and women began retarding maturation to counter men's obsession. Neoteny then evolved as a shadow of men's instinctive caricature.

Female sexual responses created boundaries around male behavior; that male reproduced best who could function within the limits. As the courting wall slowly grew from early aquatic hesitations, men faced and surmounted it as best they could. Reproduction depended not on rape but on repeated successful courting, always seeking the weakest resistance. Males focussed on the least-resistant females and the most-effective persuasion, always facing forward as soldiers in battle.

The FOCA created pedophilia, as a similar gap at an advanced age would have made old women beautiful. The FOCA's existence meant that some men would find it, and would pull all men toward it. The absence of any other preference, as in a bucket of white paint, gave incompetent males the power to taint us all with their drop of color.

Our pre-aquatic Miocene ancestors didn't rape. The most brutal males needed no force; sex was too easily gained to make assault worth the trouble. Willing and insatiable females assuaged the entire spectrum of male lusts.

With the first faint aquatic hesitations, the boundary of female eagerness began tightening. Promiscuity declined; courting and acclimation became necessary. Men could not simply push back the courting wall (creating more-accessible women) when female doubts removed promiscuity's extremes. Yet hesitations remained low in the water, making sex only a bit harder to get, and elaborate courting sufficed. So men responded with singing and language, waves thrown up by the advancing barrier of female hesitations, a hopeful rear-guard parry by men's retreating army. Back on land the courting wall advanced and grew higher, built by motherly vigilance and mature skepticism. Men backed up further, evolving to deal with increasingly reluctant women. Men varied, with some more successful and some finding no mates. Some women were relatively promiscuous, some would accept no man.

Always female preference was male desire's sole opponent. At no time could men's evolution shift to rape. Only courting brings long-term reproductive advantage; only concentrating on the feminine emotional obstacle could teach a man how to surmount it. The slowly-advancing courting wall and its weak point trapped and mesmerized the men in front of it.

Facing forward, men edged backward until they were pressed into a corner, until they bumped unseeing into rape. Sexual assault is not a reproductive option or a direction men sought to move. It is the random impact of individual men, as the wall squeezes their crowd into the corner where once they had the whole world for maneuvering.

Trench Warfare

Rape's prohibition is the vise that holds men secure while evolution sharpens their obsession for girls. Traumatic for the individual, rape for the whole species is a trivial act but a valuable measure of evolutionary pressure, as whistling steam wastes a little water while telling us that the hot and pressurized container is otherwise sealed. Punishing rape validates the Female Optimal Courting Age while announcing that other avenues are shut.

As childbirth shows female camouflage's evolutionary limits, rape shows the limit of male courting. A woman is trapped between the need for a small pelvis and her daughter's need for a big head. With individual variation some women will die in childbirth. Their deaths reflect the competitive pressure's continual presence. Our male population, with individual variation, will contain rapists as long as courting is a challenge. Courting problems do not excuse rape, but help explain it.

Rape is a symptom of confinement. When John Calhoun penned rats in dense populations they began to rape and kill each other. Rats didn't evolve into more-violent rats in a generation or two. Instead, normal rats became violent when grouped too tightly, as do humans in cities. Healthy animals' misbehavior when confined suggests that confinement itself creates pathological reactions.

Human rape shows our psychological confinement. Men's sexual outlets once included females so insatiable that males chased them away for some rest. By a wave of the arm, infant and adult males could obtain sex from equally enthusiastic females. Now a man's only acceptable option is to court women doubtful about him and sex, women who evolved for courting styles men can no longer easily manage. Male sexual options, like rats, are trying to survive in a very small space.

Rape is not an illness or a strategy, but a symptom. Not a random outcrop on a level plain, rape is a mountain range where some violent men impact the boundary of all men's confinement. Prohibiting rape of both women and little girls makes a wedge-shaped socket, holding men firm in a vise of permissible behavior. Men press in the corner where they backed up, facing forward at the courting wall, their attention inevitably directed at the Female Optimal Courting Age.

Rape in humans parallels victims crushed in a stampeding crowd. For individual families the tragic deaths will haunt them for decades, and some will never recover. But for those who learn of it with detachment, the deaths prove simply that the crowd was too confined. Too many people, too small an exit.

Sexual assault announces that the FOCA is not a break but a weakness; not unguarded but less-guarded. Young women are not *inevitably* accessible, only *relatively* accessible. Rape's presence means even the courting wall's lowest point makes a barrier that keeps pressure in the kettle. Too many men, too small a door.

Sexual assault validates the FOCA. Long before the first rapes, the courting-wall's low point gave dull men their best avenue to sex. For several million years men evolved in tightening psychological limits, with success going to the man who persuades best and frightens least. As the courting wall slowly closed in, success continually went to men who negotiated the permissible passage, though not all men needed the easiest path. When rape became a constant danger, it reflected only how tightly the courting wall had pressed men into a corner. All that time, the Female Optimal Courting Age remained the male target. With male obtuseness and sexual fear, the FOCA sharpened and focussed.

Male obsession with young girls is the toughest artifact of our species, stronger now than the grey matter where it took shape. Women accidentally formed men's minds into this shape, and we held men there for millions of years by shunning rapists. As the metal set, the mold became dispensable, yet the mold persists. Starting as a helpful competitive eye in men, the Female Optimal Courting Age became an evolutionary force on women.

Retarded maturation came from the FOCA. Once girl-child preference took shape in men, ground to a point by thousands of generations, mature women lost reproductive chances. Women countered male obsession by retarding their own maturation and neoteny became women's physical reflection of men's psychological confinement. When men's confinement was snug enough to cause women's brains to grow, it had long been tight enough to cause rape. When we hold a big-brained hominid fossil, we have found a species that struggled against sexual violence.

Neoteny defines us. Some point to its value in forging closer family bonds, or to the benefits of long youth when we have much to learn. Many conclude we became neotenic specifically *because* of these supposed advantages, with larger brains and lifelong curiosity our most-vital property. But I repeat my earlier question: Do we really benefit from neoteny or only endure it?

I will soon describe how our retarded maturity helped create our unsuspected treasure, rescuing us from a fearful plague that now kills chimps and gorillas. But human retardation evolved beyond that of our cousins not to make us smart. Brains grew just a little in the water, and only to prevent drowning. When we returned to land and male courting skills collapsed, retarded maturation was the individual female's reproductive camouflage and not the species' route to a powerful brain. When neoteny evolved, we endured it like being slowly cut by a rough diamond.

Basic Training

Like Australians we all descend from criminals; the sins are now too dusty for us to reassign the blame. Women never intended to limit men's options; they wanted only male behavior making frontal sex tolerable. They accidentally bred males with fragile controls, easily broken and more dangerous than we know.

We live in a much smaller space than we realize. Our sexual constraints seem so reasonable when taught from birth, but would suffocate our ancestors. Women, I repeat, are not jailers, and suffer confinement along with men. Women might like a good sex life without emotional encumbrances, but on average cannot be so free. Together we suffer a tight but brittle confinement.

During Army basic training at the Viet Nam war's height, I watched normal young men beat a fellow soldier with entrenching tools. Most of them were away from home for the first time, just out of school. They'd have laughed if I'd warned them, standing in their high-school hall, that they'd shroud a cowering boy in canvas and club him until the metal rang. I never dreamed of watching such a thing. None of us knew how pliable we were.

Their victim was mildly retarded, I think, and friendly as a puppy. He had no idea what was expected of him, and in a battle he might have endangered himself and others. Terrifying him into deserting seemed easier to our sergeants than doing paperwork to discharge him. They weren't evil, but they weren't gifted.

Our instructors began punishing the whole company for his trivial infractions. So these normal young men terrorized him to make him leave, while they took up a collection for his ticket home. I tried to argue his case, short of protecting his body with my own. But young men can be aimed like a weapon, and for all my words, I might as well have been bound and gagged.

He knelt in a corner while they beat him. I was close enough to hit him if I'd been weaker, or shield him if I'd been stronger. With no effort I see again in my mind the shelter half they covered him with, the soft thud of boots on flesh, the ring of a shovel blade hitting bone. Perhaps if I'd tried harder I could have stopped it. My efforts, poor as they seem to me now, at least earned me a beating of my own, later on.

We were all closer to our confinements' boundary than we knew. My fellow soldiers weren't bad, merely flexible. Like a boy or girl on a date, they didn't know what it would feel like until they got there. They didn't become violent people in the Army. Something had been inside them all along, just under the surface, ready to beat a helpless victim. With a small change of training they could have let him be. With a small change of circumstance it could have been a rape.

Human courting is an explosive experiment where we disable already-fragile hesitations to allow procreation. Rape stays always near the surface when sexual options chafe so tightly. Sexual assault is not only the typical act of a bad person but also a bad act of a typical one. Men and women need to know that this barely caged predatory instinct always rides uninvited in the back seat, with only training holding it at bay. Men and women share nothing so intimately as their confusion and peril.

Male Bashing

Proving men's dullness took me by surprise; I began with quite a different story in mind. I also never expected to find rape so near the surface when I started looking at passion. Usually training works; few men commit the crimes that any man might do. But I must admit with a sheepish grin that men often do need short sentences and simple words.

I came upon men's handicap by accident and picked it up against my will. The path laid out here I first walked from the present going backward. Brain growth timing and our small IQ advantage both show that high intelligence can't explain our evolution. But our very large brain does something important enough to risk the mother's death. Most agree that our large brain shows retarded maturation, so retarded maturation must have value. This launches theories about interdependency and long learning times.

To me it seems more direct to say that we simply look juvenile, so perhaps *looking juvenile*, by itself, brought some evolutionary gain. That would explain why our neotenic features center in the skull, being the primate's basic recognition focus. It took me a while to consider that.

Looking young brings to mind men's preference for young women, closely paralleling the hamadryas male's harem of infant females. Collecting young and tractable females makes sense for polygamous males but less so for monogamous ones. If one chooses a single female it seems wiser *not* to choose a young one, with fertility and mothering skills both untested.   If men have sought girls long enough to force our evolution in that direction, then some advantage must come from men's interest. Some suggest men pick girls to gain more of their fertile years, but men who seek young women rarely keep them. After mulling it over, I realized they might simply be easier to court; perhaps men evolved to seek not *fecundity* but *accessibility*.

Singing and language resemble the useless but elaborate behavior many animals use in courting. We can imagine their value for an aquatic ape male, yet men are clearly less-verbal now than women. If these began in courting (and therefore probably in men) then perhaps something made men less mentally agile after our aquatic time, leading them to court the least-demanding women. Eventually it dawned on me that dull men might make more-faithful and so more-tolerable husbands. I never started with that in mind.

Many argue the rewards of intense frontal sex led to long-term pairing and other human advantages, and love evolved to help make frontal sex pleasurable. In rethinking our transition out of the water, I saw love resulting from culling; passion implies *absence* of simpler pleasures. Pairing is not a hunting-ape's superior social tool but a record of emotional dependency in an inorgasmic female. Even if it makes a poorer hunter, a little obliviousness helps because it keeps a husband home.

Dullness in men creates courting problems in sons, and explains the male inability to keep up with female conversation. This illuminates our preference for young women, and shows why brains would become large (though inefficient) as camouflage. As major pieces of the puzzle slid into place, I was enraptured as by a symphony. My friends grew heartily sick of hearing about it.

Finally I could see the courting wall and its weakest point and the hidden structure. By chance I found a prize on the way, a cosmic gift in our dusty basement rubble. Soon I can show it to you, but I first had to lead you along the path in the proper direction.

I wanted only to see how the pieces fit, not to prove that we all inherit dullness. Unhappily, my model implies that we had a potential now lost; an alternate humanity which, if realized, might have scorned us as we scorn chimps. I was sorry to see this, and wish we'd kept better brains. Intelligence is clearly a disposable asset in humans, and when it caused trouble we shed some. But, like seeing senility, it is tragic to feel what has slipped away.

Foreign Femininity

Women's sexuality similarly carries no moral message. Not a superior emotional bond, passions are hyper-evolved descendants of personal preference and fear, grown by default when other rewards quit working for us. Passion, first evolved in women, is an addiction useful to the species where free choice led to celibacy.

Passion helps because women cannot have what they want. If their ideal man actually existed, they could pick him coolly as buying vegetables in a store. Instead women must choose from the men at hand, all flawed by cumulative effects of prior women's choices. Like any other hunger, passion clears the shelves regardless of the quality of the goods.

We could avoid much turmoil by genuine and innate monogamy. If after mating we stayed together for life, male wandering would not occur and male dullness would have no value. Courting, not staying home, would have retained evolutionary emphasis and we would still be a nation of poets and singers. But promiscuity flows deep in our genes, dull men angered women less, and here we are.

To lure men, women gave our species sexual signals that remain always the same. Part of their uniformity comes from men's inability to detect ovulation, part from women suppressing cycles. The whole belongs to our species, not to either sex. Though women are unconscious of these signals and may disavow any intent to send them, females cannot escape evoking male courting. Ancient primate promiscuity affects the most chaste of women and how they dress, though they deny it or don't know it.

But real promiscuity died, and the hairy picnickers are long gone. Some men and women are wildly promiscuous by human standards, though tame compared to, say, a macaque, who may couple with four different partners in an hour. The average human cannot tolerate much sex without love, and this resistance first evolved in women.

Women, far more than men, struggle with conflicting sexual pulls. For over 70 million years women's ancestors sought dominance in males, and for eight million years that dominance has evolved into a delicious terror. Yet for their first sex women tend to choose nice boys who feel safe to be with, like the men they want eventually to marry. Women like dangerous men who are safely beyond reach, and safe partners who are not quite controllable.

Men have changed much less than women during the last few million years. Men inherit from their mothers the ability to love, but can enjoy sex without affection much more readily than women can. Women have undergone a revolution of body and soul that continually leaves men baffled. Men still dream of the lusty females their ancestors knew, insatiable and accessible. In erotic films and in men's minds she appears again, as an old man half-remembers a youthful tryst. It is not men's fault that they dream of her; it is not women's fault that she died. She vanished when sex grew awkward, leaving her hesitant younger sister for men to cope with.

__Battles Continue__

Colliding needs of men and women still shape us. Not ancient history, men's girl-child caricature is growing today. Men probe for sexual access through a wall of female doubts. Women's passion conjures male alpha-ness and intimacy where it might not exist. Emotion's impact still builds a deep structure at our points of contact, and the deep structure protects us.

Our existence makes sense only if we hold an asset we haven't yet allowed for. Knowing we have little intellectual edge over other apes, finding no real value in slow growth, we cannot explain our success in the face of our cousins' slow extinction. But we accidentally forged a gem in the deep structure, and it explains why we thrive while our relatives dwindle.

If our misbehaviors truly align, if pressure far below created outcrops on a smooth plain, then the pressure also formed the prize I am sketching. If instead hypermorphic brains came from only useful slow growth and dependency in a hunter-gatherer, then my caricature is a chimera teaching no lesson. Beauty means nothing if it is merely a face men *learn* to seek. But if a deep structure exists and I described it fairly, then men's innate image is an evolved artifact formed by and helping deform the rock.

Rape has evidentiary value by indicating pressure in the kettle, pressure that formed men's caricature. Rape itself did not evolve to help us, was not our path to marriage, is not a reproductive option. It is an individual failure, a group symptom, a valuable clue. By outlining men's walls and showing that the weak point has lured men for millions of years, pressure's symptoms trace men's best evolutionary path. As long as we have shunned rape, we have guided men to evolve beauty's image.

If beauty is an evolved image formed under great pressure over a long time, incorporeal like language but nonetheless real, then we can draw on it as on our ability to speak. We imagine and remember faces more clearly than can any other ape, because men under pressure evolved a hard-wired facial image in their brains. Because beauty's image begins in our head instead of out in the world, we can fashion in our heads other images we need. The caricature is part of our treasure because it does not depend on reality to function, while our cousins must deal with the world as it is.

Humans can measure themselves against any other animal, but men and women can measure themselves only against each other. Only by comparison to a woman can we see the difference between being a human and being a man; only a man's nature shows what femininity means. We scale ourselves against each other as once we measured horses in handspans, using the only measure we have.

Men and women meet at a mental village green, at the edge of each sex's normal behavior. There men ask women to dance, and women go to football games that bore them. We go there to court and be courted. Some live there continuously, but if it feels normal to them they are not normal for their sex. Nearly all can visit it occasionally, but most of us spend most of our days closer to our center, constructing occasional imaginary partners far removed from the real nature of the other sex.

We might all like the opposite sex to live closer to our ideal. Women's self-repression given free rein would ban pornography. Intelligent people argue that depictions of sex harm women, using theories as ingenious as men's arguments that women's bodies evolved solely for men's pleasure.   Where pornography includes violence, I agree that showing it numbs us to real trauma.   Faked violence in any form gains us nothing.

But purely sexual images are only daydreams of men's ancestral sexuality, and their relations with females long-since dead.   Men did not invent pornography to hurt women, though women may feel harmed by it. Far from being a weapon, pornography is closer in spirit to a faded picture found in a shoebox, a reminder of happy times shared by people long gone.

Many disdain romances some women read by the hundreds. Average men cannot hope to match the powerful, intimate, passionate men in those stories, any more than women can live up to the easy sexuality men like to view.   When a woman consumes idealized romantic images she mentally reduces the average man by comparison.   Her expectations may rise and his courting barrier becomes more impenetrable.   We might (but I don't propose to) ban these as emotional pornography harmful to men.

Men and women are each the only *measure* of each other, but neither gives an appropriate *standard* for the other.   Neither is the one correct form of sexuality or a yardstick for all humans, but each is correct given the evolutionary path their sex followed.   We have not finished our journey, and the opposite sex will enrage and frustrate our children as it does us.

Courting gets no easier.   Love remains the dominant force behind relationships, and abandonment still hurts.   Promiscuity continues to be a daunting challenge for men; a little dullness keeps men consistently confused about women's intentions and availability.   Male shortcomings keep men at home, more or less, but don't help them find mates.

Men have no trouble distinguishing younger women and ignoring older ones.   This indicates that men's innate preference still leads them to gullible females, penetrating the courting wall at its weakest point.   It implies that women's smoothing of their maturation curve continues to improve their camouflage. We have not finished evolving. Women smoothed their maturation to counter today's male focus; they will tomorrow need to retard maturity even more for *tomorrow's* men.

Each generation reflects what has worked up to this point. It doesn't much matter in individual cases whether a person leaves progeny or not. What got us this far is the hunger we inherited, getting us off the couch, out of the house, and into the fray.

Procreation is a hopeful catapult, blindly launching possibilities forward over time. When they conceived me, my mother and father could not know if I would have children in turn, fail to win a mate, die as an infant, or be taken to the vet and fixed. Having overcome their own barriers, my parents gave me a genetic package of love and obsession and their blessings. "Here you go, whoever you become," they each said, "this is all I can give you, but it worked for me."

We face dangers harrowing to us but trivial to the species. Tiny collisions at the boundaries of male behavior give us rape and murder. No matter how tight the confinement of men's options, most men can function in that space. No matter how bruising men's behavior or obtuse their brains, most women find someone they can love. Whatever evolutionary path one sex may follow, whatever hesitation or callousness they develop, the other sex will always evolve along with them, trapped together in a massive hidden formation beneath a level plain, ready to put up with them.

Chapter 10 - Richest Of Orphans

Deep forces hollowed an empty place in the human heart, and a gem formed there with men's obsession as one glittering facet. Hollowness came on an evolutionary path leading to loneliness and fear; a cavity grew in our mind along with a caricature of girls. If it happened quickly we could not have endured, but it came gradually and we remain largely unaware of the loss. We evolved into a paradox; fear should have crippled us, but instead we are bold. Our treasure came from the shape of the hollow, and the mind built for passion.

We can map contours for the hollow, and make it visible. Not mystical, our evolution left us with a mirage in the heart, an impression of something lost in the distance. And in a mirage we found what we needed for survival. Humans bear orphancy's scar, incorporeal but not imaginary. We dimly know that our parents are somehow missing, that we have been left too much alone. Our treasure is that our relationships can fill this cavity without our even recognizing it.

Orphancy damaged us deeply. Like viewing the earth from space, we must gain some distance to see the curve of this sweeping wound. All feel the same hurt; none can display an unscarred psyche to make clear (by comparison) the injuries of the rest. No classic hypothesis of our evolution implies such a wound, yet a feeling of orphancy logically flows from the human sexual collapse and our relatives' evolution, from men's girl-child focus and women's response.

Our emotional void stands as the monumental catastrophe and gift of our evolution. Unaware of the pain you were born with, you can't easily sense the mental salve that soothes it. We fit so close around this cavity we can hardly detect it, yet we need it like we need air. Our pain damages us but also defines us. Gorillas and chimps dwindle to extinction while humans thrive, and our survival comes partly from our wound though we never saw the hurt.

Never inevitable, the hollow came from an accident of ape evolution and human brain growth. That growth was one of many paths that might have satisfied the male caricature; the caricature was one of many instinctive responses we might have evolved when courting grew difficult. And we can't know which other path would have doomed us. We had much more luck than we ever guessed.

Orphan Humanity

For millions of years no adult humans have lived. Your mother and father were not quite the parents you needed, just as many of us suspected. Through no fault of their own they didn't finish growing, as you will be incomplete for your children. All humans function as if orphaned, though unaware of their condition. Your orphancy is not obvious.

Grown humans are not adult as mature baboons are adult, but are sexually mature children. We cannot even *define* adulthood for humans, because we have no genuine adult humans to compare to the rest of us. We cannot directly experience the presence of an adult mind, but we can guess what adulthood might be by measuring what we *feel* we lack, like detecting a moon by measuring a planet's wobble. Any evidence that we seek or invent adults implies that we feel orphaned.

Our delayed maturation pushed adulthood off the end of human life spans. Delaying maturity in water gave us flexible lips, smaller canine teeth and buoyant brains. Women delayed physical maturation for sexual camouflage on land, pushing our adult form further toward old age. With retarded growth satisfying these separate requirements, the entire human animal slowed its physical maturation far beyond any other primate.

As our maturation rate slowed, our lifespans increased, perhaps doubling over the past three million years. We now live about twice as long as chimps and gorillas. Increased lifespans probably didn't matter in themselves, but slower development slowed aging as a side-effect. Our maturation delay focussed on anatomy; mental shifts and lifespan changes went along as accidental passengers.

Barring value in disconnecting them, physical and mental maturity proceed together. For the dog, bighorn sheep, and African buffalo, delays in physical maturity extend juvenile behavior. It worked the same for us, and most anthropologists credit delayed mental growth for our lifelong inquisitiveness, breadth of knowledge, and superiority. But humans did not retard their growth *in order* to delay mental maturity; women retarded physical maturity in order to compete.

Retardation delayed mental maturity accidentally, so we never become fully adult and (like our dogs) always have a bit of puppy in us. Since we have no adults, we can't quite grasp adulthood's nature. There has never been an adult human, and we communicate with other species too poorly to ask them how it feels.

We might show our conundrum (and sketch a single adulthood facet) by subtracting curiosity, which most agree is a juvenile trait. Adulthood contains no curiosity. Adulthood does not mean a state of dull resignation after deciding that answers are beyond reach, nor smug dogma claiming that answers have been provided. Both acknowledge that questions exist and answers matter. Adults of other species have no questions; all things necessary are known.

To have nothing more to learn lies beyond our grasp. We can imagine many realities except the one that lacks imagining. In real adulthood, questions are not forbidden but nonexistent. Though it sounds constricting it must feel normal to adults, and adulthood must contain other things beyond our dreams. Though I can't say what adulthood feels like, we have evidence we don't attain it and feel something lacking.

We had adults as parents for the first ninety million years of mammal evolution. As we evolved into prosimians, then primates, then apes, from forty million to twenty million years ago, adults always knew the answers. For the whole of our brain's odyssey from reptilian beginnings, a core of stolid adults formed each social species' stable base. They might sometimes be startled but were never amazed.

For thirty million years we evolved in treetops, infants and mothers clinging to each other. A moment's inattention by either could end the baby's life. Any infant who had a passionate need to hold its mother found survival; any who had less interest was ruthlessly culled. All mammal infants want their mothers near, and for thirty million years primates evolved obsessive intensity in this bond.

We needed parents for purely pragmatic reasons, and evolved strong emotions to make us cling to them. A baby chimp at five months old may let go of its mother for the *first time*. For thirty million years, whenever we held out an arm, there was our mother at the end of it. Primates, alone of all creatures, find comfort in holding another by the hand. If distressed or left alone, primates clasp their own hands for want of a better partner. Just as clasped hands evolved for parental fur, primate minds evolved for parental protection.

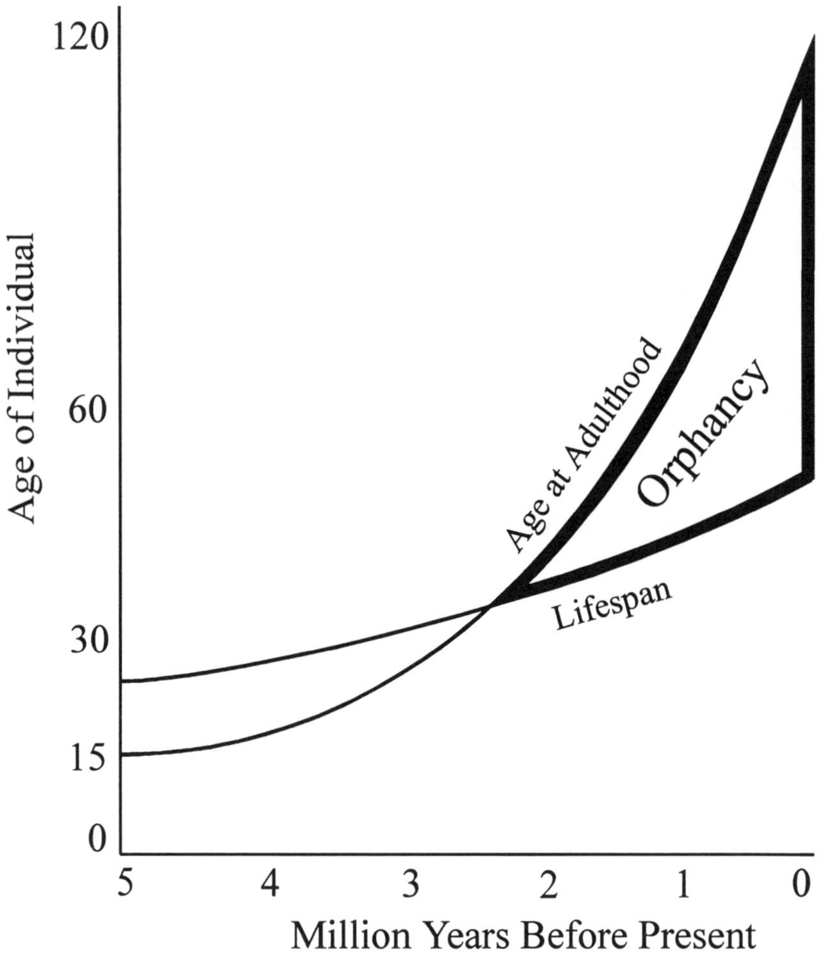

Figure 32: Lifespan and Adulthood

Brains evolve to meet their world's demands and ignore what they don't need. A baby sea turtle makes its own way from birth; it has an adult brain right from the shell, knowing nearly all it ever needs to know.

But a baby primate has parents to protect it while it learns how to live, to limit the damage ignorance can cause. Primate brains evolved to need parenting, because infants already had parents available for evolution to exploit. Watchful mothers make curiosity nonfatal.

Adults in other species do not recognize their parents because they have no more need of them. To know your parents, to find them when you need help, evolves as a trait having survival value only when parents *do* help. Adults in other species have no need a parent can fill (beyond needs any contemporary could satisfy) and so do not recognize them. Though it may live for a century, a 5-year-old tortoise knows not its mother. But humans never become adult, never outgrow their need of parents or their recognition.

Maturation delays helped women by fooling men who tried to sneak under the courting wall, generation after generation pushing our adult form further toward old age. Adult mentality followed anatomy because we gained nothing from separating them. As each individual spent a smaller part of life as a full adult, each troop had fewer adults forming a stolid core. Eventually it became rare for a troop to have any old enough to attain adult mentality, and finally adulthood vanished. We accidentally pushed adult minds off the lifespan bench while we were making room for younger bodies.

The Great Apes

Great apes share slow growth and paedomorphism and heartsickness. For unknown reasons a maturity delay appeared before our species separated, long before our aquatic time, before the Pliocene drought, and all the branches now possess it, though humans and other apes differ in degree. Juvenile minds are timid, and fear haunts our branch of the primate order. Humans carried the trend to an extreme because we had several tightly focussed evolutionary needs for a juvenile form.

Humans, gorillas, orangs and chimps all show juvenile anatomy in thickened tongues and larger brains. Primates typically have longer childhoods than most mammals and reach maturity closer to their maximum body size. Several development milestones show retardation throughout the primate order, and the hominoids (orangs, gorillas, chimps and ourselves) became the slowest-growing primates. Our Miocene ancestor probably spent a third of its life growing to adulthood, taking eight or nine years to a prosimian's six months.

Our cousins' paedomorphism shows more clearly in their minds than in their bodies. Gorillas, chimps and orangs are the three strongest primates, yet show astonishing timidity. Though we feared them before we knew them well, their gentleness now makes them our favorites. But why be timid when a little aggression coupled with their power would gain so much? They certainly didn't evolve gentleness in order that humans would love them. Neither did their timidity come from predation. Animals who often become lunch are alert to danger; shyness evolves in prey animals because it aids survival. Big cats have often preyed on humans (some of the oldest human fossils may come from predators' bone heaps), yet humans are bold. No predator commonly targets other apes, yet they are timid.

Juvenile mentality reliably causes timidity in other animals. Neotenic bighorn sheep decline fights more often than do their relatives, and grow more slowly. Dogs rarely attack their masters because adult dogs still feel like puppies inside. Adulthood brings a confidence that late-maturing animals often lack.

Our cousin ape's timidity, like our vaunted curiosity, came from juvenile mentality. They are timid because they never become adults, and so have no adult near to give support. Since they all share shyness, it came from their (and our) common Miocene ancestor. No adult human ever lived, because adulthood ebbed long before our aquatic time, before humans evolved into a separate species.

Our cousins' heartsickness comes from slow growth. Not living long enough to lose curiosity or gain full adulthood, they never become confident. They don't know this, nor can we define exactly what adulthood means. But their shyness illustrates what we'd have become had we not taken a different path.

Gorillas are pathologically timid, inbreeding to extinction because they cannot compete even for mates. Groups meet with lackluster greetings and make no sign when parting, as if numb. Though they are probably quite intelligent, their mental powers resist measurement because apathetic gorillas don't care about solving the problems primatologists present as tests for them.

Orangs are the shyest apes, though lacking natural enemies and usually left in peace by humans. ("Orangutan" means "man of the forest".) They live the most solitary existence of any primate, drifting alone through a large and stable home range. Orangs may feed for hours while never acknowledging others in the next tree, or avoid each other by sound alone.

Chimps wander in small groups within a territory, often meeting other groups and recombining. On some meetings there is great excitement, hugging and touching, but soon they lose interest and wander on. Chimps seem to suffer from unfathomable distress, hoping for solace in each new meeting and in the end finding it lacking. By the age of sixteen months, a chimp infant learns how to reassure another. The infant must be very sensitive to fear, to learn so early how to dispel it.

Great apes behave like children with hurt feelings. If they were not our closest relative we couldn't gauge their minds; we can't tell how brave an ideal koala should feel. But knowing primates, we sense that our cousins are too timid for their size. They still pose threats to each other. Rape endangers orang females, gorilla skulls often bear tooth marks and chimps sometimes beat females or kill infants. Chimps may gang up on strangers, killing them in long and violent assaults. But children can be savage, too, and our closest relatives act like children.

We great apes suffer from endemic fear and loneliness; our brains evolved for adults who no longer exist to reassure us. We all feel one hug short because neoteny pushed our parents off the end of our lifespans and now we can't find them. Chimps, orangs and gorillas delayed their lives less than we; in humans neoteny has a sharper focus. Adulthood faded in chimps, orangs and gorillas, and vanished in ourselves.

Missing our parents makes us look for them because we can't ignore their imprint in our brain. A child cries for his mother because her absence does not stop his need for her. We neotenous primates fear not only real-world hazards (made more ominous because no authentic parent reassures us), we also fear one larger than ourselves whom we cannot see but cannot stop thinking is out there, somewhere.

Jane Goodall reported watching chimps charge thunderstorms. "Charging displays" are chimps' standard dominance show, rushing noisily back and forth but rarely touching their opponent. In the storm, chimps raced up a hill making charging displays at the sky, leaving

Goodall stunned. Since one doesn't attempt to impress clouds alone, the chimps apparently felt another sort of chimp was up there making charging displays (well, noise anyway) at *them*. Fear and frustration must have overwhelmed them to cause such desperate combat against an invisible and inaccessible foe.

Fading parents left a ghost-like imprint to haunt each great ape. Chimps and gorillas are shy and submissive because each of them fails to fulfill a lifelong search. Like a blindfolded child trying to play tag, eventually frustration overcomes enthusiasm and he sits down in dismay. For ninety million years our minds evolved around adults, who fifteen million years ago began to fade away while we delayed our maturity, leaving an imprint behind. Our subconscious tells us that parents must be out there; we just can't see them.

Monkeys don't search for their parents; adults live all around them. But gorillas, orangs and chimps now slide their parents to oblivion and ours died long ago. Our cousin apes begin to feel orphaned, and just the beginning makes them shy and timid. They need the adults they evolved for, but can't find any to reassure them. Since they cannot placate or impress the chimp in the sky, he must be some *other* chimps' ally and not any friend of theirs.

We four great apes suffer orphancy together, with fragile hearts and timid souls. We humans are far more orphaned than our cousins; their parents now wither but ours died completely. Without our treetop heritage we could feel no loss, but the primate context makes relationships important, so we feel keenly the lack of parents, though we may not consciously recognize it.

Parting Company

I once spent my summers in the lake country of northern Minnesota and southern Canada. Glaciers carved there a mosaic of water and rock, leaving across the land strings of lakes like necklaces. For weeks we traversed the chains, carrying canoes and gear over land strips separating water. Beneath us lay the Canadian Shield, the oldest exposed rock on earth, surfacing in ice-cut corrugations.

We retraced routes used by French-Canadian voyageurs two centuries ago. Our canoes were built by craftsmen long dead, used by generations of earlier boys and girls now old and gone as well. I rode craft which decades earlier had carried one like me on a journey of first experiences. Canoes we paddled four decades ago still carry campers, in a continuing celebration of heritage.

On our journeys, a week's slow travel from the nearest home, on platforms of sunlit rock we sometimes found cairns. Earlier travellers had piled stones to commemorate their presence, and then moved on.

In barren northern tundra a cairn might be visible for miles, perhaps containing capped jars with messages for the rare traveller. Fewer cairns and bottles dot forests farther south, yet their mere presence sends a message. In a trackless meadow we suddenly recognize the path of another lost in time, whose eventual destination we cannot tell. We know only that they passed here and left a marker to startle us years later.

Our surprise was misplaced. Our routes intersected because a chain of causes led us to the same path others had used. We were in that part of North America (the Beautiful Land, the voyageurs called it) because glaciers pocked the land leaving highways of lakes. We chose routes where lakes chained close together with no high ridges between, as others had done. And footpaths between lakes came from centuries of travellers like ourselves, paths chosen for shortness and dry footing and steady grade, avoiding thickets and swamp.

Realizing that another had been there let us see our own presence in a new light. Our presence was a symptom of the land as well as a product of personal choice; the contours of the country will repeatedly guide travelers to choose similar routes. Glaciers long ago shaped our course as surely as they cut valleys where rivers now run. Like water, we and thousands before us in slow streams flowed over the land, gently guided by broad contours and similar preferences. Though a trail might grow over before the next traveller came by, a marker in the wilderness reveals a highway in a quiet valley.

Our cousin apes travel in another wilderness, near us but apart, illuminating another highway. In our evolution's broad valley travel three other apes, all only a little way off, still visible from here, too much like us to ignore. Over fifteen million years ago the apes separated from other primates; we came this way as one and slowly became several species close together. Yet we are bold and they are timid; we thrive and they dwindle.

If no other animal were close to us in form, we might craft a believable story of hunter-apes hairlessly sweating as they jogged in pursuit of prey. But we have three others in our immediate family, one of them an excellent and cooperative hunter. And several features making our bodies different from theirs would hinder a hunter. Weak limbs, toothlessness, and lack of smell sense are improbable adaptations in a powerful ape who specializes to hunt, but are likely in the aquatic scenario.

If no other animal were close to us in mind, we might convince ourselves of our intelligence and superiority. But there are these three again, whose intellects are so close to ours that measured differences may reflect only cultural bias. One of them regularly surpasses humans at comparable ages. And in one key aspect they are alike: They are too fearful for their size. One cowardly cousin is an embarrassment and two are a problem, but three make a pattern.

Into A Darkening Valley

Our cousins tell us that we all entered a deadly valley fifteen million years ago. None could have seen it then, but their shrinking numbers and apathy now tell us that something in this evolutionary landscape destroys our kind. Three to one they show that slowed growth is a dead end path for an ape, leading to timidity and gradual extinction. Three to one they outvote our myths, showing that humans were not destined for greatness but instead barely evaded a disaster. Three to one they show that our evolutionary valley gently leads apes toward oblivion, yet somehow humans thrive in it.

Like pandas, our cousins cling to quiet places and fill the time until extinction. Only a few thousand other apes survive where we have not yet brushed them aside. We share their genes, and our need for religion suggests we share their fears, yet we seem to have overcome orphancy's burden. More than intellect or any physical feature, boldness separates us from our timid cousins.

At least fifteen million years ago our ancestor left the central primate path and moved into a dead-end valley. Before the Pliocene drought, before any gorilla or chimp or human, there lived our precursor, closer in form to a modern chimpanzee than anything else, carrying the seed for all of us apes.

Driven by forces we can only guess, it evolved to slow life's processes. Many species have taken this route, some successful and some not. Perhaps delayed maturity gave closer family ties, a more tolerant society, a tendency to compete by impressing instead of terrorizing. Lush surroundings may have favored tolerance, giving reproductive advantage to close-packed populations. Perhaps rapidly changing environments made curiosity critical, and extending childish curiosity by delaying maturity was the evolutionary result.

Before chimps or gorillas, orangs or humans, our ancestor feared when storms raged, and shied from danger. The Pliocence drought and geographic separation sent several ape species on diverging paths within the same broad valley of slowed growth, each traveller bearing a burden of fear and timidity. Each of the cousins continually lost an imaginary contest between themselves and adults they had evolved to see but could no longer hold.

Our own hyper-retardation began with our move into the water, where aquatic physical demands urged more delays in maturation. When again back on land male courting skills collapsed, causing females to adopt increased neoteny as camouflage. Accelerating down the paedomorphic trail, we unknowingly stripped ourselves of parents' last vestiges, and left them behind. The presence of unseen beings, occasionally and vaguely felt by chimps, became constant, sharp and gnawing in ourselves.

We endured a long accident; our parents died in a slow crash. No value came from being frightened, but women found great value in growing slowly and maturing late. Our mindset of orphancy served no purpose on its own, but came partly from the common ancestor of all apes, partly as a side-effect of sexual competition's physical changes.

The human species wandered lost in the woods, burdened by a mind sensitized to emotional rewards for procreation, rushing down a path that made each generation a bit more lonely. Parents our brains had evolved to need became (by their absence) a source of pain and loneliness. As men evolved into faithful husbands, their increasing dullness pushed women's camouflage even further. As duller men became less-rewarding companions, women evolved stronger passions to make the effort worthwhile. And greater passion gave a greater sense of loss as we contemplated parents we could not find.

Our cousins' condition sketches the deadly valley around us. Their being three close together tells us the valley's effect is consistent. Retarded maturation helps an individual in the short term but can take the heart from a species over time. Lifelong childhood brings great danger when individuals have lifelong need of parents. Neoteny slowly kills ape lineages, and our three close cousins show how they die in apathy and gentle fear. Perhaps we had ten cousins a million years ago.

We should be dead. Natural selection stretched us between an increasing need for emotional involvement and a decreasing chance of satisfying our needs. As the slowest-growing, we should fear more than any other ape. Yet we are bold while they are timid, for we hold four mental assets shared by no other ape. One is language, too well known to need further discussion here. It's time to acknowledge the other three.

Braver Species

Our first asset is bravery itself. We are less timid because we endured specific evolutionary forces favoring braver individuals. Though they might have better survival chances if they had more courage, other apes passed through no similar seining and are now fearful.

Nothing in our cousins' lives demands individual courage. Even male courting competition was short-circuited in apes, though it reliably fosters aggression in other species. Female preference (instead of male dominance) grants sexual license. Male assurance comes from female approval more than from male combat, and males win dominance displays only when confident of female support. Both gorilla and chimp females will attack males who attempt to usurp an alpha role without female consent.

Females, as the backbone of ape society, rendered male combat largely moot. Male apes fight less to *establish* dominance than to *announce* the result of female votes. Political savvy instead of aggression leads to male success, and courage became discardable. Our cousins are timid because they lack selective pressure for bravery in the short term, and their society tends to weaken aggression over many generations.

Humans are brave because women needed courage to reproduce. From the first frontal sex, our record has focussed on women's fears. Men evolved desensitizers to allay fear; women evolved powerful hungers only because fear culled those with lower interest. Urged by women's need, humans have evolved fear-controlling responses and find pleasure in hazardous acts. Women often misjudge their own peril because suppressing fear has become their habit.

Courage came from short-term culling as all evolution does. A timid woman avoided possibly dangerous men, but braver women who enjoyed the fear-thrill response could relish the encounter. The more confrontational sex became, the faster timidity faded. Women's increasing courage let aquatics convert to frontal sex and later start moving back to land. Fear held the timid back when we moved, leaving survivors who could face appalling danger. We copulate in a pose other species would find terrifying, because it takes more danger to frighten us.

This has nothing to do with hunting. Hunters feel excitement and often exhaustion but rarely fear. There is danger away from camp when stalking, or getting water, for anyone can be taken by surprise. But hunters select prey for minimal risk, for the sure thing, not the glory. Few cultures chase prey that might hurt the hunters, and stalking a predator happens rarely and only for show.

Courage didn't evolve in hunters because hunters didn't need it, but women did. Animals evolve at the boundaries where chance culls them; descendants evolve faster when the slower die, a species grows agile when falls injure clumsy ones. Hunters didn't stay away out of fear, but women did. Courage evolves where fear makes failure, and women needed bravery and so evolved it.

Other apes might benefit from courage but had no path where it could evolve. Gorillas now inbreed to extinction because males are too timid to leave home. Chimps fight by waving branches because that's enough to scare another chimp unless females support him. Our courage is as foreign to them as our poetry.

Perhaps chimps puzzled, watching Jane Goodall watching them, or gorillas wondered at Dian Fossey, orangs at Birute Galdikas. Did our cousins vaguely ask what gave small and defenseless women the courage to follow their more-timid subjects? Did they grasp the vast gap in our minds that makes a crowd of them run and hide when only one of our females follows? They might wonder what makes the world look so dangerous to them yet so safe to us.

For millions of years people repeated the cycle. Though stress might make them vomit before a date, women had to confront men in order to reproduce. The smaller had to face the larger and more dangerous, and needed guts to do it. Being inheritable, bravery passed from mothers to both sons and daughters, becoming our common asset. We were bred for courage.

Orphan Individual

Passion gave our second unrecognized advantage, with obsession taking such a hold on us that we cannot feel other apes' loneliness. Though our adults are wholly gone and our illusory orphancy more profound, our passions became strong enough to bridge gaps between us. Obsessive focussing creates bonds that fill our empty places.

Our life expectancy didn't rise as maturation slowed. Infections and childbirth kept life short, and a successful first delivery didn't guarantee another. The hunter-gatherer life posed dangers for both men and women. Hairless skin was easily damaged and infected; awkward bipeds often failed to escape predators. Maximum lifespans were 70 years or more, but daily hazards kept the average closer to 35, and many died much younger than that.

With each delay in maturation, losing a parent hurt more. A woman dead at 35 would leave half a dozen children, and few teenagers would have a living parent. Children cried for mothers lost in childbirth, for parents who went to find food but never returned. No illusion here, this was a real person, their daily comfort and support, gone without warning. Ancient hominids, keyed to emotional levels not found in any other primate, were literally orphaned by the hazards facing crippled bipeds.

The death of any mammal mother often dooms her infant, for no other adult will adopt it. In primates an older sibling may take it on, but an orphan rarely grows, and commonly starves to death in glassy-eyed apathy. Jane Goodall reported how the chimp Flint built a nest near his mother Flo's body, staying there until he died.

An orphaned chimp becomes apathetic, an orphaned gorilla inbreeds, but an orphaned human (over countless generations) was evolving sharper hungers. While our cousins were slowly evolving to care less and less, we were evolving to care ever more. Unlike our cousins, we form relationships for passion more than convenience, as inorgasmic female bodies made delusion and obsession essential. When evolution made us illusory orphans, leading us to become shy and withdrawn, our passions were already strong enough to push us out of the house.

Love's obsessive focussing first evolved in aquatic women because men had little to compete with. A obsessive woman had a better chance of being pleased with a man she encountered, because her susceptibility raised his status in her eyes. This small advantage launched us on a trajectory our cousins never needed, because their females had clearer proofs of male adequacy.

We needed passion most when men came out of the water and women held back. Love evolved because emotionally susceptible women gained reproductive advantages. We might have evolved some other instinct, if we didn't already have mild love for natural selection to enhance. Passion grew from tepid love, to make a woman feel terrible when her lover stood on the beach and she in the water. Her inability to stop thinking about him was invaluably painful, for it guaranteed she would follow.

Now literal orphancy spurred passion's evolution. Dead parents left children with scarred hearts and fear of emotional contact. The slower we grew, the more it hurt. Only the people most susceptible to love's pull could overcome pain like this, passing increasing obsession to their children. Love grew to overcome the protective shyness of real and illusory orphancy, because real orphancy happened all the time. Our cousins couldn't strengthen passion to overcome illusory orphancy because they didn't have passion in the first place.

Obsession serves all our needs by giving us occasional sleepless nights. Passion is the footprint of women's fear and orphancy's pain, paralleling self-repression as the footprint of masturbation. No gravy on the rice of our lives, love is raw and clutching. No flower of heightened human intellect, love drives us to find each other despite logic and fear. It rarely works perfectly, but it works well enough.

Obsession is our heart's skin-hunger. As women evolved skin eroticism on the palm of their bodies, leading them to seek frontal sex, so love evolved as frontal-eroticism of emotions, electric hearts at the point of contact. While our doubts hold us up and out of the fray, our needs bump around our feet like a puppy on a leash, getting into mischief.

While our minds hold back, our hearts lean forward as a girl leans to reach her lover, all electric with the wanting to be touched. We cannot help but seek one who feels warm in the palm of our heart.

Love evolved not to make you feel good when you hold your lover, but to make you feel bad when apart. Love exists not to make us happy but to make us surmount obstacles to stop the pain. Human relationships exist as much to avert loneliness as to create children. For many, even a horrible mate is better than none. When we focus our hearts on another, whether parent or lover or child, we ease the gap left by long-vanished adulthood.

Monogamy is not a behavior we evolved for, though monogamy offers clear benefits. It may be good strategy to remain together until children reach (say) five years, but that doesn't mean we evolved five-year-bonds or lifetime ones. Our pairings come not from a monogamy gene but from delusions and fear-thrill and loneliness.

Sex is not monogamy's purpose; it's the friendly body nearby that gives us peaceful rest. It's easy enough to find sex; it's harder to find someone who keeps you from feeling alone in bed. A woman asks a man to stay the night less for sex in the morning than for company in the dark.

This is remarkable behavior not found in our near relatives. Even very young apes sleep in their own nests. Mighty Hunter theorists credit rampant sexuality for our liking to have our sexual partner near, but I think we sleep together because we feel lonely when we sleep apart. When we find someone who focusses our heart, we try to keep them close because their absence causes pain.

Pairing is a symptom as well as a solution. To pair is not a behavioral choice; marriage is not automatically kind and promiscuity is not automatically brutish. Our pairing is not a trick that (if they only tried it) would benefit our cousins too. Pairing comes from minds that can find comfort in it. We mate by personality not because it is superior but because we have a weakness for personality, and a tendency to obsess. We never chose to become a species who weds; we marry because we need to.

But natural selection can work with only the materials at hand, to strengthen or pare away. Other apes have preferences but lack passion, companions rather than obsessions. Only obsessive focussing can replace parents' void, and other apes had no path from companionship to obsession. No survival advantage came to them from each tiny increase in romantic interest such that obsession could result, so they stand on the far side of an emotional ravine too wide to leap. For our cousins, no mate also gives a reason for living.

We can never feel as alone as our cousins because we followed a different path to mating. When we lost adults we could fill the empty place by clinging more tightly to each other. When adults were plucked from their socket in our brains, obsessive focussing gave us a replacement to fill the gap like a prosthetic limb, imperfect but functional. Our loves take so firm a hold on our hearts that their imprints remain for a lifetime.

Vision In Bold Relief

Our imagination became the final asset separating us from our cousin apes. We can focus our obsession on an image in our mind's eye, letting us see what we cannot hold. Far from home, we can call to mind those we love. When fearful we can imagine being comforted; the dying soldier feels his mother near. Only we can find what all apes need.

If imagining absent persons were our evolutionary goal, we could hardly suit the task better. From our mothers came obsession, letting us focus on any subject long enough to make it clear in our minds. From our fathers we inherit a hard-wired sketch upon which we can draw any face. And from neoteny we gained a cavity left by adults' subtraction, an emotional hollow demanding we fill it with something.

Humans are predisposed to think about others who are either absent or never existed. Any healthy human can instantly recall a friend not seen for a year, an intimate gone for a decade. On hearing a description of race, gender and age, any normal human can envision an individual to match. To imagine another person is as easy for us as to speak.

Imagining comes quickly to us because we carry a face's sketch wired in our brains. Like a policeman's identification kit, we carry the basic parts and need only adjust the image for specific cases. In any random patterns of clouds or leaves or shadows, a human will see faces because our brains are wired that way. A cartoonist can draw a famous person by emphasizing those features that depart from our instinctive sketch; we recognize it because we each had already made the same computation when we saw that face. Embedded in our brains like the outline of language, our caricature became an asset we could draw on unthinkingly.

We don't do this merely to fill time. We imagine to fill a void, using faces as handles to fit a mental pillow into place. We imagine to recreate the feeling that the other's presence would bring. I think of one I love in order to remember how I felt in her company. I script our conversation in my head to fill a gap in my heart with her presence, and look at her picture to help my imagination play the scene. I can't really know what she's thinking, but I want to feel once more how I felt when we were last together.

At the age it learns to talk, a child also learns to imagine playmates it will likely remember for life. A child may request a place at the table for imagined buddies and demand others treat them as real people. Children imagine playmates for the same reason kittens pounce: They exercise faculties that give pleasure. Other apes can make believe; at least one chimp has been reported playing with an imaginary toy. But humans create imaginary *people* because we have a mental structure making this possible and pleasurable.

Imagining persists for life; we do it more than we care to admit. As teenagers we imagine romantic encounters earlier and more frequently than we can achieve them. As adults our dreams can fill lonely times. Our imagination is filled with reliable and accessible partners.

Being imaginary does not mean being trivial. New couples often report marathon conversations of breathtaking impact. This is an excellent sign of growing love, but the communication is half a dream. My first conversation with my wife was less with her than with the person I thought her to be. And she spoke through me to one she had already imagined meeting, who I resembled closely enough for a dialogue to begin. Only later did we start talking with each other.

Yet this may be the most honest conversation one can ever have, talking to our own hearts of the things that we feel most. To each the other is a mirror in disguise, promising to not laugh or judge, and so drawing from us all we needed to say. Sharing that event with another, helping each other to do what we cannot do alone, and giving the other an image to focus on in return, forever changes that person in our minds. In a species that needs to begin by talking, how fortunate that we dress each other in silk, if you will, for our early dialogues.

Love Is Forever

The impact another makes on us measures the importance of our empty place. Discovering a partner is a fresh thrill for an individual but an old story for the species. Filling our minds with another comes close to meeting the parents who last lived millions of years ago. We experience a rush of relief when we find love; it feels ecstatic because we need it so badly.

Discovering passion brings rapture; something about it makes us tell the world. As a young child shouts "Look at me!" balancing in a handstand, a young man seems compelled to write "John Loves Mary" on highway overpasses. Even if geese had a written language, I doubt that they (though bonded for life) would write their mates' names on roadside structures.

John didn't mean that Mary promised sex-on-demand; rather he proclaims the end of loneliness, with perfect confidence in eternity. Though surrounded by evidence of frail human pairings, we know ours will last forever. Even those who have loved and lost will stake their fate on each new arrival of passion. We feel not only that love will outlast us but that it preceded us. Lovers often feel that they've known the other all their lives, or that the two were fated to meet. I think this is no attempt to add meaning to simple lust, but a true statement of what they feel. Yet these impressions have no logical foundation.

Lovers seem foolish to one not in love. If we've never felt passion before; if it has never been in our lives; and if we rarely see it persist in others, why do we presume that in us it will never fade? Why, if we have been well-behaved, will we suddenly break laws and betray trusts to sustain passion? Where do we gain such boundless confidence in a phenomenon new to each of us?

Love fills our mental gap by focussing us on another, making ourselves feel whole. Most other species' adults don't need to fill the space, for they never feel incomplete; they cannot fall in love because they have no mental cavity. When we focus on a lover, we find all the things we can imagine there. Our imagining is a mold of what we lost by losing adulthood, though adulthood's sensations are beyond our conscious reach. When in love, we feel partly what other species' adults feel always.

Love is our heart's prosthesis, making lovers feel whole. If we see a person with an artificial leg, we assume he lost the real one but we don't laugh at the replacement. If love acts as a mental prosthesis, we must have lost a mental limb. Instead of thinking lovers foolish, we should see why seemingly foolish statements feel true for them. If love makes a person feel whole, look to see what aid the prosthesis gives.

The void in our minds is as real as any person, though we can measure it only by seeing how it affects us. It is the part of us that would have taken form if we grew to full adulthood, the part that would not feel lonely if adults were still with us. But neoteny took away adulthood and left a gap. When we fill the void with another person, we feel we've found an old friend rather than met someone new. We clothe our loves in different forms and adjust them to our individual needs, but we truly have known them all our lives. We were born with their welcome prepared in advance, a socket left by extracted parenthood.

Passion is partly a delusion where we truly fill a need by merely believing we have filled it. Delusion evolved first in women, when monogamy prevented them from each having an alpha male, giving a reproductive advantage to women who misjudged men, to those who thought inorgasmic sex was worth the risks. Passion evolved to make average men acceptable to women, and still grows stronger to meet that need. A woman will meet no perfect man; if she is to mate, she needs some skill at dressing a stranger in silk.

But passion also measures a cavity in our minds where adults were subtracted. When illusory orphancy led to loneliness and fear in chimps, gorillas and orangs, they had no mental trick to fill the space. But we were busy fighting the battle of the sexes, evolving passion to make bad sex worth the trouble. When hyper-neoteny plagued us with loneliness for missing parents, we had already built an instinct so strong and focussed that one can not feel lonely when in its grip.

First-loves feel perfect because they fill our mental cavity for the first time, like coming up for air after being underwater forever. First-love is flawless because it happens mostly in the imagination. When we first love, we feel a patch on a gaping wound we had not previously known the size of. Only from experience do we learn that our lovers are imperfect, that the cavity needs a better fit, that the next stranger needs a closer appraisal.

All lovers know they were born for passion; becoming whole needs no justification. We may spend the rest of our lives looking for truths, building certainty brick by brick, but knowledge comes instantly with passion. Love feels eternal because it is built of confidence itself; completeness born of obsession, though largely illusory, is irrefutable. All things are fair in love because we instantly know that love's value outweighs any laws or customs, though we may have never known it before.

Far from foolish, lovers' statements give their best verbal rendering of what they feel, calling to us from a place we can know only when we too feel love. When lovers say they have become one, when both say they feel whole, it is for them the literal truth. Neither liars nor insane, lovers are in a giddy free-fall of assurance and terror, reporting feelings at the very edge of what words can say, sensing for a time what once was our birthright, fearing that it feels too good and yet just right.

If imagination had only a little more maneuvering room, we could envision exactly the bond we need. Lovers and friends can never quite fill the void where parents were subtracted, though we can nearly convince ourselves they do. If human nature did not so often intrude on our obsessions, we might imagine a more-perfect union.

Our Father In Heaven

Gods express our missing parents, and probe our mind's gap from another direction. Gods show what an observer would find in an adult cut loose from a physical body. They are hand-carved replacements for what orphans seek, like stage facades that fill a real need.

Religion evolves as we whittle away what does not make us feel better and enhance what does. Each accepted aspect of a god began first in the mind of someone who said it aloud. If the thought rang true, others repeated it and taught the next generation. If an idea didn't resonate in others' hearts, they forgot or dismissed it. The human mind in contact with its god is a mind taken unawares, as when falling back we throw out our arms, showing what made our ancestors feel better and displaying instincts many deny we have.

We see in our gods what we need to see. Because gods take form in our minds, we need not adjust our view of them as we must when lovers turn out different from our imagining. Religion does not lie beyond science but is one inhabitant of science's landscape. Imagining gods exposes a place closed to other examination, probing waters deep enough to crush the scientific method. Some religions can't consider evolution, but evolutionists should surely consider religion.

I am uninterested in the reality of anyone's god. If a conscious will created the universe, that force could also shape evidence for its own existence and control what we can prove or disprove. But I am very interested in features we *attribute* to our deities. As aliens illustrate an innate human sketch, and myths show common emotional traits, so religions outline parenting we fail to get. Over thousands of generations we sculpted images of superior beings, giving gods who reflect people's readiness to see them. What we find in religion scratches a mental itch, as an amputee scratches a stump.

In religions we find parenting. In mythic birthings gods created us, and we remain their obedient children. We see gods as human in mind if not in body, angry at us when we stray, jealous if we love another, thinking like we do but on a larger scale. Our gods are perfect adults with no doubt or curiosity, knowing all that needs to be known, sure of themselves though their battles may remain undecided. Gods care for us as parents should and punish us when we err; they think of our growth for tomorrow if not our pleasure for today.

Individual religious beliefs are beside the point. Religions change by natural selection, branching and flowering as chance allows. Those that don't answer adherents' needs fade and die; those unresponsive to doubters are disbelieved. As a rule-of-thumb, any religion will have enough logic to withstand any argument comprehensible to its most demanding follower. A religion might describe one omnipotent parent or a hundred small ones, and our hearts would get the same benefit.

The believer tries to imagine his god so clearly that the image assumes living characteristics, as a child's imaginary friends seem real, as my love comes alive when I gaze at her picture. Envisioning a living god (not all can do it) is like finding a real parent possessing qualities we feel short of, like recognizing the face of a friend, impossible to describe but unmistakable. Religious rapture is a real event, not a marketing ploy. Assurance comes from completeness; a god replaces our inner void with a new structure molded to fit, and the rapture's power measures our relief. Gods betray a deep need for adults when we cannot be adults ourselves.

Passionate bonding is a multi-purpose tool, evolved first in women to allow frightening sex. Obsessive focussing was only part of women's reaction, and with frontal eroticism it lured women to endure what they would otherwise avoid. Purely by chance, women's mindset became useful when neoteny made illusory orphans of us. Religion copes with orphancy by dressing a different object in passion's silk, obsessively focussing on a more-permanent figure. When orphancy became painful, we made do with the best answer we could find. If mating satisfied all our needs, religion would not have evolved as another way to fill them. Love for another person fills our mind and gives a warm body to embrace but gives little parenting; religion gives perfect parenting but nothing to hold.

To assuage neoteny's loneliness our minds need only a focal point. Whether we think gods are capricious or cruel they still give structure and meaning, because any obsessive focussing provides the mind's prosthetic part. Our gods never surprise us, because before we attribute any event to a god we have already fitted that event into the god's motives. Belief in a god gives coherence to a world that formed without our comprehension as its goal. Filling the mind with a god completes our mental structure, gives us assurance even if we imagine difficult gods.

A group of believers, by agreeing on their view of a god, creates a deity who can outlive any member of the flock. Their god is indeed immortal. Religion cuts from whole cloth the thing it demands to see, as we create loves on shaky foundations or as we by simple agreement create the scientific method. Religion is real though gods begin in imagination. And religion has impact because it affects the faithful *whether or not* their supreme being has an existence apart from their imaginations. It might not matter whether gods exist, but it matters very much that people think so.

Gods are what we once saw in adults of our troop, knowing more and having more power. We bargain with them, trading sacrifices or simple allegiance for protection. As in any primate troop, the followers gain power by making friends with the alpha. A straight-laced church praying for world peace and a wild-eyed cult bent on bloody mayhem both gain confidence from believing that their god helps them, that the super alpha male chose them alone as his special allies.

Religion became our mental lives' infrastructure whether we each feel devout or not, affecting us daily though we ignore it. A fascist living in a democracy scorns democracy, yet draws upon democracy's liberty to speak against it. Though religion may seem trivial or evil to some, none of us can avoid living among religious people. We may not share a god, but as social animals we cannot avoid sharing confidence though it pass between us as secretly as radiation, softly as pollen, continuously as sitting near a low fire.

Living in a meaningful world gives us an anchor, and religion gives meaning. Atheists could not imagine a secular view of the world without religion's example. To construct a coherent all-encompassing scheme for the cosmos (accurate or not) entirely without evidence is our most-significant mental achievement. No other invention comes close.

Imagining a god affects a human group's chances of survival. Tribes with gods were more confident than tribes with none. People who imagined omnipotent gods trod on people who followed complaisant deities. God-linked dogma supplies answers and removes doubt, where lack of dogma may let us question our purpose. That religion often leads to mayhem is beside the point; like any artificial limb, religion can function without being pretty.

Our filling the world owes much to imagining a friendly giant. It was not language or weapons that gave us the courage to advance where life was hard. It was our willingness to bargain with unseen fate, making a deal with gods we thought might oppose us, giving ourselves license to go. Whatever we can imagine doing, we can envision our gods supporting.

Here we left our cousin apes seeking their quiet places in forest shadows. With no friendly giant on their side and no clue to what it all means, gorillas, chimps and orangs see danger in all unknown things. They spend hazardous and buffeted lives, hiding and watching the world pass by. But we see ourselves as members of the chosen troop, finding meaning and order no other animal can see. While our cousins watch the world go by, we believe the world watches us.

<u>Marker In The Wilderness</u>

Understanding our heart's genesis doesn't lessen its magic or power. It is both a gem in our hands and a cairn in the wilderness, a treasure showing where we and our cousins parted. Like any marker on a trail, it may clarify our course but need not change it. I found the trail while on other business and am leaving a note in the jar. We've followed a larger path than we suspected, and I think we need to see it.

Our marker and uniqueness came from language, courage, passion and imagination. All began in our sexual competition and collapse, and all make us different from other apes. Like wood or paint in a sign, each has no intrinsic meaning. Yet our cousins are dying in shadows, haunted by ghosts because they lack what we have.

Another animal with different needs might evolve one or all of these without any significance larger than the feature itself. Nothing demands that a species use songs for more than courting, or that visual preference lead to more than long tail feathers. We acquired other features that did not build our treasure. Crafting weapons using opposable thumbs makes for stirring stories, but wasn't what saved us from extinction.

Our treasure, a gem formed by long and deep pressure in our hearts, comes from language, courage, passion and our caricature combined in an animal who needed their total effect. In our basement rubble we found a way to face danger by knowing we can, find solutions by focussing on problems, ease loneliness by recalling a loved one, touch hearts by sharing a thought, shape supportive gods by thinking of them, and create a meaningful universe by talking it over. Our hearts evolved passionate incandescence, brightest where they touch others, and now light our way through a valley shadowed by death.

We are bold and our cousins are timid. Both our courage and their fear come from misunderstanding; the meaning we see exists only because we think of it, and they fear a giant who lives only in their minds. We love, and our loves guide our lives; they only like, and drift aimlessly. We are bold without good cause, and for no reason they fear everything.

We parted from them by nothing more than being on the coast when the Pliocene drought began. Aquatic life gave us language and embraces. Orgasm's death gave birth to passion, women's rage, and the courting collapse. Dull men used innate sketches to locate vulnerable females. Female competition then retarded our maturation and exacerbated the illusory orphancy we share with other apes.

By bare chance, our male sexual competition created a foundation image on which we can sculpt what we seek, gods or lovers. Lovers we build alone and in silence. Gods we shape together, as language and obsessive focussing let us compare, adjust, and pass on those images. Over thousands of generations we have clothed the foundation in imagined faces and bodies, given it histories, personalities and thousands of names.

We weave our lives of reality and imagination and memory. Once we fill our empty place, that person lives on in us and fades only slowly. Our loves warm our hearts when we recall them, like opening a scrapbook where each image glows with heat. In his wallet the soldier carries a picture of his love, to contemplate in a quiet moment and bring her close again. For any other animal a photograph is only paper. For humans it can be a talisman, an identity, a companion, a reason to continue.

A human who has loved will never again be completely alone. Absent or dead, the beloved still lives in the mind and fills our empty space. Whether imagining a god or a loved one, we fill an aching need as a chimp never can. When holding this image, each of us has a perfect companion and gains confidence other hominoids can't know. Chimps and gorillas never needed to focus on each other. Other apes needed no language to court and now have no tool for whittling gods to shape. Never evolved for obsession and imagining, our cousins remain lonely and afraid.

Confidence is contagious in a social animal. When you are lost, anyone who knows the way is a wonderful companion. Anyone with a fulfilled inner image gives confidence to his neighbors; the evangelist enthralled by his god gains followers, and all the world loves a lover. Each such person is a spring of confidence from which the community draws; in every crowd some are in such a state. We all share the confidence of those in rapture, just as shoulder to shoulder we share courage. By language and deed we communicate the mood; written words and gods make it firm and clear. We give ourselves license to go by merely agreeing that we can, while our cousins wistfully watch from the shadows.

Human evolution had a lucky accident. Fighting the battle of the sexes saved us from a melancholic plague at home; struggling in the trenches raised meaningful contours on a featureless plain. Our courting collapse pushed us further along the hominoid route of illusory orphancy, yet the same collapse gave us mental tools to conquer pain. Our good fortune came by pure chance; three of the four Great Apes failed to win the prize. Of three aquatic primates, we know only that the one who survived possessed it. Caught by surprise in desperate need, we dug in our pockets and found we'd been carrying the necessities. Such fortune leaves me breathless.

Sign In Our Heads

Human uniqueness lies in emotions, not tools. Our advantage is in our instinctive and obsessive focussing, not our intellect. Intimate attachments are no chance product of a powerful human brain, but are instead our evolution's main channel.

Many remain unable to acknowledge any instinct in humans, or passion in a fossil, but instinctive passions guided our evolution. Anthropologists weigh alternate scavenging strategies, but our ancestors' feelings counts for much more. They are beyond interviewing, but the couple who left muddy footprints at Laetoli was very close to us. The man loved the woman he walked with but did not understand her, and her youthful face and slim shape first drew him to speak to her. And she thought he was special.

The scientific method has trouble with us and with our brothers. Experiments with chimpanzee languages have largely failed because our science cannot probe an ape's brain without our own feelings intruding. Experiments in the classic style require that we place chimps in solitary confinement from birth, where they would respond to mechanical cues by depersonalized examiners. Such an environment would destroy the process by which language is learned and would drive the subject insane. We refuse to cause such pain, so chimp language talents lie beyond the grasp of a researcher with feelings. We can more easily wound species that are farther from ourselves.

Things closest to our hearts are hardest to study. We compute a planet's orbit, not an astronomer's love. Language is both product and symptom of social interaction; we cannot reproduce it in isolation. Passion is similarly elusive, for we have barely touched the surface of what our brains do, and we have much to unlearn before we will see clearly. Our inability to apply scientific rigor to our hearts shows a lack in our science, not a flaw in our hearts. To ignore a species' vital property because it resists laboratory probing is intellectual dereliction. Only the foolishly unfeeling can think that the Laetoli couple felt nothing.

Some find it disturbing to contemplate a cave man squatting in some dark corner of their minds. They want to feel free of primitive ties, yet I'm pleased that our ancestors inhabit us. Instinct does not cage us in dark corners, but floats us in a warm bath where we (like whales) can breathe. Instinct makes vital acts easier, and gives our minds time to appreciate the magic.

Instinct guides us through unknown territory, and much of what I am was in me before I was born. Instincts told me when I was about to confront trouble and how to avoid it. I possessed instinctive mechanisms for making babies, though as a baby I didn't know where to fit things. I knew at birth how to nurse and how to cuddle; I had an inborn hunger to touch others of my kind. Instinct guides most of us to find a mate and rewards us for the effort.

We are missiles of reproduction's hopeful catapult, messengers our parents launched blindly into the future. We who scorn trial and error all came from trial and error. If not the fittest, our parents were at least the better ones, the ones who survived. We carry them in our blood; the rules they instinctively followed worked for them, so they passed them on to us. My father died before I knew him well, but he lives in me. I couldn't let him go if I tried.

What we learn is not trivial, but we err in thinking that only learning matters. Our passions' depth and uniformity mark them as instinctive; their reliable effect defines our evolution's central channel. They show where we stand and illuminate the way we've come. Awkward and cumbersome, sometimes painful, our emotions are necessary baggage.

The Path Ahead

We can intentionally improve our situation only by choosing what to accomplish and knowing what to overcome. I don't mean to *justify* how things work, but to appraise forces we must deal with. We got this far by natural selection, and in the process inherited unruly instincts. To amend our behavior and injure each other less often, we need to see the path and the obstacles.

As long as we believe that passions are mysterious and beyond understanding, they will remain mysterious and misunderstood. If we see ourselves in the primate context, seeing the various forces shaping us, each with respect to its own source, then we might feel less bewildered.

We need not fear that such knowledge will remove the magic or the power.

Our sexual ills shape both our evolution and our present lives. Hungers that bring us together are the same hungers causing obsession, rape, and madness. To be pursued may bring joy or terror to a person, depending on whether they *want* to be pursued. One who causes harm may be a terrible person, or a good person gripped by a terrifying instinct.

Some speak of sexual faults as a cycle to be broken. If we can raise one perfect generation, some say, we will see no more injury. Our sexual ills are no virus to eradicate, but an unruly beast we must domesticate in each generation, like teaching a buffalo to use a litter box. If we must share rooms with this instinct-beast for the next few million years, while we evolve in ways we can't predict, we should at least understand it.

I don't suggest we try to breed out our problems. Though evolving got us here, planning our future evolution seems ludicrous. If I am wrong, my recommendations for eugenics would be meaningless. If I am right, then our survival came from our failures. Speech announces crippled dominance; our caricature and finally our confidence came from dull men's attempts to cheat. The flaw a eugenicist might weed out could be our next rescuer if left to flower.

When their burdens seem too great, the devout ask their gods for help. In their minds they hand over part of their load, acknowledging their own lack of control over fate. They trust their religion to have an inner coherence, though they may feel personally traumatized by what they imagine their god has done. In mixed fear and trust they bargain with gods, offering allegiance in return for help. Able then to imagine their god at their side, they feel lightened and encouraged. Yet this does not absolve them of responsibility for their own acts, and may instead hold them to a higher standard.

We should model our relationship with love along similar lines. We can acknowledge passion's power without relieving ourselves of responsibility for what we do. We can trust our emotions to have an inner coherence though their purpose is sometimes hidden, their effect sometimes painful. Our hearts demand what they need without giving reasons. Clearly seeing our flaws, knowing the peril of an unthinking act, we can share the burden of emotions with our instinctive core where they evolved. By feeding them measured sacrifices, we can enlist their better aspects on our side while harming no one.

Message In The Jar

What about the eternal verities? The ground where my lover walks was supposed to be holy; passion was supposed to move the stars and planets. Where does love fit into the cosmos if it's just something happening in our imaginations?

All species focus inward, seeking what benefits them, ignoring how another species might perceive the same rewards. Our finest wines start as poisonous body wastes of yeasts, generating alcohol as sewage until they die in its saturation. They are bound to a chemical treadmill by genetic coding, both consuming and producing as evolution compels them to do. Busy yeasts do not become convivial, don't see us eagerly waiting for them to reproduce until they suffocate.

Our intoxicating emotional life matches that of no other animal. We clearly understand anger and fear, seeing them in other species, but love's uniqueness hinders our probing it. Our lack of parallels leaves us oblivious to our predicament. Where should we place passion in our scheme of understanding, if we can't take notes while another species displays it?

We generate emotion from within, swimming in a wine of passionate products. We are innately bound to a mental treadmill, built by many million years' evolution for reproduction against great odds. Like the yeast, we cannot help doing what we evolved to do. Hungry for relationships, we clothe our inner images with the forms of people we meet or imagine, and convince ourselves that we find what we seek.

We generate not poison but nectar; our minds' products thrill us beyond measure. We thrive on what we imagine and would die without imagining. We nourish our own vital passions using no more than our natural abilities. If we find no other with such desperate need we also find no other so able to sustain itself. From thin air we conjure love, for passion gives us life.

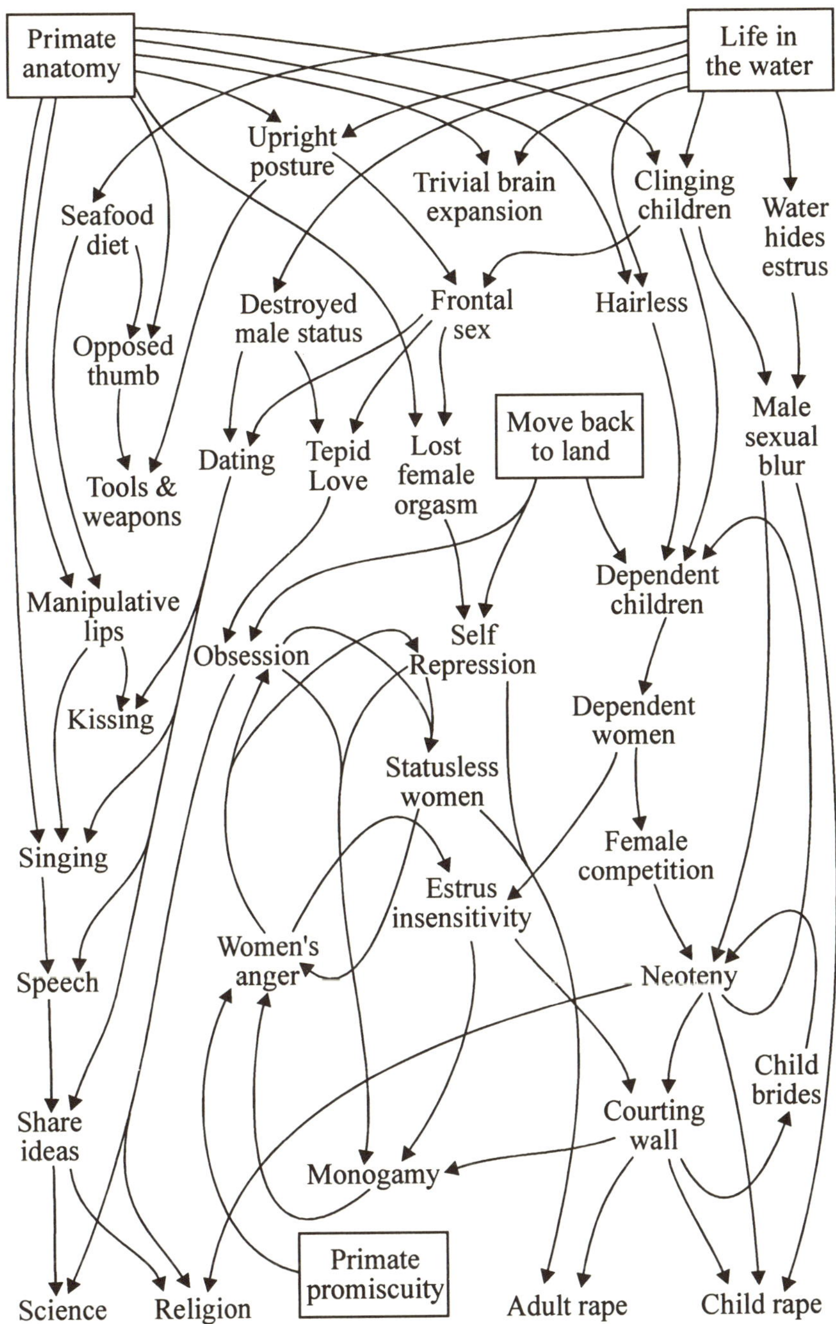

Figure 33: Human Evolution

Index

Bibliography

Ardrey, Robert, <u>African Genesis</u>, 1961

Ardrey, Robert, <u>The Territorial Imperative</u>, 1966 Atheneum

Ardrey, Robert, <u>The Social Contract</u>, 1970 Atheneum

Ardrey, Robert, <u>The Hunting Hypothesis</u>, 1976 Atheneum

Armbruster, David, Allen, Robert and Billingsly, Hubert, <u>Swimming and Diving</u>, 1963 C. V. Mosby Co

Bakker, Robert T., <u>The Dinosaur Heresies</u>, 1986 Kensington Pub. Co

Barash, David P., <u>The Hare and the Tortoise</u>, 1986 Viking

Barnouw, Victor, <u>Physical Anthropology and Archeology</u>, 1971 Dorsey Press

Birdsell, J. B., <u>Human Evolution</u>, 1972 Rand McNally Co

Brace, C. Loring and Montagu, M. F. Ashley, <u>Man's Evolution</u>, 1965 Macmillan Pub. Co

Brewer, Stella, <u>The Chimps of Mt. Asserik</u>, 1978 Alfred A. Knopf

Brown, Michael, <u>Search for Eve</u>, 1990 Harper & Row

Brownmiller, Susan, <u>Against Our Will</u>, 1975 Simon & Shuster

Buettner-Janusch, John, <u>Physical Anthropology: A Perspective</u>, 1973 Vail-Ballou Press

Bullough, Vern, <u>Sexual Variance in Society and History</u>, 1976 Univ. of Chicago Press

Bygott, J. D., <u>Cannibalism among Wild Chimpanzees</u>, 1972 Nature

Calvin, William, <u>The Throwing Madonna</u>, 1983 Bantam Books

Calvin, William, <u>The River That Flows Uphill</u>, 1986

Calvin, William, <u>The Cerebral Symphony</u>, 1990 Bantam Books

Campbell, Bernard, <u>Human Evolution</u>, 1974 Aldine Pub. Co.

Cassell, Carol, <u>Swept Away</u>, 1984 Simon & Shuster

Chartwell Books, Inc, <u>Illustrator's Figure Reference Manual</u>, 1987 Chartwell Books, Inc

Crawford, Michael and Marsh, David, <u>The Driving Force</u>, 1989 Harper & Row

Darwin, Charles, <u>The Origin of Species</u>, 1979 Avenel Books

Darwin, Charles, <u>The Descent Of Man</u>, Unkn Random House

Dawkins, Richard, <u>The Blind Watchmaker</u>, 1987 W. W. Norton

De Waal, Frans, <u>Chimpanzee Politics</u>, 1982 Harper & Row

Desmond, Adrian, <u>The Ape's Reflexion</u>, 1979 Blond & Briggs, Ltd

Deutsch, Ronald, <u>The Key to Feminine Response in Marriage</u>, 1968 Random House

Edey, Maitland and Johanson, Donald, <u>Blueprints</u>, 1989 Little, Brown and Co

Eiger, Marvin and Olds, Sally, The Complete Book of Breastfeeding, 1972 Workman Pub. Co

Eiseley, Loren, The Unexpected Universe, 1964 Harcourt, Brace & World

Eiseley, Loren, All The Strange Hours, 1975 Harcourt, Brace & World

Eiseley, Loren, The Star Thrower, 1978 Harcourt Brace Jovanovich

Eltringham, S. Keith, Elephants, 1982 Blandford Press

Fenton, Carrol Lane & Fenton, Mildred Adams, The Fossil Book, 1958 Doubleday & Co.

Fisher, Helen, The Sex Contract, 1982 William Morrow & Co

Fossey, Dian, Gorillas in the Mist, 1983 Houghton Mifflin Co

Freeman, Derek, Margaret Mead and Samoa, 1983 Harvard University Press

Freud, Sigmund, The Basic Writings of Sigmund Freud, 1938 Random House

Friday, Nancy, My Mother/My Self, 1977 Delacorte Press

Friday, Nancy, Men In Love, 1980 Delacorte Press

Friday, Nancy, Jealousy, 1985 William Morrow & Co

Gaylin, Willard, Rediscovering Love, 1986 Penguin Books

Ghileri, Michael, East of the Mountains of the Moon, 1988

Van Lawick-Goodall, Jane, My Friends the Wild Chimpanzees, 1967 National Geographic Society

Van Lawick-Goodall, Jane, In The Shadow Of Man, 1971 Houghton Mifflin Co

Van Lawick-Goodall, Jane, The Chimpanzees of Gombe, 1986 Belknap Press

Gould, Stephen Jay, Ontogeny & Phylogeny, 1977 Belknap Press

Gould, Stephen Jay, The Panda's Thumb, 1980 W. W. Norton

Gould, Stephen Jay, The Mismeasure of Man, 1981 W. W. Norton

Gould, Stephen Jay, Hen's Teeth and Horse's Toes, 1984 W. W. Norton

Gould, Stephen Jay, The Flamingo's Smile, 1985 W. W. Norton

Gould, Stephen Jay, Wonderful Life, 1989 W. W. Norton

Gribbin, John & Cherfas, Jeremy, The Monkey Puzzle, 1982 McGraw-Hill

Hammer, Signe, Passionate Attachments, 1982 Rawson Associates

Hardy, Alister, Was Man More Aquatic In The Past, 1960 The New Scientist

Hardy, Alister, Has Man an Aquatic Past?, 1960 The Listener

Hardy, Alister, Was There a Homo aquaticus?, 1977 Zenith

Harnad, Stevan, Horst Steklis and Jane Lancaster, Eds, Origins and Evolution of Language and Speech, 1976 Annals of the New York Academy of Sciences

Harrison, R. J. et al, ed, <u>Behavior & Physiology of Pinnipeds</u>, 1968 Appleton-Century-Crofts

Harrison, R. J. and Kooyman, G. L., <u>General Physiology of the Pinnipedia</u>, 1968 Appleton-Century-Crofts

Hesse, Hermann, <u>Magister Ludi</u>, or <u>Das Glasperlenspiel</u>, 1943 Bantam

Highwater, Jamake, <u>Myth & Sexuality</u>, 1990 New American Library

Hill, Justine, <u>Women Talking</u>, 1976 Lyle Stuart, Inc

Hite, Shere, <u>The Hite Report</u>, 1976 Macmillan Pub. Co

Hite, Shere, <u>The Hite Report On Male Sexuality</u>, 1981 Alfred A. Knopf

Hite, Shere, <u>Women and Love</u>, 1987 Alfred A. Knopf

Hopkins, Budd, <u>Intruders,</u> 1987 Random House

Horner, John and Gorman, James, <u>Digging Dinosaurs</u>, 1988 Workman Pub. Co

Howells, William, <u>Mankind in the Making</u>, 1967 Doubleday & Co.

Hrdy, Sarah Blaffer, <u>The Woman That Never Evolved</u>, 1981 Harvard University Press

Hulse, Frederick, <u>The Human Species</u>, 1971 Random House

Hyde, Janet, <u>Understanding Human Sexuality</u>, 1982 McGraw-Hill

Johanson, Donald & Edey, Maitland, <u>Lucy</u>, 1981 Simon & Shuster

Johanson, Donald & Shreve, James, <u>Lucy's Child</u>, 1989 William Morrow & Co

Kern, Stephen, <u>Anatomy & Destiny</u>, 1975 Grove Press

Kinsey, Alfred et al, <u>Sexual Behavior in the Human Male</u>, 1948 W. B. Saunders Co

Kinsey, Alfred et al, <u>Sexual Behavior in the Human Female</u>, 1953 W. B. Saunders Co

Kohler, Wolfgang, <u>Mentality of Apes</u>

Korn, Noel et al, Ed, <u>Human Evolution</u>, 1959 Henry Holt

Kurten, Bjorn, <u>Pleistocene Mammals of Europe</u>, 1968 Weidenfeld & Nicolson

Kurten, Bjorn, <u>The Ice Age</u>, 1972 G. P. Putnam

La Lumiere, Leon, <u>The Evolution of Human Bipedalism</u>

LeVay, Simon, <u>The Sexual Brain</u>, 1993 MIT Press

Lee, Richard and Irven DeVore, <u>Man the Hunter</u>, 1968

Leonard, Linda Schierse, <u>Witness To The Fire</u>, 1989 Shambhala Publications

Lieberman, Philip, <u>On the Origins of Language</u>, 1975

Linden, Eugene, <u>Apes, Men and Language</u>, 1975

Linden, Eugene, <u>Silent Partners</u>, 1986 Times Books

Loomis, Andrew, <u>Figure Drawing For All It's Worth</u>, 1943 The Viking Press

Lorenz, Konrad, <u>On Aggression</u>, 1963 Harcourt, Brace & World

MacKinnon, John, In Search Of The Red Ape, 1974 Holt, Rinehart And Winston

Maccoby, Eleanor and Jacklin, Carol, The Psychology of Sex Differences, 1974 Stanford University Press

Margulis, Lynn and Sagan, Dorion, Mystery Dance, 1991 Summit Books

Marshall, Donald and Suggs, Robert, Human Sexual Behavior, 1971 Basic Books, Inc

Mason, Jeffrey, and McCarthy, Susan, When Elephants Weep, 1995, Delacorte Press

Masters, R. E. L., Patterns of Incest, 1963 Julian Press

Masters, William and Johnson, Virginia, Human Sexual Response, 1966 Little, Brown and Co

Masters, William and Johnson, Virginia, The Pleasure Bond, 1970 Little, Brown and Co

Masters, William and Johnson, Virginia, Human Sexual Inadequacy, 1970 Little, Brown and Co

Masters, W., Johnson, V. & Kolodny, R., Masters and Johnson on Sex and Human Loving, 1986 Little, Brown and Co

McFarland, David, The Oxford Companion to Animal Behavior, 1987 Oxford U. Press

McMinn, R. M. H. and Hutchings, R. T., A Colour Atlas of Human Anatomy, 1977 Wolfe Medical Books

Mead, Margaret, Sex and Temperament in Three Primitive Societies, 1935 William Morrow & Co

Mead, Margaret, Male And Female, 1949 William Morrow & Co

Money, John, Love and Love Sickness, 1980

Montagu, Ashley, The Human Revolution, 1965 World Publishing Co

Montagu, Ashley and Matson, Floyd, The Human Connection, 1979 McGraw-Hill

Moore, Ruth, Man, Time and Fossils, 1953 Alfred A. Knopf

Morgan, Elaine, The Descent of Woman, 1972 Stein & Day

Morgan, Elaine, The Aquatic Ape, 1982 Stein & Day

Morris, Desmond, The Naked Ape, 1967 McGraw-Hill

Morris, Desmond, The Human Zoo, 1969

Morris, Desmond, Intimate Behaviour, 1971 Random House

Moss, Cynthia, Elephant Memories, 1988 Fawcett Columbine

Napier, John, The Antiquity of Human Walking, 1967 W. H. Freeman & Co

Ornstein, Robert, The Evolution of Consciousness, 1991 Prentice Hall Press

Perper, Timothy, Sex Signals: The Biology of Love, 1985

Peterson, Richard, Social Behavior in Pinnipeds, 1968 Appleton-Century-Crofts

Pilbeam, David, The Evolution of Man, 1970 Funk & Wagnalls

Poirier, Frank, In Search Of Ourselves, 1974 Burgess Publishing Co

Premack, David, Intelligence in Ape and Man, 1976 Lawrence Earlbaum Associates

Prudden, Bonnie, Your Baby Can Swim, 1974 Reader's Digest Press

Raup, David, Extinction - Bad Genes or Bad Luck, 1991 W. W. Norton

Reynolds, Vernon, The Apes, 1967 E. P. Dutton & Co.

Richter, Paul, Artistic Anatomy, 1971 Watson-Guptill Publications

Robinson, John, Early Hominid Posture & Locomotion, 1972 Univ. of Chicago Press

Rosen, Julius, Going Ape, 1988 Contemporary Books

Rubin, Isadore & Kirkendall, Lester, Sex in the Adolescent Years, 1968 Association Press

Rubin, Lillian B., Intimate Strangers, 1983 Harper & Row

Sagan, Carl, Broca's Brain, 1974 Random House

Sagan, Carl, Dragons of Eden, 1977 Random House

Sagan, Eli, Freud, Women and Morality, 1988 Basic Books, Inc

Sarnoff, Suzanne and Sarnoff, Irving, Masturbation and Adult Sexuality, 1979 M. Evans and Co

Scarf, Maggie, Intimate Partners, 1987 Random House

Schwartz, Jeffrey, The Red Ape, 1987 Houghton Mifflin Co

Sherfey, Mary Jane, The Nature & Evolution of Female Sexuality, 1966 Random House

Stewart, George, Man: an Autobiography, 1946 Random House

Strum, Shirley, Almost Human, 1987 Random House

Sutcliffe, Antony J., On The Track of Ice Age Mammals, 1985 Harvard University Press

Swindler, Daris and Wood, Charles, An Atlas of Primate Gross Anatomy, 1973 U. of Washington Press

Tannehill, Reay, Sex In History, 1980 Stein & Day

Tattersall, Ian, The Fossil Trail, 1995, Oxford University Press

Teleki, Geza, The Predatory Behavior of Wild Chimpanzees, 1973

Tennov, Dorothy, Love and Limerence, 1979 Stein & Day

Thomas, Lewis, Lives of a Cell, 1974 Penguin Books

Tierney, Patrick, The Highest Altar, 1989 Viking

Tiger, Lionel and Fox, Robin, The Imperial Animal, 1971 Holt, Rinehart And Winston

Tinbergen, Nikolaas, The Study of Instinct, 1989 Oxford U. Press

Twain, Mark, The Adventures of Tom Sawyer, 1876 Alfred Knopf

Washburn, S. L. and Lancaster, C. S., The Evolution of Hunting, 1968

Weinrich, James, <u>Sexual Landscapes</u>, 1987 Charles Scribner's Sons
Wendt, Herbert, <u>In Search of Adam</u>, 1956 Houghton Mifflin Co
Wendt, Herbert, <u>The Sex Life of the Animals</u>, 1965 Simon & Shuster
Willis, Delta, <u>The Hominid Gang</u>, 1989 Viking
Wills, Christopher, <u>The Wisdom of the Genes</u>, 1989 Basic Books, Inc
Wilson, Edward, <u>Sociobiology: The New Synthesis</u>, 1975
Wilson, Edward, <u>On Human Nature</u>, 1978
Young, Wayland, <u>Eros Denied</u>, 1964 Grove Press
Young-Bruehl, Elisabeth, <u>Freud On Women</u>, 1990 W. W. Norton
Zuckerman, Solly, <u>The Social Life of Monkeys and Apes</u>, 1981 Routledge
    & Kegan Paul

www.PassionateApe.com

The Book

Where to find it
Overview of the thesis
Excepts and chapter outline

The Publisher

Addresses - eMail and SnailMail
Wholesalers and quantity discounts
Other projects as they develop

The Larger Perspective

Passionate Ape discussion forum
The Aquatic Ape Hypothesis and more
Other useful links and discussion groups

The Author and Mrs. Author

Sherry's side of the story
Addresses - eMail and SnailMail
Further musings on unrelated subjects

 # Order Form

If your local bookseller does not carry this title,
you can order it from the publisher.

Want to give it as a gift? Want it autographed?
Tell us how you'd like it inscribed (within reason).

Name: ____________________________

Address: ____________________________

City: ____________________________

State: ____________________________

Zip: ____________________________

eMail: ____________________________

Number of copies: _______

Price each: $21.95

Subtotal: _______

Washington State orders add 8% sales tax: _______

Shipping and Handling add $3 per book: _______

Total: _______

Enclose check or money order, no stamps, no cash.
Prices subject to change without notice.

Mail to: RiverForest Press
P.O. Box 11694
Bainbridge Island, WA
98110-5694

www.PassionateApe.com

The Book

Where to find it
Overview of the thesis
Excepts and chapter outline

The Publisher

Addresses - eMail and SnailMail
Wholesalers and quantity discounts
Other projects as they develop

The Larger Perspective

Passionate Ape discussion forum
The Aquatic Ape Hypothesis and more
Other useful links and discussion groups

The Author and Mrs. Author

Sherry's side of the story
Addresses - eMail and SnailMail
Further musings on unrelated subjects

# Order Form

If your local bookseller does not carry this title,
you can order it from the publisher.
Want to give it as a gift? Want it autographed?
Tell us how you'd like it inscribed (within reason).

Name: ___________________________________

Address: _________________________________

City: ____________________________________

State: ___________________________________

Zip: ____________________________________

eMail: ___________________________________

Number of copies: _______

Price each: $21.95

Subtotal: _______

Washington State orders add 8% sales tax: _______

Shipping and Handling add $3 per book: _______

Total: _______

Enclose check or money order, no stamps, no cash.
Prices subject to change without notice.

Mail to: RiverForest Press
P.O. Box 11694
Bainbridge Island, WA
98110-5694